# Laboratory Experiments for

# General, Organic & Biochemistry

## FIFTH EDITION

**FREDERICK A. BETTELHEIM**
**JOSEPH M. LANDESBERG**

Adelphi University

THOMSON

BROOKS/COLE

Australia • Canada • Mexico • Singapore
Spain • United Kingdom • United States

**THOMSON**

**BROOKS/COLE**

This book is dedicated to our wives: Vera S. Bettelheim and Lucy G. Landesberg,
whose help, understanding, and patience enabled us to write this book.

**CHEMISTRY EDITOR:** John Holdcroft
**ASSISTANT EDITOR:** Alyssa White
**MARKETING MANAGER:** Jeffrey L. Ward
**MARKETING ASSISTANT:** Mona Weltmer
**ADVERTISING PROJECT MANAGER:** Stacey Purviance
**PROJECT MANAGER, EDITORIAL PRODUCTION:** Belinda Krohmer
**PRINT/MEDIA BUYER:** Judy Inouye
**PRODUCTION SERVICE:** Thompson Steele, Inc.
**COPY EDITOR:** Thompson Steele, Inc.

**ILLUSTRATOR:** J/B Woolsey and Thompson Steele, Inc.
**COVER DESIGNER:** Larry Didona
**COVER IMAGE:** Crystal: Cholesterol.© Michael W. Davidson/Photo Researchers, Inc.
**COVER PRINTER:** Darby Printing Company
**COMPOSITOR:** Thompson Steele, Inc.
**PRINTER:** Darby Printing Company

**Library of Congress Control Number:** 2003104151

ISBN 0-534-40192-9

**Brooks/Cole—Thomson Learning**
**10 Davis Drive**
**Belmont, CA 94002**
**USA**

**Asia**
Thomson Learning
5 Shenton Way #01-01
UIC Building
Singapore 068808

**Australia/New Zealand**
Thomson Learning
102 Dodds Street
Southbank, Victoria 3006
Australia

**Canada**
Nelson
1120 Birchmount Road
Toronto, Ontario M1K 5G4
Canada

**Europe/Middle East/Africa**
Thomson Learning
High Holborn House
50/51 Bedford Row
London WC1R 4LR
United Kingdom

**Latin America**
Thomson Learning
Seneca, 53
Colonia Polanco
11560 Mexico D.F.
Mexico

**Spain/Portugal**
Paraninfo
Calle/Magallanes, 25
28015 Madrid, Spain

## Warnings About Safety Precautions

Some of the experiments contained in this Laboratory Manual involve a degree of risk on the part of the instructor and student. Although performing the experiments is generally safe for the college laboratory, unanticipated and potentially dangerous reactions are possible for a number of reasons, such as improper measuring or handling of chemicals, improper use of laboratory equipment, failure to follow laboratory safety procedures, and other causes. Neither the publisher nor the authors can accept any responsibility for personal injury or property damage resulting from the use of this publication.

# Preface

In preparing the fifth edition of this Laboratory Manual, we wish to thank our colleagues who made this new edition possible by adopting our manual for their courses. This fifth edition coincides with the publication of the seventh edition of the textbook: *Introduction to General, Organic and Biochemistry* by Bettelheim, Brown and March. The textbook shares the outline and pedagogical philosophy with this book. As in previous editions, we have strived for the clearest possible writing in the procedures. The experiments give the student a meaningful, reliable laboratory experience that consistently work, while covering the basic principles of general, organic and biochemistry. Throughout the years, feedback from Colleges and Universities made us aware that we have managed to achieve a manual that not only eases the student's task in performing experiments, but also is student friendly. Our new edition maintains this standard and improves upon it.

The major changes in this new edition are as follows: (1) We added two new experiments and eliminated two of the old ones. This still keeps a total of 52 experiments and still provides a wide selection of experiments from which the instructor can choose. One of the new experiments demonstrates the fermentation of sucrose to ethanol; the second shows the action of a neurotransmission enzyme. (2) We improved the procedures of a number of experiments as a result of our observations of how our students carried out these experiments in our laboratories at Adelphi. (3) Safety issues and waste disposal are emphasized throughout this edition. (4) We further improved our aim to minimize the use of hazardous chemicals where possible and to design experiments that work either on a small scale or on a semi-micro scale. (5) Most Pre-Lab and Post-Lab Questions have been changed or modified.

As in previous editions, three basic goals were followed in all the experiments: (1) the experiments should illustrate the concepts learned in the classroom; (2) the experiments should be clearly and concisely written so that students will easily understand the task at hand, will work with minimal supervision because the manual provides enough information on experimental procedures, and will be able to perform the experiments in a 2-1/2 hour laboratory period; (3) the experiments should not only be simple demonstrations, but also should contain a sense of discovery.

It did not escape our attention that in adopting this manual of Laboratory Experiments, the instructor must pay attention to budgetary constraints. All experiments in this manual generally use only inexpensive pieces of equipment and glassware. A few spectrophotometers and pH meters are necessary in a number of experiments. A few experiments may require more specialized, albeit inexpensive, equipment, for example, a few viscometers; for these, the instructor might wish to do them as a demonstration.

The 52 experiments in this book will provide suitable choice for the instructor to select about 25 experiments for a two-semester or three-quarter course. The following are the principle features of this book:

1.  The **Table of Contents** is organized so that the first 23 experiments illustrate the principles of general chemistry, the next 12 those of organic chemistry, and remaining 17 those of biochemistry.

2. Each experiment starts out with a **Background** which contains information beyond the textbook material. All the relevant principles and their applications are reviewed in this **Background** section.

3. The **Procedure** part provides a step-by-step description of the experiments. Clarity of writing in this section is of the utmost importance for successful execution of the experiments. **Caution!** signs alert students when dealing with dangerous chemicals, such as strong acids or bases.

4. **Pre-Lab Questions** are provided to familiarize students with the concepts and procedures before they start the experiments. By requiring students to answer these questions and by grading their answers, we accomplish the task of preparing the students for the experiments.

5. In the **Report Sheet** we not only ask for the recording of raw data, but we also request some calculations to yield secondary data.

6. The **Post-Lab Questions** are designed so that students should be able to reflect upon the results, interpret them, and relate their significance.

7. At the end of the book in **Appendix 3,** we provide Stockroom Personnel with detailed instructions on preparation of solutions and other chemicals for each experiment. We also give detailed instructions as to how much material is needed for a class of 25 students.

An **Instructor's Manual** that accompanies this book is **solely for the use of the Instructor.** It helps in the grading process by providing ranges of the experimental results we obtained from class use. In addition it alerts the instructor to some of the difficulties that may be encountered in certain experiments. The disposal of waste material is discussed for each experiment. For further information, see *Prudent Practices in the Laboratory*, National Academy Press, Washington, DC (1995).

We hope that you will find our book of Laboratory Experiments helpful in instructing your students. We anticipate that students will find it inspiring in studying different aspects of chemistry.

Garden City, NY
April 2003

Frederick A. Bettelheim
Joseph M. Landesberg

# Acknowledgments

These experiments have been used by our colleagues over the years and their criticism and expertise were instrumental in refinement of the experiments. We thank Stephen Goldberg, Robert Halliday (dec.), Cathy Ireland, Mahadevappa Kumbar, Robert Lippman, Jerry March (dec.), Sung Moon, Donald Opalecky, Reuben Rudman, Charles Shopsis, Kevin Terrance, and Stanley Windwer. We acknowledge the contributions of Dr. Jessie Lee, Community College of Philadelphia.

We also thank the following reviewers for their thoughtful comments and suggestions: R. E. Bozak, California State University, Hayward; Robert Bruner, Contra Costa College; Bridget Dube, San Antonio College; Katherine Jimison, Cuesta College; Jo Kohn, Olympic College; Eli C. Minkoff, Bates College; Margaret Sequin, San Francisco State University; Conrad Stanitski, University of Central Arkansas ; and Mona Wahby, Macomb Community College.

We extend our appreciation to the entire staff at Thomson Learning-Brooks/Cole, especially to John Holdcroft, Chemistry Editor, and Alyssa White, Assistant Chemistry Editor, for their encouragement and excellent efforts in producing this book.

# Table of Contents

**Figure 1** • Common laboratory equipment

Crucible tongs

Test tube holder

Bunsen burner
(Tirrill type)

Ring support

Utility clamp

Clay triangle

Ring stand with
support

Buret clamp

Wire gauze

Evaporating dish

Watch glass

Tripod

**Figure 1** • Continued

# Practice Safe Laboratory

A few precautions can make the laboratory experience relatively hazard-free and safe. These experiments are on a small scale and as such, many of the dangers found in the chemistry laboratory have been minimized. In addition to specific regulations that you may have for your laboratory, the following **DO** and **DON'T RULES** should be observed at all times.

## DO RULES

❑ **Do wear approved safety glasses or goggles at all times.**

The first thing you should do after you enter the laboratory is to put on your safety eyewear. The last thing you should do before you leave the laboratory is to remove them. Contact lens wearers must wear additional safety goggles; prescription glasses can be used instead.

❑ **Do wear protective clothing.**

Wear sensible clothing in the laboratory: e.g., no shorts, no tank tops, no sandals. Be covered from the neck to the feet. Laboratory coats or aprons are recommended. Tie back long hair, out of the way of flames.

❑ **Do know the location and use of all safety equipment.**

This includes eyewash facilities, fire extinguishers, fire showers, and fire blankets. In case of fire, do not panic, clear out of the immediate area, and call your instructor for help.

❑ **Do use proper techniques and procedures.**

Closely follow the instructions given in this laboratory manual. These experiments have been student tested; however, accidents do occur but can be avoided if the steps for an experiment are followed. Pay heed to the **Caution!** signs in a procedure.

❑ **Do discard waste material properly.**

Organic chemical waste should be collected in appropriate waste containers and *not flushed down sink drains*. Dilute, nontoxic solutions may be washed down the sink with plenty of water. Insoluble and toxic waste chemicals should be collected in properly labeled waste containers. Follow the directions of your instructor for alternative or special procedures.

❑ **Do be alert, serious, and responsible.**

The best way you can prepare for an experiment is to read the procedure carefully and be aware of the hazards before stepping foot into the laboratory.

# DON'T RULES

❏ **Do not eat or drink in the laboratory.**

Consume any food or drink before entering the laboratory. Chemicals could get into food or drinks, causing illness. If you must take a break, wash your hands thoroughly before leaving.

❏ **Do not smoke in the laboratory.**

Smoke only in designated smoking areas outside the laboratory. Flammable gases and volatile flammable reagents could easily explode.

❏ **Do not taste any chemicals or breathe any vapors given off by a reaction.**

If there is a need to smell a chemical, you will be shown how to do it safely.

❏ **Do not get any chemicals on your skin.**

Wash off the exposed area with plenty of water should this happen. Notify your instructor at once. Wear gloves as indicated by your instructor.

❏ **Do not clutter your work area.**

Your laboratory manual and the necessary chemicals, glassware, and hardware are all that should be on your benchtop. This will avoid spilling chemicals and breaking glassware.

❏ **Do not enter the chemical storage area or remove chemicals from the supply area.**

Everyone must have access to the chemicals for the day's experiment. Removal of a chemical from the storage or supply area only complicates the proper execution of the experiment for the other students.

❏ **Do not perform unauthorized experiments.**

Any experiment not authorized presents a hazard to any person in the immediate area.

❏ **Do not take unnecessary risks.**

These **DO** and **DON'T RULES** for a safe laboratory are not an exhaustive list, but are a minimum list of precautions that will make the laboratory a safe and fun activity. Should you have any questions about a hazard, ask your instructor *first*—not your laboratory partner. Finally, if you wish to know about the dangers of any chemical you work with, read the Material Safety Data Sheet (MSDS). These sheets should be on file in the chemistry department office.

# Laboratory techniques: use of the laboratory gas burner; basic glassworking

## Background

### The Laboratory Gas Burner

Tirrill or Bunsen burners provide a ready source of heat in the chemistry laboratory. In general, since chemical reactions proceed faster at elevated temperatures, the use of heat enables the experimenter to accomplish many experiments more quickly than would be possible at room temperature. The burner illustrated in Fig. 1.1 is typical of the burners used in most general chemistry laboratories.

**Figure 1.1**
The Bunsen burner.

Violet outer cone
Pale-blue middle cone
Dark-blue inner cone
Hottest part of the flame (>800°C)
Barrel
Air vents
Gas inlet
Gas control valve
Main gas valve
Base

A burner is designed to allow gas and air to mix in a controlled manner. The gas often used is "natural gas," mostly the highly flammable and odorless hydrocarbon methane, $CH_4$. When ignited, the flame's temperature can be adjusted by altering the various proportions of gas and air. The gas flow can be controlled either at the main gas valve or at the gas control valve at the base of the burner. Manipulation of the air vents at the bottom of the barrel allows air to enter and mix with the gas. The hottest flame has a nonluminous violet outer cone, a pale-blue middle cone, and a dark-blue inner cone; the air vents, in this case, are opened sufficiently to assure complete combustion of the gas. Combustion of the gas yields carbon dioxide and water.

$$CH_{4(g)} + 2O_{2(g)} \longrightarrow CO_{2(g)} + 2H_2O_{(g)}$$

Lack of air produces a cooler, luminous yellow flame. This flame lacks the inner cone, most likely is smokey, and often deposits soot on objects it contacts. The soot is due to small particles of carbon from unburned fuel, which luminesce and give the yellow flame. Too much air blows out the flame.

## Basic Glassworking

In the chemistry laboratory, it is often necessary to modify apparatus made from glass or to connect pieces of equipment with glass tubing. Following correct procedures for working with glass, especially glass tubing, is important.

Glass is a super-cooled liquid. Unlike crystalline solids which have sharp melting points, glass softens when heated, flows, and thus can be worked. Bending, molding, and blowing are standard operations in glassworking.

Not all glass is the same; there are different grades and compositions. Most laboratory glassware is made from borosilicate glass (containing silica and borax compounds). Commercially, this type of glass is known as *Pyrex* (made by Corning Glass) or *Kimax* (made by Kimble glass). This glass does not soften very much below 800°C and, therefore, requires a very hot flame in order to work it. A Bunsen burner flame provides a hot enough temperature for general glassworking. In addition, borosilicate glass has a low thermal coefficient of expansion. This refers to the material's change in volume per degree change in temperature. Borosilicate glass expands or contracts slowly when heated or cooled. Thus, glassware composed of this material can withstand rapid changes in temperature and can resist cracking.

Soft glass consists primarily of silica sand, $SiO_2$. Glass of this type softens in the region of 300–400°C, and because of this low softening temperature is not suitable for most laboratory work. It has another unfortunate property that makes it a poor material for laboratory glassware. Soft glass has a high thermal coefficient of expansion. This means that soft glass expands or contracts very rapidly when heated or cooled; sudden, rapid changes in temperature introduce too much stress into the material, and the glass cracks. While soft glass can be worked easily using a Bunsen burner, care must be taken to prevent breakage; with annealing, by first mildly reheating and then uniformly, gradually cooling, stresses and strains can be controlled.

### Objectives

1. To learn how to use a Bunsen burner.
2. To learn basic glassworking by bending and fire-polishing glass tubing.

## Procedure

### The Laboratory Gas Burner; Use of the Bunsen Burner

1. Before connecting the Bunsen burner to the gas source, examine the burner and compare it to Fig. 1.1. Be sure to locate the gas control valve and the air vents and see how they work.

2. Connect the gas inlet of your burner to the main gas valve by means of a short piece of thin-walled rubber tubing. Be sure the tubing is long enough to provide some slack for movement on the bench top. Close the gas control valve. If your burner has a screw-needle valve, turn the knob clockwise. Close the air vents. This can be done by rotating the barrel of the burner (or sliding the ring over the air vents if your burner is built this way).

3. Turn the main gas valve to the open position. Slowly open the gas control valve counterclockwise until you hear the hiss of gas. Quickly strike a match or use a gas striker to light the burner. With a lighted match, hold the flame to the top of the barrel. The gas should light. How would you describe the color of the flame? Hold a Pyrex test tube in this flame. What do you observe?

4. Carefully turn the gas control valve, first clockwise and then counterclockwise. What happens to the flame size? (If the flame should go out, or if the flame did not light initially, shut off the main gas valve and start over, as described above.)

5. With the flame on, adjust the air vents by rotating the barrel (or sliding the ring). What happens to the flame as the air vents open? Adjust the gas control valve and the air vents until you obtain a flame about 3 or 4 in. high, with an inner cone of blue (Fig. 1.1). The tip of the pale blue inner cone is the hottest part of the flame.

6. Too much air will blow out the flame. Should this occur, close the main gas valve immediately. Relight following the procedure in step 3.

7. Too much gas pressure will cause the flame to rise away from the burner and "roar" (Fig. 1.2). If this happens, reduce the gas flow by closing the gas control valve until a proper flame results.

**Figure 1.2**
The flame rises away
from the burner.

8. "Flashback" sometimes occurs. If so, the burner will have a flame at the bottom of the barrel. Quickly close the main gas valve. Allow the barrel to cool. Relight following the procedures in step no. 3.

## Basic Glassworking; Working with Glass Tubing

### Cutting glass tubing

1. Obtain a length of glass tubing (6 or 8 mm in diameter). Place the tubing flat on the bench top, and with a grease pencil mark off a length of 30 cm. Grasp a triangular file with one hand, placing your index finger on a flat side of the file. With your other hand, hold the tubing firmly in place against the bench top. At the mark, press the edge of the file down firmly on the glass, and in one continuous motion scratch the glass (Fig. 1.3).

**Figure 1.3**
Cutting glass tubing with
a triangular file.

**2.** Place a drop of water on the scratch (this seems to help the glass break). Wrap the tubing with cloth or paper towels and grasp with both hands, as shown in Fig. 1.4. Place your thumbs on the unscratched side of the tubing, one thumb on either side of the scratch. Position the scratch away from your body and face. Snap the tubing by simultaneously pushing with both thumbs and pulling with both hands toward your body. The tubing should break cleanly where the glass was scratched. Should the tubing not break, repeat the procedure described above.

**Figure 1.4**
Breaking glass tubing.

## Glass bends

**1.** Turn off the Bunsen burner and place a wing top on the barrel. The wing top will spread out the flame so that a longer section of glass will be heated to softness. Relight the burner and adjust the flame until the blue inner cone appears along the width of the wing top (Fig. 1.5).

**Figure 1.5**
Wing top on the Bunsen burner.

Bright blue flame

**2.** Hold the midsection of the newly cut glass tubing in the flame. Keep the tubing in the hottest part of the flame, just above the spread-out blue cone (Fig. 1.6). Rotate the tubing continuously to obtain uniform heating. As the glass gets hot, the flame should become yellow; this color is due to sodium ions, which are present in the glass.

**Figure 1.6**
Holding the glass tubing
in the flame.

When the glass gets soft and feels as if it is going to sag, remove the glass from the flame. Hold it steady without twisting or pulling (Fig. 1.7), and quickly, but gently, bend it to the desired angle (Fig. 1.8).

**Figure 1.7** • Hold before bending.

**Figure 1.8** • Quickly bend.

A good bend has a smooth curve with no constrictions (Fig. 1.9).

**Figure 1.9**

**A good bend.**

**Poor bends.**

## CAUTION!

Hot glass looks like cold glass. When finished with a piece of hot glass, place it out of the way on your bench top, on a piece of wire gauze. Glass cools slowly, so do not attempt to pick up any piece until you test it. Hold your hand above the glass without touching; you will be able to sense any heat. If your fingers get burnt by touching hot glass, immediately cool them with cold water and notify your instructor.

### *Fire polishing*

1. To remove sharp edges from cut glass, a hot flame is needed to melt and thereby smooth out the glass.

2. If the wing top is on the burner, turn off the gas and carefully remove the wing top from the barrel with a pair of crucible tongs. The wing top may be hot.

3. Relight the gas and adjust to the hottest flame. Hold one end of the cooled tubing in the hottest part of the flame (just above the blue inner cone). Slowly rotate the tube (Fig. 1.10).

**Figure 1.10**
Fire polishing.

The flame above the glass tubing should become yellow as the glass gets hot and melts. Be careful not to overmelt the glass, in order to prevent the end from closing. After a short time (approx. 1 min.), remove the glass from the flame and examine the end; fire polishing will round the edges. Reheat if necessary to complete the polishing. When the end is completely smooth, lay the hot glass on a piece of wire gauze to cool. Be sure the glass is completely cooled before you attempt to polish the other end.

4. Show your instructor your glass bend with the ends completely fire polished.

### Making stirring rods

Cut some solid glass rods (supplied by the instructor) into 20-cm lengths. Fire polish the ends.

### Drawing capillary tubes

1. Cut a piece of glass tubing about 20 cm in length.

2. Heat the middle of the glass tubing in the flame just above the inner blue cone. Don't use a wing top. Rotate the tube in the flame until it softens (Fig. 1.11 A).

Step A                                          Step B

**Figure 1.11** • Techniques for drawing capillary tubes.

3. As the glass sags, remove the tubing from the flame. Gently pull on each end, as straight as possible, until the capillary is as small as desired (Fig. 1.11 B).

4. Carefully place the tubing on the bench top and allow the glass to cool.

5. With a triangular file, carefully cut a piece of the drawn-out capillary tube (approx. 10 cm). Seal one end by placing it in the flame. Show your instructor your sealed capillary tube.

| Chemicals and Equipment |
| --- |
| 1. Glass tubing (6-mm and 8-mm OD) |
| 2. Glass rod (6-mm OD) |
| 3. Bunsen burner |
| 4. Wing top |
| 5. Wire gauze |
| 6. Crucible tongs |

## Experiment 1

### PRE-LAB QUESTIONS

**1.** Laboratory experiments often use heat for their chemical reactions. Explain why this is done.

**2.** What does the valve at the base of a Bunsen burner do?

**3.** Describe the appearance of the hottest flame that a Bunsen burner can produce.

**4.** Glass is a solid material but unlike a crystalline solid, it does not have a sharp melting point. Explain.

**5.** Why can borosilicate glass be used for standard operations in glassblowing?

## Experiment 1

## REPORT SHEET

### *Bunsen burner*

**1.** What is the color of the flame when the air vents are closed?


**2.** What happened to the Pyrex test tube in this flame?


**3.** What happens to the flame when the gas control valve is turned?


**4.** Describe the effect on the flame as the air vents were opened.


### *Glassworking*

Let the instructor comment on your glass experiments.


**1.** 90° angle bend:


**2.** Fire polishing:


**3.** Glass stirring rod:


**4.** Capillary tube:

## POST-LAB QUESTIONS

**1.** What are the products of complete combustion of gas in a Bunsen flame?

**2.** A student's Bunsen flame is yellow and smokey. What causes the yellow flame?

**3.** How can a student change a luminous yellow flame into a nonluminous blue flame?

**4.** Why must a freshly cut piece of glass tubing be fire polished before it is used?

**5.** The diagram below is of a nonluminous Bunsen flame. Indicate the approximate region for the hottest part of the flame.

# Experiment 2

## Laboratory measurements

### Background

#### Units of Measurement

The metric system of weights and measures is used by scientists of all fields, including chemists. This system uses the base 10 for measurements; for conversions, measurements may be multiplied or divided by 10. Table 2.1 lists the most frequently used factors in the laboratory which are based on powers of 10.

| Table 2.1 | Frequently Used Factors | | |
|---|---|---|---|
| **Prefix** | **Power of 10** | **Decimal Equivalent** | **Abbreviation** |
| Micro | $10^{-6}$ | 0.000001 | $\mu$ |
| Milli | $10^{-3}$ | 0.001 | m |
| Centi | $10^{-2}$ | 0.01 | c |
| Kilo | $10^{3}$ | 1000 | k |

The measures of length, volume, mass, energy, and temperature are used to evaluate our physical and chemical environment. Table 2.2 compares the metric system with the more recently accepted SI system (International System of Units). The laboratory equipment associated with obtaining these measures is also listed.

| Table 2.2 | Units and Equipment | | |
|---|---|---|---|
| **Measure** | **SI Unit** | **Metric Unit** | **Equipment** |
| Length | Meter (m) | Meter (m) | Meterstick |
| Volume | Cubic meter ($m^3$) | Liter (L) | Pipet, graduated cylinder, Erlenmeyer flask, beaker |
| Mass | Kilogram (kg) | Gram (g) | Balance |
| Energy | Joule (J) | Calorie (cal) | Calorimeter |
| Temperature | Kelvin (K) | Degree Celsius (°C) | Thermometer |

#### Accuracy, precision, and significant figures

Chemistry is a science that depends on experience and observation for data. It is an empirical science. An experiment that yields data requires the appropriate measuring devices in order to get accurate measurements. Once the data is in hand, calculations are done with the numbers obtained. How good the calculations are depends on a number of factors: (1) how careful you are in taking the measurements (laboratory techniques), (2) how good your measuring device is in getting a true measure (accuracy), and (3) how reproducible the measurement is (precision).

The measuring device usually contains a scale. The scale, with its subdivisions or graduations, tells the limits of the device's accuracy. You cannot expect to obtain a measurement better than your instrument is capable of reading. Consider the portion of the ruler shown in Fig. 2.1.

**Figure 2.1** • Reading a metric ruler.

There are major divisions labeled at intervals of 1 cm and subdivisions of 0.1 cm or 1 mm. The accuracy of the ruler is to 0.1 cm (or 1 mm); that is the measurement that is known for certain. However, it is possible to estimate to 0.01 cm (or 0.1 mm) by reading in between the subdivisions; this number is less accurate and of course, is less certain. In general, you should be able to record the measured value to one more place than the scale is marked. For example, in Fig. 2.1 there is a reading marked on the ruler. This value is 8.75 cm: two numbers are known with certainty, *8.7*, and one number, *0.05*, is uncertain since it is the *best estimate* of the fractional part of the subdivision. The number recorded, 8.75, contains 3 significant figures, 2 certain plus 1 uncertain. When dealing with *significant figures*, remember: (1) the uncertainty is in the last recorded digit, and (2) the number of significant figures contains the number of digits definitely known, plus one more that is estimated.

The manipulation of significant figures in multiplication, division, addition, and subtraction is important. It is particularly important when using electronic calculators which give many more digits than are useful or significant. If you keep in mind the principle that the final answer can be no more accurate than the least accurate measurement, you should not go wrong. A few examples will demonstrate this.

---

**EXAMPLE 1**

Divide 9.3 by 4.05. If this calculation is done by a calculator, the answer found is 2.296296296. However, *a division should have as an answer the same number of significant figures as the least accurately known (fewest significant figures) of the numbers being divided.* One of the numbers, 9.3, contains only 2 significant figures. Therefore, the answer can only have 2 significant figures, i.e., 2.3 (rounded off).

---

## EXAMPLE 2

Multiply 0.31 by 2.563. Using a calculator, the answer is 0.79453. *As in division, a multiplication can have as an answer the same number of significant figures as the least accurately known (fewest significant figures) of the numbers being multiplied.* The number 0.31 has 2 significant figures (the zero fixes the decimal point), therefore, the answer can only have 2 significant figures, i.e., 0.79 (rounded off).

## EXAMPLE 3

Add 3.56 + 4.321 + 5.9436. A calculator gives 13.8246. *With addition (or subtraction), the answer is significant to the least number of decimal places of the numbers added (or subtracted).* The least accurate number is 3.56, measured only to the hundredth's place. The answer should be to this accuracy, i.e., 13.82 (rounded off to the hundredth's place).

Finally, how do precision and accuracy compare? *Precision* is a determination of the reproducibility of a measurement. It tells you how closely several measurements agree with one another. Several measurements of the same quantity showing high precision will cluster together with little or no variation in value; however, if the measurements show a wide variation, the precision is low. *Random errors* are errors which lead to differences in successive values of a measurement and affect precision; some values will be off in one direction or another. One can estimate the precision for a set of values for a given quantity as follows: estimate $= \pm\Delta/2$, where $\Delta$ is the difference between the highest and lowest values.

*Accuracy* is a measure of how closely the value determined agrees with a known or accepted value. Accuracy is subject to *systematic errors*. These errors cause measurements to vary from the known value and will be off in the same direction, either too high or too low. A consistent error in a measuring device will affect the accuracy, but always in the same direction. It is important to use properly calibrated measuring devices. If a measuring device is not properly calibrated, it may give high precision, but with none of the measurements being accurate. However, a properly calibrated measuring device will be both precise and accurate. (See Fig. 2.2.) A systematic error is expressed as the difference between the known value and the average of the values obtained by measurement in a number of trials.

**Figure 2.2** • Precision and accuracy illustrated by a target.

High precision and high accuracy

High precision and poor accuracy

Poor precision and poor accuracy

## Procedure

### Length: use of the meterstick (or metric ruler)

1. The meterstick is used to measure length. Examine the meterstick in your kit. You will notice that one side has its divisions in inches (in.) with subdivisions in sixteenths of an inch; the other side is in centimeters (cm) with subdivisions in millimeters (mm). Some useful conversion factors are listed below.

| | | | |
|---|---|---|---|
| 1 km | = 1000 m | 1 in. | = 2.54 cm |
| 1 m | = 100 cm | 1 ft. | = 30.48 cm |
| 1 cm | = 10 mm | 1 yd. | = 91.44 cm |
| 1 m | = 1000 mm | 1 mi. | = 1.6 km |

A meterstick that is calibrated to 0.1 cm can be read to the hundredth's place; however, only a 0 (0.00) or a 5 (0.05) may appear. A measurement falling directly on a subdividision is read as a 0 in the hundredth's place. A measurement falling anywhere between adjacent subdivisions is read as a 5 in the hundredth's place.

2. With your meterstick (or metric ruler), measure the length and width of this laboratory manual. Take the measurements in inches (to the nearest sixteenth of an inch) and in centimeters (to the nearest 0.5 cm). Record your response on the Report Sheet (1).

3. Convert the readings in cm to mm and m (2).

4. Calculate the area of the manual in $in^2$, $cm^2$, and $mm^2$ (3). Be sure to express your answers to the proper number of significant figures.

---

**EXAMPLE 4**

A student measured a piece of paper and found it to be 20.30 cm by 29.25 cm. The area was found to be
$$20.30 \text{ cm} \times 29.25 \text{ cm} = 593.8 \text{ cm}^2$$

---

### Volume: use of a graduated cylinder, an Erlenmeyer flask, and a beaker

1. Volume in the metric system is expressed in liters (L) and milliliters (mL). Another way of expressing milliliters is in cubic centimeters ($cm^3$ or cc). Several conversion factors for volume measurements are listed below.

| | | | |
|---|---|---|---|
| 1 L | = 1000 mL | 1 qt. | = 0.96 L |
| 1 mL | = 1 $cm^3$ = 1 cc | 1 gal. | = 3.79 L |
| 1 L | = 0.26 gal. | 1 fl. oz. | = 29.6 mL |

**2.** The graduated cylinder is a piece of glassware used for measuring the volume of a liquid. Graduated cylinders come in various sizes with different degrees of accuracy. A convenient size for this experiment is the 100-mL graduated cylinder. Note that this cylinder is marked in units of 1 mL; major divisions are of 10 mL and subdivisions are of 1 mL. Estimates can be made to the nearest 0.1 mL. When a liquid is in the graduated cylinder, you will see that the level in the cylinder is curved with the lowest point at the center. This is the *meniscus*, or the dividing line between liquid and air. When reading the meniscus for the volume, be sure to read the *lowest* point on the curve and not the upper edge. To avoid errors in reading the meniscus, the eye's line of sight must be perpendicular to the scale (Fig. 2.3). In steps 3 and 4, use the graduated cylinder to see how well the marks on an Erlenmeyer flask and a beaker measure the indicated volume.

**Figure 2.3**
Reading the meniscus
on a graduated cylinder.

82.58 mL – Incorrect
82    mL – Incorrect
82.5  mL – Correct

**3.** Take a 50-mL graduated Erlenmeyer flask (Fig. 2.4) and fill with water to the 50 mL mark. Transfer the water, completely and without spilling, to a 100-mL graduated cylinder. Record the volume on the Report Sheet (4) to the nearest 0.1 mL; convert to L.

**Figure 2.4**
A 50-mL graduated
Erlenmeyer flask.

**4.** Take a 50-mL graduated beaker (Fig. 2.5), and fill with water to the 40-mL mark. Transfer the water, completely and without spilling, to a dry 100-mL graduated cylinder. Record the volume on the Report Sheet (5) to the nearest 0.1 mL; convert to L.

**Figure 2.5**
A 50-mL graduated beaker.

5. What is the error in mL and in percent for obtaining 50.0 mL for the Erlenmeyer flask and 40.0 mL for the beaker (6)? Calculate the % error in the following way: [(volume by graduated cylinder) minus (volume using the mark of the Erlenmeyer or the beaker) divided by (volume by graduated cylinder)] times 100.

6. Which piece of glassware will give you a more accurate measure of liquid: the graduated cylinder, the Erlenmeyer flask, or the beaker (7)?

## Mass: use of the laboratory balance

1. Mass measurements of objects are carried out with the laboratory balance. Many types of balances are available for laboratory use. The proper choice of a balance depends upon what degree of accuracy is needed for a measurement. The standard units of mass are the kilogram (kg) in the SI system and the gram (g) in the metric system. Some conversion factors are listed below.

| | |
|---|---|
| 1 kg = 1000 g | 1 lb. = 454 g |
| 1 g = 1000 mg | 1 oz. = 28.35 g |

Four types of balances are illustrated in Figs. 2.6, 2.8, 2.10, and 2.12. A platform triple beam balance is shown in Fig. 2.6. This balance can weigh objects up to 2610 g. Since the scale is marked in 0.1-g divisions, it is mostly used for rough weighing; weights to 0.01g can be estimated. Figure 2.7 illustrates how to take a reading on this balance.

**Figure 2.6**
A platform triple beam balance.

**Figure 2.7**
Reading on a platform triple beam balance.

461.75 g

(estimated figure)

The single pan, triple beam (Centogram) balance is shown in Fig. 2.8. This Centogram balance has a higher degree of accuracy since the divisions are marked in 0.01-g (estimates can be made to 0.001 g) increments.

**Figure 2.8**
A single pan, triple beam balance (Centogram).

Smaller quantities of material can be weighed on this balance (to a maximum of 311 g). Figure 2.9 illustrates how a reading on this balance would be taken.

**Figure 2.9**
Reading on a single pan, triple beam balance.

Top loading balances show the highest accuracy (Figs. 2.10 and 2.12). Objects can be weighed very rapidly with these balances because the total weight, to the nearest 0.001 g, can be read directly off either an optical scale (Fig. 2.11) or a digital readout (Fig. 2.12). Balances of this type are very expensive and one should be used only after the instructor has demonstrated their use.

**Figure 2.10**
A top loading balance.

**Figure 2.11**
Reading on a top loading balance.

**Figure 2.12**
A top loading balance with a digital readout.

---

**CAUTION!**

When using any balance, never drop an object onto the pan; place it gently in the center of the pan. Never place chemicals directly on the pan; use either a glass container (watch glass, beaker, weighing bottle) or weighing paper. Never weigh a hot object; hot objects may mar the pan. Buoyancy effects will cause incorrect weights. Clean up any chemical spills in the balance area to prevent damage to the balance.

---

2. Weigh a quarter, a test tube (100 × 13 mm), and a 125-mL Erlenmeyer flask. Express each weight to the proper number of significant figures. Use a platform triple beam balance, a single pan, triple beam balance (Centogram), and a top loading balance for these measurements. Use the table on the Report Sheet to record each weight.

3. The single pan, triple beam balance (Centogram) (Fig. 2.8) is operated in the following way.

a. Place the balance on a level surface; use the leveling foot to level.

b. Move all the weights to the zero position at left.

c. Release the beam lock.

d. The pointer should swing freely in an equal distance up and down from the zero or center mark on the scale. Use the zero adjustment to make any correction to the swing.

e. Place the object on the pan (remember the caution).

f. Move the weight on the middle beam until the pointer drops; make sure the weight falls into the "V" notch. Move the weight back one notch until the pointer swings up. This beam weighs up to 10 g, in 1-g increments.

g. Now move the weights on the back beam until the pointer drops; again be sure the weight falls into the "V" notch. Move the weight back one notch until the pointer swings up. This beam weighs up to 1 g, in 0.1-g increments.

h. Lastly, move the smallest weight (the cursor) on the front beam until the pointer balances, that is, swings up and down an equal distance from the zero or center mark on the scale. This last beam weighs up to 0.1 g, in 0.01-g increments.

i. The weight of the object on the pan is equal to the weights shown on each of the three beams (Fig. 2.8). Weights to 0.005 g may be estimated.

j. Repeat the movement of the cursor to check your precision.

k. When finished, move the weights to the left, back to zero, and arrest the balance with the beam lock.

### Temperature: use of the thermometer

1. Routine measurements of temperature are done with a thermometer. Thermometers found in chemistry laboratories may use either mercury or a colored fluid as the liquid, and degrees Celsius (°C) as the units of measurement. The fixed reference points on this scale are the freezing point of water, 0°C, and the boiling point of water, 100°C. Between these two reference points, the scale is divided into 100 units, with each unit equal to 1°C. Temperature can be estimated to 0.1°C. Other thermometers use either the Fahrenheit (°F) or the Kelvin (K) temperature scale and use the same reference points, that is, the freezing and boiling points of water. Conversion between the scales can be accomplished using the formulas below.

$$°F = \frac{9}{5}°C + 32.0 \qquad °C = \frac{5}{9}(°F - 32.0) \qquad K = °C + 273.15$$

**EXAMPLE 5**

Convert 37.0°C to °F and K.

$$°F = \frac{9}{5}(37.0°C) + 32.0 = 98.6°F$$
$$K = 37.0°C + 273.15 = 310.2 \text{ K}$$

2. Use the thermometer in your kit and record to the nearest 0.1°C the temperature of the laboratory at *room temperature*. Use the Report Sheet to record your results.

3. Record the temperature of boiling water. Set up a 250-mL beaker containing 100 mL water, and heat on a hot plate until boiling. Hold the thermometer in the boiling water for at least 1 min. before reading the temperature (*be sure not to touch the sides of the beaker*). Using the Report Sheet, record your results to the nearest 0.1°C.

4. Record the temperature of ice water. Into a 250-mL beaker, add enough crushed ice to fill halfway. Add distilled water to the level of the ice. Stir the ice water gently with a glass rod for 1 min. before reading the thermometer. Hold the thermometer in the ice water for at least 1 min. before reading the temperature. *Use caution; be careful not to touch the walls of the beaker with the thermometer or to hit the thermometer with the glass rod.* Read the thermometer to the nearest 0.1°C. Record your results on the Report Sheet.

---

**CAUTION!**

When reading the thermometer, do not hold the thermometer by the bulb. Body temperature will give an incorrect reading. If you are using a mercury thermometer and the thermometer should break accidentally, call the instructor for proper disposal of the mercury. Mercury is toxic and very hazardous to your health. Do not handle the liquid or breathe its vapor.

---

5. Convert your answers to questions 2, 3, and 4 into °F and K.

---

**Chemicals and Equipment**

1. 50-mL graduated beaker
2. 50-mL graduated Erlenmeyer flask
3. 100-mL graduated cylinder
4. Meterstick or ruler
5. Quarter
6. Balances
7. Hot plates

---

## Experiment 2

### PRE-LAB QUESTIONS

1. A standardized weight obtained from the National Bureau of Standards has the value of 20.000 g. When it was used to calibrate a student's top loading balance, the balance showed the following readings: 19.995, 20.005, 20.001. Comment on the accuracy and precision of the balance.

2. Can one get an accurate determination of the weight of a hot beaker? Explain.

3. Name the type of balance you would use for the following determinations:

   a. A rough weight of 65 g _____.

   b. An accurate weight of 65.356 g _____.

4. Name the piece of laboratory equipment you would use to find

   a. the length of a table _____.

   b. the mass (or weight) of a glass marble _____.

   c. the volume of water in a drinking glass _____.

5. Solve the following problems and record the answers to the proper number of significant figures.

   a. $11.10 - 3.5 =$

   b. $1.04 + 5.438 + 8.102 =$

   c. $6.15 \div 1.2 =$

   d. $2.62 \times 21.43 =$

## Experiment 2

## REPORT SHEET

### *Length*

1. Length _____ in.     _____ cm

   Width _____ in.     _____ cm

2. Length _____ mm     _____ m

   Width _____ mm     _____ m

3. Area _____ in$^2$     _____ cm$^2$     _____ mm$^2$
   (Show calculations)

### *Volume*

4. Erlenmeyer flask _____ mL     _____ L

5. Beaker _____ mL     _____ L

6. Error in volume:
   Erlenmeyer flask _____ mL     _____ %

   Beaker _____ mL     _____ %

$$\% \text{ Error} = \frac{\text{Error in volume}}{\text{Total volume}} \times 100$$

## Mass

| OBJECT | Platform g | Platform mg | Centogram g | Centogram mg | Top Loading g | Top Loading mg |
|---|---|---|---|---|---|---|
| Quarter | | | | | | |
| Test tube | | | | | | |
| 125-mL Erlenmeyer | | | | | | |

## Temperature

| | °C | °F | K |
|---|---|---|---|
| Room temperature | | | |
| Ice water | | | |
| Boiling water | | | |

How well do your thermometer readings agree with the accepted values for the freezing point and boiling point of water? Express any discrepancy as a deviation in degrees.

## POST-LAB QUESTIONS

1. A student measured a book's cover and recorded the length as 28.1 cm and the width as 22.7 cm. The area, using a calculator, was reported to be 637.87 cm$^2$. What should have been reported by the student? Explain your answer.

2. A man weighed 152 lbs. Express the quantity in kilograms. Show your work.

3. Pike's Peak in Colorado is the 32nd highest peak in the world at 14,110 ft. in height. Express this height in meters (m) and kilometers (km). Show your work.

4. A 3.456 g sample of iron filings was added to a beaker weighing 83.65 g. What is the combined weight of the paper and sample in grams (g) and milligrams (mg). Show your work.

5. Two students each weighed a 125-ml Erlenmeyer flask that had a true weight of 80.562 g. Each student recorded three weights for the flask and took an average. Below are the results.

Which set of results is more accurate? _____.

Which set of results is more precise? _____.

|  | Student A | Student B |
|---|---|---|
|  | 80.560 | 80.400 |
|  | 80.555 | 79.551 |
|  | 80.565 | 80.793 |
| Average |  |  |

What can be said of the results from Student A and Student B?

# Density determination

Samples of matter can be identified by using characteristic physical properties. A substance may have a unique color, odor, melting point, or boiling point. These properties do not depend on the quantity of the substance and are called *intensive properties*. Density also is an intensive property and may serve as a means for identification.

The *density* of a substance is the *ratio of its mass per unit volume*. Density can be found mathematically by dividing the mass of a substance by its volume. The formula is $\mathbf{d} = \frac{\mathbf{m}}{\mathbf{V}}$, where $\mathbf{d}$ is density, $\mathbf{m}$ is mass, and $\mathbf{V}$ is volume. While mass and volume do depend on the quantity of a substance (these are *extensive properties*), the ratio is constant at a given temperature. The units of density, reported in standard references, is in terms of g/mL (or g/cc or g/cm$^3$) at 20°C. The temperature is reported since the volume of a sample will change with temperature and, thus, so does the density.

---

**EXAMPLE**

A bank received a yellow bar, marked gold, of mass 453.6 g, and volume 23.5 cm$^3$. Is it gold? (Density of gold = 19.3 g/cm$^3$ at 20°C.)

$$d = \frac{m}{V} = \frac{453.6 \text{ g}}{23.5 \text{ cm}^3} = 19.3 \text{ g/cm}^3$$

Yes, it is gold.

---

**Objectives**

1. To determine the densities of regular- and irregular-shaped objects and use them as a means of identification.
2. To determine the density of water.
3. To determine the density of a small irregular-shaped object by flotation technique.

### Density of a Regular-Shaped Object

1. Obtain a solid block from the instructor. Record the code number.

2. Using your metric ruler, determine the dimensions of the block (length, width, height) and record the values to the nearest 0.05 cm (1). Calculate the volume of the block (2). Repeat the measurements for a second trial.

3. Using a single pan, triple beam balance (Centogram) or a top loading balance (if available), determine the mass of the block (3). Record the mass to the nearest 0.001 g. Calculate the density of the block (4). Repeat the measurements for a second trial.

### Density of an Irregular-Shaped Object

1. Obtain a sample of unknown metal from your instructor. Record the code number.

2. Obtain a mass of the sample of approximately 5 g. Be sure to record the exact quantity to the nearest 0.001 g (5).

3. Choose either a 10-mL or a 25-mL graduated cylinder. For small metal pieces, you should use the 10-mL graduated cylinder. For large metal pieces, you should use the 25-mL graduated cylinder. Fill the graduated cylinder approximately halfway with water. Record the exact volume to the nearest 0.05 mL (6). [Depending on the number of subdivisions between milliliter divisions, you can read to the hundredth's place. For example, you can read the 10-mL graduated cylinder to the hundredth's place where only a 0 (the reading is directly on a subdivision, e.g. 0.00) or a 5 (anywhere between two adjacent subdivisions, e.g. 0.05) can be used.]

4. Place the metal sample into the graduated cylinder. (If the pieces of metal are too large for the opening of the 10-mL graduated cylinder, use the larger graduated cylinder.) Be sure all of the metal is below the water line. Gently tap the sides of the cylinder with your fingers to ensure that no air bubbles are trapped in the metal. Read the new level of the water in the graduated cylinder to the nearest 0.05 mL (7). Assuming that the metal does not dissolve or react with the water, the difference between the two levels represents the volume of the metal sample (8) (Fig. 3.1).

**Figure 3.1**
Measurement of volume of
an irregular-shaped object.

**5.** Carefully recover the metal sample and dry it with a paper towel. Repeat the experiment.

**6.** Calculate the density of the metal sample from your data (9). Determine the average density from your trials, reporting to the proper number of significant figures.

**7.** Determine the identity of your metal sample by comparing its density to the densities listed in Table 3.1 (10).

| Table **3.1** | Densities of Selected Metals | |
|---|---|---|
| **Sample** | **Formula** | **Density (g/cm$^3$)** |
| Aluminum | Al | 2.70 |
| Iron | Fe | 7.86 |
| Tin (white) | Sn | 7.29 |
| Zinc | Zn | 7.13 |
| Lead | Pb | 11.30 |

**8.** Recover your metal sample and return it as directed by your instructor.

## Use of the Spectroline Pipet Filler

**1.** Examine the Spectroline pipet filler and locate the valves marked "A," "S," and "E" (Fig. 3.2). These operate by pressing the flat surfaces between the thumb and forefinger.

**2.** Squeeze the bulb with one hand while you press valve "A" with two fingers of the other hand. The bulb flattens as air is expelled. If you release your fingers when the bulb is flattened, the bulb remains collapsed.

**Figure 3.2**
The Spectroline pipet filler.

Aspirate (air)

Suction

Empty

**3.** Carefully insert the pipet end into the Spectroline pipet filler (Fig. 3.3). The end should insert easily and not be forced.

**4.** Place the tip of the pipet into the liquid to be pipetted. Make sure that the tip is below the surface of the liquid at all times.

**Figure 3.3**
Using the Spectroline pipet
filler to pipet.

Etched mark

5. With your thumb and forefinger, press valve "S." Liquid will be drawn up into the pipet. By varying the pressure applied by your fingers, the rise of the liquid into the pipet can be controlled. Allow the liquid to fill the pipet to a level slightly above the etched mark on the stem. Release the valve; the liquid should remain in the pipet.

6. Withdraw the pipet from the liquid. Draw the tip of the pipet lightly along the wall of the beaker to remove excess liquid.

7. Adjust the level of the meniscus of the liquid by carefully pressing valve "E." The level should lower until the curved meniscus touches the etched mark (Fig. 3.4). Carefully draw the tip of the pipet lightly along the wall of the beaker to remove excess liquid.

**Figure 3.4**
Adjusting the curved meniscus
of the liquid to the etched mark.

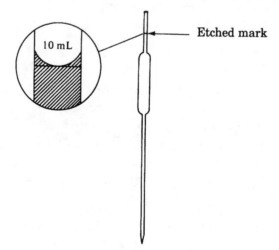

10 mL

Etched mark

8. Drain the liquid from the pipet into a collection flask by pressing valve "E." Remove any drops on the tip by touching the tip of the pipet against the inside walls of the collection flask. Liquid should remain inside the tip; the pipet is calibrated with this liquid in the tip.

## Density of Water

1. Obtain approximately 50 mL of distilled water from your instructor. Record the temperature of the water (11).

2. Take a clean, dry 50-mL beaker; weigh to the nearest 0.001 g (12).

3. With a 10-mL volumetric pipet, transfer 10.00 mL of distilled water into the preweighed beaker using a Spectroline pipet filler (Fig. 3.3). (Before transferring the distilled water, be sure there are no air bubbles trapped in the volumetric pipet. If there are, gently tap the pipet to dislodge the air bubbles, and then refill to the line.) Immediately weigh the beaker and water and record the weight to the nearest 0.001 g (13). Calculate the weight of the water by subtraction (14). Calculate the density of the water at the temperature recorded (15).

---

### CAUTION!

Never use your mouth when pipetting.

---

4. Repeat step no. 3 for a second trial. Be sure all the glassware used is clean and dry.

5. Calculate the average density (16). Compare your average value at the recorded temperature to the value reported for that temperature in a standard reference.

## Density of a Small Irregular-Shaped Object by Flotation Technique

1. Obtain two small (2-mm) plastic chips from your instructor.

2. Place a 100-mL graduated cylinder containing a small magnetic spin-bar on a magnetic stirrer. Add 30 mL of acetone and begin to stir the liquid slowly. Add the plastic chips to the liquid. Stop the stirring and note that the chips will sink to the bottom.

3. With slow intermittent stirring, add water with an eye dropper. Watch the plastic chips as you add the water; see if they rise or stay on the bottom. (If air bubbles appear, take a glass stirring rod and stir slowly, with the magnetic stirrer off, to dislodge bubbles from the chips.) If the chips stay on the bottom, keep adding more drops of water until the chips rise and float in the middle of the liquid. At this point, when the chips are floating in the middle of the liquid, *the liquid has the same density as that of the plastic chips.* (If you add too much water, the chips will float on the surface. If this happens, add acetone back to the solution, dropwise, until the chips float in the middle.)

4. Weigh a clean and dry 50-mL beaker to the nearest 0.001 g. Record the weight on your Report Sheet (17).

5. Using a Spectroline pipet filler (Fig. 3.3), transfer exactly 10.00 mL of liquid from the graduated cylinder to the beaker. Weigh to the nearest 0.001 g (18), and by subtraction determine the weight of the liquid. Record it on your Report Sheet (19).

**6.** Repeat step 5 for a second trial. Be sure all the glassware used is clean and dry.

**7.** Calculate the density of the liquid, and hence the density of the plastic chips (20). Determine the average density of the plastic chips.

---

### Chemicals and Equipment

1. Magnetic spin-bar
2. Magnetic stirrer
3. Spectroline pipet filler
4. 10-mL volumetric pipet
5. Solid wood block
6. Aluminum
7. Lead
8. Tin
9. Zinc
10. Polyethylene plastic chips
11. Acetone

---

## Experiment 3

### PRE-LAB QUESTIONS

1. Identify the following characteristics as either an *intensive* property or an *extensive* property.

    **a.** Melting point _____.

    **b.** Color _____.

    **c.** Volume _____.

    **d.** Mass _____.

    **e.** Density _____.

2. What are the units of density?

3. Why is it necessary to state the temperature with a density value?

4. If exactly 100.0 mL of a liquid has a mass of 80.60 g, what is its density? Show your work.

5. A prospector panned some 'yellow sand' from a stream. The sand weighed 405 g and had a volume of 21.0 cm$^3$. Was the sand pure gold or only "fool's gold" (iron pyrite)? (The density of gold is 19.3 g/cm$^3$ and that of pyrite is 5.0 g/cm$^3$ at 20°C.) Show your work to justify your answer.

## Experiment 3

## REPORT SHEET

Report all measurements and calculations to the correct number of significant figures.

| *Density of a regular-shaped object* | Trial 1 | Trial 2 |
|---|---|---|
| Unknown code number _____ | | |
| 1. Length | _____ cm | _____ cm |
| Width | _____ cm | _____ cm |
| Height | _____ cm | _____ cm |
| 2. Volume (L × W × H) | _____ cm$^3$ | _____ cm$^3$ |
| 3. Mass | _____ g | _____ g |
| 4. Density: (3)/(2) | _____ g/cm$^3$ | _____ g/cm$^3$ |
| Average density of block | | _____ g/cm$^3$ |

| *Density of an irregular-shaped object* | Trial 1 | Trial 2 |
|---|---|---|
| Unknown code number _____ | | |
| 5. Mass of metal sample | _____ g | _____ g |
| 6. Initial volume of water | _____ mL | _____ mL |
| 7. Final volume of water | _____ mL | _____ mL |
| 8. Volume of metal: (7) − (6) | _____ mL | _____ mL |
| 9. Density of metal: (5)/(8) | _____ g/mL | _____ g/mL |
| Average density of metal | | _____ g/mL |

10. Identity of unknown metal _____

**Density of water**

| | Trial 1 | Trial 2 |
|---|---|---|
| **11.** Temperature of water | _____ °C | _____ °C |
| **12.** Weight of 50-mL beaker | _____ g | _____ g |
| Volume of water | 10.00 mL | 10.00 mL |
| **13.** Weight of beaker and water | _____ g | _____ g |
| **14.** Weight of water: (13) − (12) | _____ g | _____ g |
| **15.** Density of water: (14)/10.00 mL | _____ g/mL | _____ g/mL |
| **16.** Average density of water | | _____ g/mL |
| Density found in literature | | _____ g/mL |

**Density by flotation technique**

| | Trial 1 | Trial 2 |
|---|---|---|
| **17.** Weight of 50-mL beaker | _____ g | _____ g |
| Volume of liquid | 10.00 mL | 10.00 mL |
| **18.** Weight of beaker and liquid | _____ g | _____ g |
| **19.** Weight of liquid: (18) − (17) | _____ g | _____ g |
| **20.** Density of liquid: (19)/10.00 mL | _____ g/mL | _____ g/mL |
| Average density of plastic chips | | _____ g/mL |

## POST-LAB QUESTIONS

**1.** All the metals listed in Table 3.1 can have their densities determined by the 'displacement-of-a-liquid' method, as in the **Procedure.** Why can you use this method?

**2.** Olive oil (density 0.918 g/cm$^3$) separates from the water-vinegar in a salad dressing and is visible as a layer. Is the olive oil the top or the bottom layer? Explain your answer.

3. Ethyl alcohol has a density of 0.791 g/cm³ at 20°C. How many milliliters (mL) are needed to have 30.0 g of liquid? Show your work.

4. A student uses a volumetric pipet calibrated to deliver exactly 50.0 mL of a liquid and then measures the mass of the liquid. How is the density determination of the liquid affected in the situations which follow? Would the density increase, decrease, or not be affected? Explain your answer.

   a. A dirty pipet is used and water droplets adhere to the inner walls of the pipet.

   b. The student did not allow sufficient time for all the liquid to empty from the pipet.

   c. The student blew out all the liquid from the pipet, including the small amount from the tip.

   d. The student did not remove air bubbles from the pipet before delivering the liquid.

   e. At the mark on the pipet, the student read the upper edge of the meniscus, not the lowest point on the curve.

5. Iron should sink in water since its density is greater than that of water (see Table 3.1). However, ships (for example, tankers and cruise ships) float even though their hulls are built of steel, an iron alloy. Explain why this is possible.

# The separation of the components of a mixture

## Background

Mixtures are not unique to chemistry; we use and consume them on a daily basis. The beverages we drink each morning, the fuel we use in our automobiles, and the ground we walk on are mixtures. Very few materials we encounter are pure. Any material made up of two or more substances that are not chemically combined is a mixture.

The isolation of pure components of a mixture requires the separation of one component from another. Chemists have developed techniques for doing this. These methods take advantage of the differences in physical properties of the components. The techniques to be demonstrated in this laboratory are the following:

1. *Sublimation*. This involves heating a solid until it passes directly from the solid phase into the gaseous phase. The reverse process, when the vapor goes back to the solid phase without a liquid state in between, is called condensation or deposition. Some solids which sublime are iodine, caffeine, and *para*-dichlorobenzene (mothballs).

2. *Extraction*. This uses a solvent to selectively dissolve one component of the solid mixture. With this technique, a soluble solid can be separated from an insoluble solid.

3. *Decantation*. This separates a liquid from an insoluble solid sediment by carefully pouring the liquid from the solid without disturbing the solid (Fig. 4.1).

**Figure 4.1**
Decantation.

4. *Filtration*. This separates a solid from a liquid through the use of a porous material as a filter. Paper, charcoal, or sand can serve as a filter. These materials allow the liquid to pass through but not the solid (see Fig. 4.4 in the **Procedure**).

5. *Evaporation*. This is the process of heating a mixture in order to drive off, in the form of vapor, a volatile liquid, so as to make the remaining component dry.

The mixture that will be separated in this experiment contains three components: naphthalene, $C_{10}H_8$, common table salt, NaCl, and sea sand, $SiO_2$. The separation will be done according to the scheme in Fig. 4.2 by

1. heating the mixture to sublime the naphthalene,

2. dissolving the table salt with water to extract, and

3. evaporating water to recover dry NaCl and sand.

**Figure 4.2** • Separation scheme.

## Procedure

1. Obtain a clean, dry 150-mL beaker and carefully weigh it to the nearest 0.001 g. Record this weight for beaker 1 on the Report Sheet (1). Obtain a sample of the unknown mixture from your instructor; use a mortar and pestle to grind the mixture into a fine powder. With the beaker still on the balance, carefully transfer

approximately 2 g of the unknown mixture into the beaker. Record the weight of the beaker with the contents to the nearest 0.001 g (2). Calculate the exact sample weight by subtraction (3).

2. Place an evaporating dish on top of the beaker containing the mixture. Place the beaker and evaporating dish on a wire gauze with an iron ring and ring stand assembly as shown in Fig. 4.3. Place ice in the evaporating dish, being careful not to get any water on the underside of the evaporating dish or inside the beaker.

**Figure 4.3**
Assembly for sublimation.

3. Carefully heat the beaker with a Bunsen burner, increasing the intensity of the flame until vapors appear in the beaker. A solid should collect on the underside of the evaporating dish. After 10 min. of heating, remove the Bunsen burner from under the beaker. Carefully remove the evaporating dish from the beaker and collect the solid by scraping it off the dish with a spatula. Drain away any water from the evaporating dish and add ice to it, if necessary. Stir the contents of the beaker with a glass rod. Return the evaporating dish to the beaker and apply the heat again. Continue heating and scraping off solid until no more solid collects. Discard the naphthalene into a special container provided by your instructor.

4. Allow the beaker to cool until it reaches room temperature. Weigh the beaker with the contained solid (4). Calculate the weight of the naphthalene that sublimed (5).

5. Add 25 mL of distilled water to the solid in the beaker. Heat and stir for 5 min.

6. Weigh a second clean, dry 150-mL beaker with 2 or 3 boiling chips, to the nearest 0.001 g (6).

7. Assemble the apparatus for gravity filtration as shown in Fig. 4.4.

**Figure 4.4**
Gravity filtration.

Residue

Funnel tip should touch the
beaker in such a way that
filtrate will run down the
wall of the beaker

Filtrate

8. Fold a piece of filter paper following the technique shown in Fig. 4.5.

**Figure 4.5**
Steps for folding a filter
paper for gravity filtration.

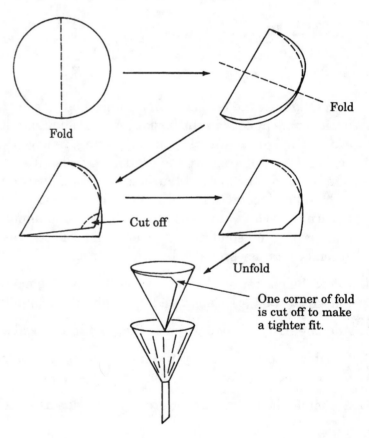

Fold

Fold

Cut off

Unfold

One corner of fold
is cut off to make
a tighter fit.

9. Wet the filter paper with water and adjust the paper so that it lies flat on the glass of the funnel.

10. Position the second beaker under the funnel.

11. Pour the mixture through the filter, first decanting most of the liquid into beaker 2, and then carefully transferring the wet solid into the funnel with a rubber policeman. Collect all the liquid (called the filtrate) in beaker 2.

12. Rinse beaker 1 with 5–10 mL of water, pour over the residue in the funnel, and add the liquid to the filtrate; repeat with an additional 5–10 mL of water.

13. Place beaker 2 and its contents on a wire gauze with an iron ring and ring stand assembly as shown in Fig. 4.6a. Begin to heat gently with a Bunsen burner. Control the flame in order to prevent boiling over. As the volume of liquid is reduced, solid sodium chloride will appear. Reduce the flame to avoid bumping of the solution and spattering of the solid. When all of the liquid is gone, cool the beaker to room temperature. Weigh the beaker, chips, and the solid residue to the nearest 0.001 g (7). Calculate the weight of the recovered NaCl by subtraction (8).

a. Evaporation of a volatile
liquid from a solution.

b. Heating a solid
to dryness.

**Figure 4.6** • Assembly for evaporation.

14. Carefully weigh a third clean, dry 150-mL beaker to the nearest 0.001 g (9). Transfer the sand from the filter paper to beaker 3. Heat the sand to dryness in the beaker with a burner, using the ring stand and assembly shown in Fig. 4.6b (or use an oven at T = 90–100°C, if available). Heat carefully to avoid spattering; when dry, the sand should be freely flowing. Allow the sand to cool to room temperature. Weigh the beaker and the sand to the nearest 0.001 g (10). Calculate the weight of the recovered sand by subtraction (11).

15. Calculate

   **a.** Percentage yield using the formula:

   $$\% \text{ yield} = \frac{\text{grams of solid recovered}}{\text{grams of initial sample}} \times 100$$

**b.** Percentage of each component in the mixture by using the formula:

$$\% \text{ component} = \frac{\text{grams of component isolated}}{\text{grams of initial sample}} \times 100$$

---

### EXAMPLE

A student isolated the following from a sample of 1.132 g:

> 0.170 g of naphthalene
> 0.443 g of NaCl
> 0.499 g of sand
> _____
> 1.112 g solid recovered

The student calculated the percentage yield and percentage of each component as follows:

$$\% \text{ yield} = \frac{1.112 \text{ g (solid recovered)}}{1.132 \text{ g (original sample)}} \times 100 = 98.2\%$$

$$\% \text{ C}_{10}\text{H}_8 = \frac{0.170 \text{ g (naphthalene)}}{1.132 \text{ g (original sample)}} \times 100 = 15.0\%$$

$$\% \text{ NaCl} = \frac{0.443 \text{ g (NaCl)}}{1.132 \text{ g (original sample)}} \times 100 = 39.1\%$$

$$\% \text{ sand} = \frac{0.499 \text{ g (sand)}}{1.132 \text{ g (original sample)}} \times 100 = 44.1\%$$

---

### Chemicals and Equipment

1. Unknown mixture
2. Balances
3. Boiling chips
4. Evaporating dish, 6 cm
5. Filter paper, 15 cm
6. Mortar and pestle
7. Oven (if available)
8. Ring stands (3)
9. Rubber policeman

## Experiment 4

### PRE-LAB QUESTIONS

**1.** A coffee drink is made by passing hot water through ground coffee beans in a coffee maker. Is the drink a mixture? What separation technique was used in the preparation of the drink?

**2.** Both sublimation and evaporation bring material into the vapor phase. How do the two methods of separation differ?

**3.** Do the five methods described in the **Background** involve physical changes or chemical changes of matter? Explain.

**4.** Both naphthalene and *para*-dichlorobenzene are effective moth repellents. What property of these chemicals allow them to be used in mothballs for the protection of woollen clothing?

## Experiment 4

**REPORT SHEET**

1. Weight of beaker 1                             _____ g

2. Weight of beaker 1 and mixture           _____ g

3. Weight of mixture: (2) − (1)              _____ g

4. Weight of beaker 1 and solid after sublimation    _____ g

5. Weight of naphthalene: (2) − (4)         _____ g

6. Weight of beaker 2                             _____ g

7. Weight of beaker 2 and NaCl             _____ g

8. Weight of NaCl: (7) − (6)                _____ g

9. Weight of beaker 3                             _____ g

10. Weight of beaker 3 and sand            _____ g

11. Weight of sand: (10) − (9)             _____ g

*Calculations*

12. Weight of recovered solids:

   (5) + (8) + (11)                      _____ g

13. Percentage yield (percentage of solids **recovered**):

   % = [(12)/(3)] × 100             _____ %

14. Percentage of naphthalene:

   % = [(5)/(3)] × 100              _____ %

15. Percentage of NaCl:

   % = [(8)/(3)] × 100              _____ %

16. Percentage of sand:

   % = [(11)/(3)] × 100            _____ %

## POST-LAB QUESTIONS

1. The weight of naphthalene was obtained by weighing the collected solid. Was this an accurate weight of the naphthalene present in the original sample? Explain your answer.

2. In step 14 the sand was not completely free of water when the weight was reported. How does this affect the percent composition of the sand in the mixture?

3. In your experiment did you recover 100% of the original mixture? What was your percent recovery?

4. Snow in the winter can slowly disappear even when the temperature is freezing. How do you account for this observation?

5. A sample of French fried potatoes weighing 150 g was extracted with the volatile organic solvent hexane. The recovered cooking oil weighed 12.3 g. What methods from this experiment could be used to separate and isolate the cooking oil after the extraction? What was the percent oil in the potatoes? Show your calculations.

# Experiment 5

## Resolution of a mixture by distillation

Distillation is one of the most common methods of purifying a liquid. It is a very simple method: a liquid is brought to a boil, the liquid becomes a gas, the gas condenses and returns to the liquid state, and the liquid is collected.

Everyone has had an opportunity to heat water to a boil. As heat is applied, water molecules increase their kinetic energy. Some molecules acquire sufficient energy to escape from the liquid phase and enter into the vapor phase. The vapor above the liquid exerts a pressure, called the vapor pressure. As more and more molecules obtain enough energy to escape into the vapor phase, the vapor pressure of these molecules increases. Eventually the vapor pressure equals the pressure exerted externally on the liquid (this external pressure usually is caused by the atmosphere). Boiling occurs when this condition is met, and the temperature where this occurs is called the boiling point.

In distillation, the process described is carried out in an enclosed system, such as is illustrated in Fig 5.2. The liquid in the boiling flask is heated to a boil, and the vapor rises through tubing. The vapor then travels into a tube cooled by water, which serves as a condenser, where the vapor returns to the liquid state. If the mixture has a low-boiling component (a volatile substance with a high vapor pressure), it will distill over first and can be collected. Higher-boiling and nonvolatile components (substances with low vapor pressure) remain in the boiling flask. Only by applying more heat will the higher-boiling component be distilled. Nonvolatile substances will not distill. For example, pure or "distilled" water for steam irons or car batteries is prepared this way.

Normal distillations, procedures carried out at atmospheric pressure, require "normal" boiling points. However, when boiling takes place in a closed system, it is possible to change the boiling point of the liquid by changing the pressure in the closed system. If the external pressure is reduced, usually by using a vacuum pump or a water aspirator, the boiling point of the liquid is reduced. Thus, heat-sensitive liquids, some of which decompose when boiled at atmospheric pressure, distill with minimum decomposition at reduced pressure and temperature. The relation of temperature to vapor pressure for the organic compound aniline can be shown by the curve in Fig 5.1. The organic liquid aniline, $C_6H_5NH_2$, a compound used to make synthetic dyes, can be distilled at 184°C (760 mm Hg) or at 68°C (10 mm Hg).

## Figure 5.1
Temperature–vapor pressure curve for aniline.

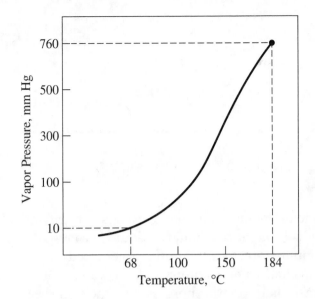

## Objectives

1. To use distillation to separate a mixture.
2. To show that distillation can purify a liquid.

## Procedure

1. In this experiment a salt–water mixture will be separated by distillation. The volatile water will be separated from the nonvolatile salt (sodium chloride, NaCl). The purity of the collected distilled water will be demonstrated by chemical tests specific for sodium ions ($Na^+$) and chloride ions ($Cl^-$).

2. Assemble an apparatus as illustrated in Fig 5.2. A kit containing the necessary glassware can be obtained from your instructor. The glassware contains standard taper joints, which allow for quick assembly and disassembly. Before fitting the pieces together, apply a light coating of silicone grease to each joint to prevent the joints from sticking.

3. Use 100-mL round-bottom flasks for the boiling flask and the receiving flask. Fill the boiling flask with 50 mL of the prepared salt–water mixture. Add two boiling chips to the boiling flask to ensure smooth boiling of the mixture and to prevent bumping. Be sure that the rubber tubing to the condenser enters the lower opening and empties out of the upper opening. Turn on the water faucet and allow the water to fill the jacket of the condenser slowly, so as not to trap air. Take care not to provide too much flow, otherwise the hoses will disconnect from the condenser. Adjust the bulb of the thermometer to below the junction of the condenser and the distillation column. **Be sure that the opening of the vacuum adapter is open to the atmosphere.**

**Figure 5.2** • A distillation apparatus.

4. Gently heat the boiling flask with a Bunsen burner. Eventually the liquid will boil, vapors will rise and enter the condenser, and liquid will recondense and be collected in the receiving flask.

5. Discard the first 1 mL of water collected. Record the temperature of the vapors as soon as the 1 mL of water has been collected. Continue collection of the distilled water until approximately one-half of the mixture has distilled. Record the temperature of the vapors at this point. Turn off the Bunsen burner and allow the system to return to room temperature.

6. The distilled water and the liquid in the boiling flask will be tested.

7. Place in separate clean, dry test tubes (100 × 13 mm) 2 mL of distilled water and 2 mL of the residue liquid from the boiling flask. Add to each sample 5 drops of silver nitrate solution. Look for the appearance of a white precipitate. Record your observations. Silver ions combine with chloride ions to form a white precipitate of silver chloride.

$$Ag^+ + Cl^- \longrightarrow AgCl(s) \text{ (White precipitate)}$$

8. Place in separate clean, dry test tubes (100 × 13 mm) 2 mL of distilled water and 2 mL of the residue liquid from the boiling flask. Obtain a clean nickel wire from your instructor. In the hood, dip the wire into concentrated nitric acid and hold the wire in a Bunsen burner flame until the yellow color in the flame disappears. Dip the wire into the distilled water sample. Put the wire into the Bunsen burner flame. Record the color of the flame. Repeat the above procedure, cleaning the wire, dipping the wire into the liquid from the boiling flask, and observing the color of the Bunsen burner flame. Record your observations. Sodium ions produce a bright yellow flame with a Bunsen burner.

9. Make sure you wipe the grease from the joints before washing the glassware used in the distillation.

### Chemicals and Equipment

1. Boiling chips
2. Bunsen burner
3. Clamps
4. Distillation kit
5. Silicone grease
6. Thermometer
7. Nickel wire
8. Concentrated nitric acid, $HNO_3$
9. Salt–water mixture
10. 0.5 M silver nitrate, 0.5 M $AgNO_3$

## Experiment 5

### PRE-LAB QUESTIONS

1. When does boiling of a liquid occur?



2. What happens to water molecules as heat is applied?



3. In a distillation, how are the vapors collected?



4. Is it possible for a liquid to boil at a temperature other than the "normal" boiling point? How could you do this?



5. Water boils at a lower temperature in Denver, Colorado (the 'Mile-High City') than in New York City. Why?

## Experiment 5

## REPORT SHEET

1. Barometric pressure               _____

2. Boiling point of water at measured pressure    _____

3. Temperature of vapor after collecting 1 mL    _____

4. Temperature of vapor at end of distillation    _____

| Solution | Observation with 0.5 M AgNO$_3$ | Color in Flame Test |
|---|---|---|
| Distilled water | | |
| Liquid in boiling flask | | |

## POST-LAB QUESTIONS

1. What would happen if the opening to air on the vacuum adapter was closed during distillation?

**2.** Why could you not detect chloride ion, $Cl^-$, in the liquid collected in the distilling flask?

**3.** How could you get aniline to distill at 100°C ("normal" b. p. is 184°C)?

**4.** What is the advantage of carrying out a distillation at reduced pressure?

# Experiment 6

## The empirical formula of a compound: the Law of Constant Composition

## Background

One of the most important fundamental observations in chemistry is summarized as the **Law of Constant (or Definite) Composition:** any pure chemical compound is made up of two or more elements in the same proportion by mass. In addition, it does not matter where the compound is found. Consider water from your kitchen tap and water from the Pacific Ocean: both are composed of the same elements, hydrogen and oxygen, and are found in exactly the same proportion—89% oxygen and 11% hydrogen—by weight. We also know that the compound water is composed of 2 atoms of hydrogen and 1 atom of oxygen and has the formula $H_2O$. If we consider that the mass of oxygen is 16 times the mass of hydrogen, water will always be found to contain 89% oxygen and 11% hydrogen.

For example, we can find the exact percentages for water and verify the above numbers. Using the **molecular formula** (the actual number of atoms in each molecule of a compound) of water, $H_2O$, the **gram molecular weight** or **molar mass** (the weight in grams of 1 mole of a compound) can be calculated:

$$2H = 2 \times 1.008 = 2.016$$
$$1O = 1 \times 15.999 = \underline{15.999}$$
$$18.015 = 18.015 \text{ g/mole}$$

The percent composition of each element in water can then be calculated:

$$\%H = \frac{2.016}{18.015} \times 100 = 11.19 = 11\%$$

$$\%O = \frac{15.999}{18.015} \times 100 = 88.81 = 89\%$$

These values are constant and are never found in any other proportion!

The **empirical formula** (the simplest whole number ratio of atoms in a compound) is experimentally the simplest formula of a compound that can be found. For water, the formula, $H_2O$, is both the empirical and the molecular formula. Some other examples are carbon dioxide gas, $CO_2$; methane gas, $CH_4$; and hydrogen chloride gas, HCl. However, for the compound benzene, while the molecular formula is $C_6H_6$, the empirical formula is CH. Another example is the sugar found in honey, fructose: the molecular formula is $C_6H_{12}O_6$ and the empirical formula is $CH_2O$.

The empirical formula of a compound can be determined in a laboratory experiment by finding the ratio between the number of moles of the elements in the compound. The number of moles of each element can be calculated from the experimental values of the weights in which the elements combine by dividing by their corresponding atomic weights. If the molecular weight and the empirical formula of the compound are known, then the molecular formula of the compound can be determined.

## Method

In this experiment we will verify that the empirical formula of copper(II) chloride is $CuCl_2$, and in so doing, demonstrate the Law of Constant Composition. We will do this by reducing a known weight of copper(II) chloride with aluminum to elemental copper. The reaction is shown by the following equation:

$$3CuCl_2 + 2Al \rightarrow 3Cu + 2AlCl_3 \tag{1}$$

From the weight of $CuCl_2$, and the weight of Cu, subtraction will give the weight of Cl. From these weights, the mole ratio of copper to chlorine, the empirical formula, and the percent composition of $CuCl_2$ can then be calculated.

---

### EXAMPLE

Copper(II) chloride, 5.503 g, is reduced by excess aluminum and gives elemental copper, 2.603 g, according to equation (1). Using this data, the following calculations can be made:

1. Weight of chlorine in $CuCl_2$: (5.503 g $CuCl_2$) − (2.603 g Cu) = 2.900 g Cl.

2. Moles of Cu: (2.603 g Cu) $\times \left( \dfrac{1 \text{ mole Cu}}{63.55 \text{ g Cu}} \right)$ = 0.04100 mole Cu.

3. Moles of Cl: (2.900 g Cl) $\times \left( \dfrac{1 \text{ mole Cl}}{35.45 \text{ g Cl}} \right)$ = 0.08181 mole Cl.

4. Mole ratio of Cu to Cl: 0.04100 : 0.08181.

5. Simplest whole number ratio of Cu to Cl:

$$\frac{0.04100}{0.04100} : \frac{0.08181}{0.04100} = 1 : 2$$

6. The empirical formula for copper(II) chloride is $CuCl_2$.

7. %Cu in sample from data: $\dfrac{2.603 \text{ g}}{5.503 \text{ g}} \times 100$ = 47.30%.

The theoretical calculated value of %Cu in $CuCl_2$, using the atomic masses, is 47.27%.

---

### Objectives

1. To calculate the percent composition of an element in a compound.
2. To verify the empirical formula of copper(II) chloride.
3. To illustrate the Law of Constant Composition.

---

## Procedure

1. Weigh out between 5 and 6 g of $CuCl_2$; record the weight to the nearest 0.001 g on your Report Sheet (1). Do not weigh directly on the balance pan, but be sure to use a container or weighing paper.

2. Transfer the $CuCl_2$ to a 250-mL beaker. Add 60 mL of distilled water and stir the contents with a glass stirring rod until the solid is completely dissolved.

3. Obtain a 45-cm length of aluminum wire (approx. 1.5 g). Make a flat coil on one end of the wire, and a handle at the other end. Make the handle long enough so that the wire can be hung over the side of the beaker. The coil must be covered by the solution and should reach the bottom of the beaker (Fig. 6.1).

**Figure 6.1**
Suspension of aluminum wire in $CuCl_2$ solution.

4. As the reaction proceeds, you will see flakes of brown copper accumulating on the wire. Occasionally shake the wire to loosen the copper. The disappearance of the initial blue color of the copper(II) ions indicates that the reaction is complete.

5. Test for the completion of the reaction.

   **a.** With a clean Pasteur pipet, place 10 drops of the supernatant solution into a clean test tube (100 × 13 mm).

   **b.** Add 3 drops of 6 M aqueous ammonia to the test tube. If a dark blue solution appears, copper(II) ions are still present, and the solution should be heated to 60°C for 15 min. (Use a hot plate.)

6. When the supernatant no longer tests for $Cu^{2+}$ ions, the reaction is complete. Shake the aluminum wire so that all the copper clinging to it will fall into the solution. With a wash bottle filled with distilled water, wash the aluminum wire to remove any remaining residual copper. Remove the unreacted aluminum wire from the solution and discard into a solid waste container provided by your instructor.

7. Set up a vacuum filtration apparatus as shown in Fig. 6.2.

**Figure 6.2** • Vacuum filtration setup using the Büchner funnel.

8. Weigh a filter paper that fits into the Büchner funnel to the nearest 0.001 g; record on your Report Sheet (2).

9. Moisten the filter paper with distilled water, turn on the water aspirator, and filter the copper through the Büchner funnel. With a rubber policeman move any residue left in the beaker to the Büchner funnel; then rinse down all the copper in the beaker with water from a wash bottle and transfer to the Büchner funnel. If filtrate is cloudy, refilter, slowly. Finally, wash the copper in the funnel with 30 mL of acetone (to speed up the drying process). Let the copper remain on the filter paper for 10 min. with the water running to further the drying process.

10. Carefully remove the filter paper from the Büchner funnel so as not to tear the paper or lose any copper. Weigh the filter paper and the copper to the nearest 0.001 g and record on your Report Sheet (3). By subtraction obtain the weight of copper (4). From the weight of copper(II) chloride (1) and the weight of copper (4), the weight of chlorine can be calculated in the sample by subtraction (5).

11. From the experimental data, determine the empirical formula of copper(II) chloride, and the error in determining the percent of copper.

---

### Chemicals and Equipment

1. Aluminum wire (no. 18)
2. Acetone
3. 6 M aqueous ammonia, 6 M $NH_3$
4. Copper(II) chloride
5. Filter paper (Whatman no. 2, 7.0 cm)
6. Hot plate
7. Rubber policeman
8. Test tube (100 × 13 mm)
9. Pasteur pipets
10. Vacuum filtration setup
11. Wash bottle

# Experiment 6

## PRE-LAB QUESTIONS

**1.** How is the molar mass calculated?

**2.** How does the empirical formula differ from the molecular formula?

**3.** Given the following molecular formulas, write the empirical formulas.

| Molecular Formula | Empirical Formula |
|---|---|
| $C_2H_6$ | _____ |
| $C_3H_6O_3$ | _____ |
| $B_2H_6$ | _____ |

**4.** Calculate the percentage by weight of Cl in $AlCl_3$. Show your work.

## Experiment 6

**REPORT SHEET**

1. Weight of copper(II) chloride $CuCl_2$ _____ g

2. Weight of filter paper _____ g

3. Weight of filter paper and copper, Cu _____ g

4. Weight of Cu: (3) − (2) _____ g

5. Weight of Cl in sample: (1) − (4) _____ g

6. Molar mass of Cu _____ g

7. Molar mass of Cl _____ g

8. Number of moles of Cu atoms in sample: (4)/(6) _____ moles

9. Number of moles of Cl atoms in sample: (5)/(7) _____ moles

10. Mole ratio of Cu atoms to Cl atoms: (8):(9) _____

11. Simple whole number mole ratio of Cu atoms to Cl atoms _____

12. Empirical formula for copper(II) chloride _____

13. Percentage of Cu in sample: % = [(4)/(1)] × 100 _____ %

14. Actual percentage of Cu in $CuCl_2$:

$$\% = \frac{(6)}{(6) + [2 \times (7)]} \times 100$$
_____ %

15. Percentage error:

$$\% = \frac{(14) - (13)}{(14)} \times 100$$
_____ %

## POST-LAB QUESTIONS

1. How would the following affect the accuracy of your determination of the percentage composition of copper?

   a. The aluminum wire broke into pieces and was not separated from the copper metal.

   b. Although the solution gave an intense blue color when treated with $NH_3$, a student continued the experiment without heating but just filtered the solution.

2. Does the experiment enable you to determine the weight of chlorine in $CuCl_2$ by direct measurement?

3. Copper(I) chloride has the molecular formula $Cu_2Cl_2$. What is the empirical formula? Calculate the weight of one mole of material.

4. In an experiment 12.072 g of tin(IV) chloride, $SnCl_4$, yielded 5.35 g of tin metal. Assuming that the original sample was pure, how much should be obtained? Find the percent error. (Your answer should be to the correct number of significant figures.)

# Experiment 7

## Determination of the formula of a metal oxide

### Background

Through the use of chemical symbols and numerical subscripts, the formula of a compound can be written. The simplest formula that may be written is the *empirical formula*. In this formula, the subscripts are in the form of the simplest whole number ratio of the atoms in a molecule or of the ions in a formula unit. The *molecular formula*, however, represents the actual number of atoms in a molecule. For example, although $CH_2O$ represents the empirical formula of the sugar, glucose, $C_6H_{12}O_6$, represents the molecular formula. For water, $H_2O$, and carbon dioxide, $CO_2$, the empirical and the molecular formulas are the same. Ionic compounds are generally written as empirical formulas only; for example, common table salt is NaCl.

The formation of a compound from pure components is independent of the source of the material or of the method of preparation. If elements chemically react to form a compound, they always combine in definite proportions by weight. This concept is known as the *Law of Constant Composition*.

If the weight of each element that combines in an experiment is known, then the number of moles of each element can be determined. The empirical formula of the compound formed is the ratio between the number of moles of elements in the compound. This can be illustrated by the following example. If 32.06 grams of sulfur is burned in the presence of 32.00 grams of oxygen, then 64.06 grams of sulfur dioxide results. Thus

$$\frac{32.06 \text{ g S}}{32.06 \text{ g/mole S}} = 1 \text{ mole of sulfur}$$

$$\frac{32.00 \text{ g O}}{16.00 \text{ g/mole O}} = 2 \text{ moles of oxygen}$$

and the mole ratio of sulfur : oxygen is 1 : 2. The empirical formula of sulfur dioxide is $SO_2$. This also is the molecular formula.

In this experiment, the moderately reactive metal, magnesium, is combined with oxygen. The oxide, magnesium oxide, is formed. The equation for this reaction, based on the known chemical behavior, is

$$2Mg(s) + O_2(g) \xrightarrow{\text{heat}} 2MgO(s)$$

If the mass of the magnesium is known and the mass of the oxide is found in the experiment, the mass of the oxygen in the oxide can be calculated:

$$\begin{array}{l} \text{mass of magnesium oxide} \\ \underline{- \text{ mass of magnesium}} \\ \phantom{-} \text{mass of oxygen} \end{array}$$

As soon as the masses are known, the moles of each component can be calculated. The moles can then be expressed in a simple whole number ratio and an empirical formula written.

When 2.43 g of magnesium was burned in oxygen, 4.03 g of magnesium oxide was produced.

$$
\begin{array}{rl}
\text{mass of magnesium oxide} & = 4.03 \text{ g} \\
\underline{- \text{ mass of magnesium}} & \underline{= 2.43 \text{ g}} \\
\text{mass of oxygen} & = 1.60 \text{ g}
\end{array}
$$

$$\text{No. of moles of magnesium} = \frac{2.43 \text{ g}}{24.31 \text{ g/mole}} = 0.100 \text{ moles}$$

$$\text{No. of moles of oxygen} = \frac{1.60 \text{ g}}{16.00 \text{ g/mole}} = 0.100 \text{ moles}$$

The molar ratio is 0.100 : 0.100 = 1 : 1
The empirical formula is $Mg_1 O_1$ or MgO.

$$\%Mg = \frac{2.43 \text{ g}}{4.03 \text{ g}} \times 100 = 60.3\%$$

In the present experiment, magnesium metal is heated in air. Air is composed of approximately 78% nitrogen and 21% oxygen. A side reaction occurs between some of the magnesium and the nitrogen gas:

$$3Mg(s) + N_2(g) \xrightarrow{\text{heat}} Mg_3N_2(s)$$

Not all of the magnesium is converted into magnesium oxide; some becomes magnesium nitride. However, the magnesium nitride can be converted to magnesium oxide by the addition of water:

$$Mg_3N_2(s) + 3H_2O(l) \xrightarrow{\text{heat}} 3MgO(s) + 2NH_3(g)$$

As a result, all of the magnesium is transformed into magnesium oxide.

## Objectives

1. To prepare a metal oxide.
2. To verify the empirical formula of a metal oxide.
3. To demonstrate the Law of Constant Composition.

## Procedure

### CAUTION!

A hot crucible can cause severe burns if handled improperly. Be sure to allow the crucible to cool sufficiently before handling. Always handle a hot crucible with crucible tongs.

## Cleaning the Crucible

1. Obtain a porcelain crucible and cover. Carefully clean the crucible *in the hood* by adding 10 mL of 6 M HCl to the crucible; allow the crucible to stand for 5 min. with the acid. With crucible tongs, pick up the crucible, discard the HCl, and rinse the crucible with distilled water from a plastic squeeze bottle.

2. Place the crucible in a clay triangle, which is mounted on an iron ring and attached to a ring stand. Be sure the crucible is firmly in place in the triangle. Place the crucible cover on the crucible slightly ajar (Fig. 7.1a).

**Figure 7.1**
(a) Heating the crucible.
(b) Picking up the crucible with crucible tongs.

3. Begin to heat the crucible with the aid of a Bunsen burner in order to evaporate water. Increase the heat, and, with the most intense flame (the tip of the inner blue cone), heat the crucible and cover for 5 min.; a cherry red color should appear when the bottom is heated strongly. Remove the flame. With tongs, remove the crucible to a heat-resistant surface and allow the crucible and cover to reach room temperature.

4. When cool, weigh the crucible and cover to 0.001 g (1). (Be sure to handle with tongs since fingerprints leave a residue.)

5. Place the crucible and cover in the clay triangle again. Reheat the crucible to the cherry red color for 5 min. Allow the crucible and cover to cool to room temperature. Reweigh when cool (2). Compare weight (1) and weight (2). If the weight differs by more than 0.005 g, heat the crucible and cover again for 5 min. and reweigh when cool. Continue heating, cooling, and weighing until the weight of the crucible and cover are constant to within 0.005 g.

## Forming the Oxide

1. Using forceps to handle the magnesium ribbon, cut a piece approximately 12 cm in length and fold the metal into a ball; transfer to the crucible. Weigh the crucible, cover, and magnesium to 0.001 g (3). Determine the weight of magnesium metal (4) by subtraction.

2. Transfer the crucible to the clay triangle; the cover should be slightly ajar (Fig. 7.1a).

3. Using a small flame, gently apply heat to the crucible. Should fumes begin to appear, remove the heat and cover the crucible immediately. Again place the cover ajar and continue to gently heat for 10 min. (If fumes appear, cover as before.) Remove the flame and allow the assembly to cool for 2 min. With tongs, remove the cover. If the magnesium has been fully oxidized, the contents should be a dull gray. Shiny metal means there is still free metal present. The cover should be replaced as before and the crucible heated for an additional 5 min. Reexamine the metal and continue heating until no shiny metal surfaces are present.

4. When all the metal appears as the dull gray oxide, half-cover the crucible and gently heat with a small Bunsen flame. Over a period of 5 min., gradually adjust the intensity of the flame until it is at its hottest, then heat the crucible to the cherry red color for 5 min.

## Completing the Reaction

1. Discontinue heating and allow the crucible assembly to cool to room temperature. Remove the cover and, with a glass rod, *carefully* break up the solid in the crucible. With 0.5 mL (10 drops) of distilled water dispensed from an eye dropper, wash the glass rod, adding the water to the crucible.

2. Set the cover ajar on the crucible and *gently* heat with a small Bunsen flame to evaporate the water. (Be careful to avoid spattering while heating; if spattering occurs, remove the heat and quickly cover the crucible completely.)

3. When all the water has been evaporated, half-cover the crucible and gradually increase to the hottest flame. Heat the crucible and the contents with the hottest flame for 10 min.

4. Allow the crucible assembly to cool to room temperature. Weigh the cool crucible, cover, and magnesium oxide to 0.001 g (5).

5. Return the crucible, cover, and magnesium oxide to the clay triangle. Heat at full heat of the Bunsen flame for 5 min. Allow to cool and then reweigh (6). The two weights, (5) and (6), must agree to within 0.005 g; if not, the crucible assembly must be heated for 5 min., cooled, and reweighed until two successive weights are within 0.005 g.

## Calculations

1. Determine the weight of magnesium oxide (7) by subtraction.

2. Determine the weight of oxygen (8) by subtraction.

3. From the data obtained in the experiment, calculate the empirical formula of magnesium oxide.

---

### Chemicals and Equipment

1. Clay triangle
2. Porcelain crucible and cover
3. Crucible tongs
4. Magnesium ribbon
5. Eye dropper
6. 6 M HCl

---

## Experiment 7

**PRE-LAB QUESTIONS**

1. Below are molecular formulas of selected organic compounds. Write the empirical formula for each:

   **a.** $C_8H_8$ (Styrene)

   **b.** $C_6H_{12}O_6$ (Glucose)

   **c.** $C_3H_8O_3$ (Glycerol)

2. Calculate the molar mass of aluminum oxide, $Al_2O_3$, in grams. Show your work.

3. Calculate the percentage, by weight, of each of the elements (Al, O) in aluminum oxide. Show your work.

## Experiment 7

## REPORT SHEET

**1.** Weight of crucible and cover (1) _____ g

**2.** Weight of crucible and cover (2) _____ g

**3.** Weight of crucible, cover, and Mg (3) _____ g

**4.** Weight of Mg metal (4): (3) − (2) _____ g

**5.** Weight of crucible, cover, and oxide (5) _____ g

**6.** Weight of crucible, cover, and oxide (6) _____ g

**7.** Weight of magnesium oxide (7): (6) − (2) _____ g

**8.** Weight of oxygen (8): (7) − (4) _____ g

**9.** Number of moles of magnesium
(4)/24.31 g/mole _____ moles

**10.** Number of moles of oxygen
(8)/16.00 g/mole _____ moles

**11.** Simplest whole number ratio
of Mg atoms to O atoms = (9):(10) _____ : _____

**12.** Empirical formula for magnesium oxide _____

**13.** % Mg in the oxide from data
% = [(4)/(7)] × 100 _____ %

**14.** % Mg calculated from the formula MgO
% = [24.31 g/40.31 g] × 100 _____ %

**15.** Error
$\% = \dfrac{(14) - (13)}{(14)} \times 100$ _____ %

## POST-LAB QUESTIONS

1. Write the balanced equation for the formation of iron(II) oxide, FeO, from the elements of iron, Fe, and oxygen, $O_2$.

2. Write the two chemical equations that describe the conversion of magnesium into magnesium oxide.

3. What error in calculation would result if, in the procedure for forming the magnesium oxide, the fumes in the initial heating were allowed to escape?

4. Mercuric oxide decomposes when heated and forms free mercury metal and releases oxygen gas. If a sample of the oxide, 2.096 g, is decomposed, 1.942 g of mercury is obtained. Determine the empirical formula for mercuric oxide. (*Hint:* first determine the number of moles of mercury and oxygen in the sample.)

# Experiment 8

## Classes of chemical reactions

The Periodic Table shows over 100 elements. The chemical literature describes millions of compounds that are known—some isolated from natural sources, some synthesized by laboratory workers. The combination of chemicals, in the natural environment or the laboratory setting, involves chemical reactions. The change in the way that matter is composed is a *chemical reaction*, a process wherein reactants (or starting materials) are converted into products. The new products often have properties and characteristics that are entirely different from those of the starting materials.

Four ways in which chemical reactions may be classified are combination, decomposition, single replacement (substitution), and double replacement (metathesis).

Two elements reacting to form a compound is a *combination reaction*. This process may be described by the general formula:

$$A + B \longrightarrow AB$$

The rusting of iron or the combination of iron and sulfur are good examples.

$$4Fe(s) + 3O_2(g) \longrightarrow 2Fe_2O_3(s) \text{ (rust)}$$
$$Fe(s) + S(s) \longrightarrow FeS(s)$$

Two compounds reacting together as in the example below also is a combination reaction.

$$CaO(s) + CO_2(g) \longrightarrow CaCO_3(s)$$

A compound which breaks down into elements or simpler components typifies the *decomposition reaction*. This reaction has the general formula:

$$AB \longrightarrow A + B$$

Some examples of this type of reaction are the electrolysis of water into hydrogen and oxygen:

$$2H_2O(l) \longrightarrow 2H_2(g) + O_2(g)$$

and the decomposition of potassium iodate into potassium iodide and oxygen:

$$2KIO_3(s) \longrightarrow 2KI(s) + 3O_2(g)$$

The replacement of one component in a compound by another describes the *single replacement* (or *substitution*) reaction. This reaction has the general formula:

$$AB + C \rightarrow CB + A$$

Processes which involve oxidation (the loss of electrons or the gain of relative positive charge) and reduction (the gain of electrons or the loss of relative positive charge) are typical of these reactions. Use of Table 8.1, the activity series of common metals, enables chemists to predict which oxidation-reduction reactions are possible. A more active metal, one higher in the table, is able to displace a less active metal, one listed lower in the table, from its aqueous salt. Thus aluminum metal displaces copper metal from an aqueous solution of copper(II) chloride; but copper metal will not displace aluminum from an aqueous solution of aluminum(III) chloride.

$$2Al(s) + 3CuCl_2(aq) \longrightarrow 3Cu(s) + 2AlCl_3(aq)$$

$$Cu(s) + AlCl_3(aq) \longrightarrow No\ Reaction$$

(Note that Al is oxidized to $Al^{3+}$ and $Cu^{2+}$ is reduced to Cu.)

Hydrogen may be displaced from water by a very active metal. Alkali metals are particularly reactive with water, and the reaction of sodium with water often is exothermic enough to ignite the hydrogen gas released.

$$2Na(s) + 2HOH(l) \longrightarrow 2NaOH(aq) + H_2(g) + heat$$

(Note that Na is oxidized to $Na^+$ and $H^+$ is reduced to $H_2$.)

Active metals, those above hydrogen in the series, are capable of displacing hydrogen from aqueous mineral acids such as HCl or $H_2SO_4$; however, metals below hydrogen will not replace hydrogen. Thus zinc reacts with aqueous solutions of HCl and $H_2SO_4$ to release hydrogen gas, but copper will not.

$$Zn(s) + 2HCl(aq) \longrightarrow ZnCl_2(aq) + H_2(g)$$

$$Cu(s) + H_2SO_4(aq) \longrightarrow No\ reaction$$

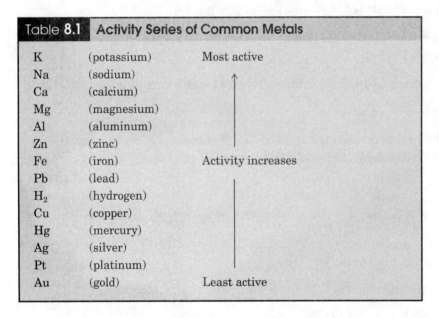

| Table 8.1 | Activity Series of Common Metals | | |
|---|---|---|---|
| K | (potassium) | Most active | |
| Na | (sodium) | | |
| Ca | (calcium) | | |
| Mg | (magnesium) | | |
| Al | (aluminum) | | |
| Zn | (zinc) | | |
| Fe | (iron) | Activity increases | |
| Pb | (lead) | | |
| $H_2$ | (hydrogen) | | |
| Cu | (copper) | | |
| Hg | (mercury) | | |
| Ag | (silver) | | |
| Pt | (platinum) | | |
| Au | (gold) | Least active | |

Two compounds reacting with each other to form two different compounds describes *double replacement* (or *metathesis*). This process has the general formula:

$$AB + CD \longrightarrow AD + CB$$

There are two replacements in the sense that A replaces C in CD and C replaces A in AB. This type of reaction generally involves ions which form in solution either from the dissociation of ionic compounds or the ionization of molecular compounds. The reaction of an aqueous solution of silver nitrate with an aqueous solution of sodium chloride is a good example. The products are sodium nitrate and silver chloride. We know a reaction has taken place since the insoluble precipitate silver chloride forms and separates from solution.

$$AgNO_3(aq) + NaCl(aq) \longrightarrow NaNO_3(aq) + AgCl(s) \text{ (White precipitate)}$$

In general, a double replacement results if one combination of ions leads to a precipitate, a gas or an un-ionized or very slightly ionized species such as water. In all of these reaction classes, it is very often possible to use your physical senses to observe whether a chemical reaction has occurred. The qualitative criteria may involve the formation of a gaseous product, the formation of a precipitate, a change in color, or a transfer of energy.

## Objectives

1. To demonstrate the different types of chemical reactions.
2. To be able to observe whether a chemical reaction has taken place.
3. To use chemical equations to describe a chemical reaction.

## Procedure

### Combination Reactions

1. Obtain a piece of aluminum foil approximately 2 × 0.5 in. Hold the foil at one end with a pair of forceps or crucible tongs and hold the other end in the hottest part of the flame of a Bunsen burner. Observe what happens to the foil. Record your observation and complete a balanced equation if you see that a reaction has occurred (1). Place the foil on a wire gauze to cool.

2. Obtain a piece of copper foil approximately 2 × 0.5 in. (A copper penny, one minted before 1982, may be substituted.) Hold the foil at one end with a pair of forceps or crucible tongs and hold the other end in the hottest part of the flame of a Bunsen burner. Observe what happens to the metal. Record your observation and complete a balanced equation if you see that a reaction has occurred (2). Place the foil on a wire gauze to cool.

3. Scrape some of the gray solid from the surface of the aluminum obtained in step no. 1 into a test tube (100 × 13 mm). Add 1 mL of water and shake the test tube. Is the solid soluble? Record your observation (3).

4. Scrape some of the black solid from the surface of the copper obtained in step no. 2 into a test tube (100 × 13 mm). Add 1 mL of water and shake the test tube. Is the solid soluble? Record your observation (4).

### Decomposition Reactions

1. *Decomposition of ammonium carbonate.* Place 0.5 g of ammonium carbonate into a clean, dry test tube (100 × 13 mm). Gently heat the test tube in the flame of a Bunsen burner (Fig. 8.1). As you heat, hold a piece of wet red litmus paper with forceps at the mouth of the test tube. What happens to the solid? Are any gases produced? What

happens to the color of the litmus paper? Ammonia gas acts as a base and turns moist red litmus paper blue. Record your observations and complete a balanced equation if you see that a reaction has occurred (5).

---

**CAUTION!**

When heating the contents of a solid in a test tube, do not point the open end towards anyone.

---

**Figure 8.1**
Position for holding a test tube in a Bunsen burner flame.

2. *Decomposition of potassium iodate.*

   **a.** Obtain three clean, dry test tubes (100 × 13 mm). Label them and add 0.5 g of compound according to the table below.

   | Test Tube No. | Compound |
   |---------------|----------|
   | 1 | $KIO_3$ |
   | 2 | $KIO_3$ |
   | 3 | $KI$ |

   **b.** Heat test tube no. 1 with the hottest flame of the Bunsen burner as shown in Fig. 8.2. Keep the test tube holder at the upper end of the test tube. While test tube no. 1 is being heated, hold a glowing wooden splint just inside the opening of the test tube. (The splint should not be flaming but should be glowing with sparks after the flame has been blown out. *Do not drop the glowing splint into the hot $KIO_3$.*) Oxygen supports combustion. The glowing splint should glow brighter or may burst into flame in the presence of oxygen. Record what happens to the glowing splint (6).

   **c.** Remove the test tube from the flame and set it aside to cool.

**Figure 8.2**
Testing for oxygen gas.

**d.** Add 5 mL of distilled water to each of the three test tubes and mix thoroughly to ensure that the solids are completely dissolved. Add 10 drops of 0.1 M $AgNO_3$ solution to each test tube. Observe what happens to each solution. Record the colors of the precipitates and write complete balanced equations for the reactions taking place in test tube no. 2 and test tube no. 3. (The $KIO_3$ and KI solids can be distinguished by the test results with $AgNO_3$: AgI is a yellow precipitate; $AgIO_3$ is a white precipitate.) (7)

**e.** Does a precipitate result when the residue in test tube no. 1 is mixed with $AgNO_3$ solution (8)? What compound is present in test tube no. 1 after heating $KIO_3$ (9)?

**f.** Write a complete balanced equation for the decomposition reaction (10).

## Single Replacement Reactions

**1.** In a test tube rack, set up labeled test tubes (100 × 13 mm) numbered from 1 through 9. Place 1 mL (approx. 20 drops) of the appropriate solution in the test tube with a small piece of metal as outlined in the table below. Use forceps to handle the metal.

| Test Tube No. | Solution | Metal |
|---|---|---|
| 1 | $H_2O$ | Ca |
| 2 | $H_2O$ | Mg |
| 3 | $H_2O$ | Al |
| 4 | 3 M HCl | Zn |
| 5 | 6 M HCl | Pb |
| 6 | 6 M HCl | Cu |
| 7 | 0.1 M $NaNO_3$ | Al |
| 8 | 0.1 M $CuCl_2$ | Al |
| 9 | 0.1 M $AgNO_3$ | Cu |

**2.** Observe the mixtures over a 20-min. period of time. Note any color changes, any evolution of gases, any formation of precipitates, or any energy changes (hold each test tube in your hand and note whether the solution becomes warmer or colder) that occur during each reaction; record your observations in the appropriate spaces on the Report Sheet (9). Write a complete and balanced equation for each reaction that occurred. For those cases where no reaction took place, write "No Reaction."

**3.** Dispose of the unreacted metals as directed by your instructor. *Do not discard into the sink.*

## Double Replacement Reaction

1. Each experiment in this part requires mixing equal volumes of two solutions in a test tube (100 × 13 mm). Use about 10 drops of each solution. Record your observation at the time of mixing (10). When there appears to be no evidence of a reaction, feel the test tube for an energy change (exothermic or endothermic). The solutions to be mixed are outlined in the table below.

| Test Tube No. | Solution No. 1 | Solution No. 2 |
|---|---|---|
| 1 | 0.1 M NaCl | 0.1 M KNO$_3$ |
| 2 | 0.1 M NaCl | 0.1 M AgNO$_3$ |
| 3 | 0.1 M Na$_2$CO$_3$ | 3 M HCl |
| 4 | 3 M NaOH | 3 M HCl |
| 5 | 0.1 M BaCl$_2$ | 3 M H$_2$SO$_4$ |
| 6 | 0.1 M Pb(NO$_3$)$_2$ | 0.1 M K$_2$CrO$_4$ |
| 7 | 0.1 M Fe(NO$_3$)$_3$ | 3 M NaOH |
| 8 | 0.1 M Cu(NO$_3$)$_2$ | 3 M NaOH |

2. For those cases where a reaction occurred, write a complete and balanced equation. Indicate precipitates, gases, and color changes. Table 8.2 lists some insoluble salts. For those cases where no reaction took place, write "No Reaction."

3. Discard the solutions as directed by your instructor. *Do not discard into the sink.*

| Table 8.2 | Some Insoluble Salts |
|---|---|
| AgCl | Silver chloride (white) |
| Ag$_2$CrO$_4$ | Silver chromate (red) |
| AgIO$_3$ | Silver iodate (white) |
| AgI | Silver iodide (yellow) |
| BaSO$_4$ | Barium sulfate (white) |
| Cu(OH)$_2$ | Copper(II) hydroxide (blue) |
| Fe(OH)$_3$ | Iron(III) hydroxide (red) |
| PbCrO$_4$ | Lead(II) chromate (yellow) |
| PbI$_2$ | Lead(II) iodide (yellow) |
| PbSO$_4$ | Lead(II) sulfate (white) |

## Chemicals and Equipment

1. Aluminum foil
2. Aluminum wire
3. Copper foil
4. Copper wire
5. Ammonium carbonate, $(NH_4)_2CO_3$
6. Potassium iodate, $KIO_3$
7. Potassium iodide, KI
8. Calcium turnings
9. Magnesium ribbon
10. Mossy zinc
11. Lead shot
12. 3 M HCl
13. 6 M HCl
14. 3 M $H_2SO_4$
15. 3 M NaOH
16. 0.1 M $AgNO_3$
17. 0.1 M NaCl
18. 0.1 M $NaNO_3$
19. 0.1 M $Na_2CO_3$
20. 0.1 M $KNO_3$
21. 0.1 M $K_2CrO_4$
22. 0.1 M $BaCl_2$
23. 0.1 M $Cu(NO_3)_2$
24. 0.1 M $CuCl_2$
25. 0.1 M $Pb(NO_3)_2$
26. 0.1 M $Fe(NO_3)_3$

## Experiment 8

**PRE-LAB QUESTIONS**

For each of the reactions below, classify as a combination, decomposition, single replacement, or double replacement.

**1.** $Ca(s) + Cl_2(g) \longrightarrow CaCl_2(s)$ _____

**2.** $2Cu(s) + O_2(g) \longrightarrow 2CuO(s)$ _____

**3.** $Ca(NO_3)_2(aq) + H_2SO_4(aq) \longrightarrow 2HNO_3(aq) + CaSO_4(s)$ _____

**4.** $NH_3(aq) + HCl(aq) \longrightarrow NH_4Cl(aq)$ _____

**5.** $Hg(NO_3)_2(aq) + 2NaI(aq) \longrightarrow HgI_2(s) + 2NaNO_3(aq)$ _____

**6.** $AgNO_3(aq) + NaCl(aq) \longrightarrow AgCl(s) + NaNO_3(aq)$ _____

**7.** $Zn(s) + H_2SO_4(aq) \longrightarrow ZnSO_4(aq) + H_2(g)$ _____

**8.** $H_2CO_3(aq) \longrightarrow CO_2(g) + H_2O(l)$ _____

**9.** $2H_2O(l) \longrightarrow 2H_2(g) + 2O_2(g)$ _____

**10.** $2Li(s) + 2H_2O(l) \longrightarrow 2LiOH(aq) + H_2(g)$ _____

## Experiment 8

### REPORT SHEET

Write *complete, balanced equations* for all cases that a reaction takes place. Your observation that a reaction occurred would be by a color change, by the formation of a gas, by the formation of a precipitate, or by an energy change (exothermic or endothermic). Those cases showing no evidence of a reaction, write "No Reaction."

*Classes of chemical reactions*                                    *Observation*

**Combination reactions**

1. _____ Al(s) + _____ $O_2$(g) ⟶                            _____

2. _____ Cu(s) + _____ $O_2$(g) ⟶                            _____

3. Solubility of aluminum oxide

4. Solubility of copper oxide

**Decomposition reactions**

5. _____ $(NH_4)_2CO_3$(s) ⟶                                   _____

6. What happens to the glowing splint?                           _____

7. _____ $KIO_3$(aq) + _____ $AgNO_3$(aq) ⟶                  _____

   _____ KI(aq) + _____ $AgNO_3$(aq) ⟶                       _____

8. Results when the residue of $KIO_3$ is mixed with $AgNO_3$ solution.  _____

9. The formula of the residue present after heating $KIO_3$.     _____

10. _____ $KIO_3$(s) $\xrightarrow{\text{Heat}}$

## Classes of chemical reactions

### Single replacement reactions
Observation

**9.** *Test tube no.*

1. _____ $Ca(s) + $ _____ $H_2O(l)$ _____

2. _____ $Mg(s) + $ _____ $H_2O(l)$ _____

3. _____ $Al(s) + $ _____ $H_2O(l)$ _____

4. _____ $Zn(s) + $ _____ $HCl(l)$ _____

5. _____ $Pb(s) + $ _____ $HCl(l)$ _____

6. _____ $Cu(s) + $ _____ $HCl(l)$ _____

7. _____ $Al(s) + $ _____ $NaNO_3(aq)$ _____

8. _____ $Al(s) + $ _____ $CuCl_2(aq)$ _____

9. _____ $Cu(s) + $ _____ $AgNO_3(aq)$ _____

### Double replacement reactions

**10.** *Test tube no.*

1. _____ $NaCl(aq) + $ _____ $KNO_3(aq)$ _____

2. _____ $NaCl(aq) + $ _____ $AgNO_3(aq)$ _____

3. _____ $Na_2CO_3(aq) + $ _____ $HCl(aq)$ _____

4. _____ $NaOH(aq) + $ _____ $HCl(aq)$ _____

5. _____ $BaCl_2(aq) + $ _____ $H_2SO_4(aq)$ _____

6. _____ $Pb(NO_3)_2(aq) + $ _____ $K_2CrO_4(aq)$ _____

7. _____ $Fe(NO_3)_3(aq) + $ _____ $NaOH(aq)$ _____

8. _____ $Cu(NO_3)_2(aq) + $ _____ $NaOH(aq)$ _____

## POST-LAB QUESTIONS

1. Zinc metal, Zn, reacts with 0.5 M HCl, but mercury metal, Hg, does not. Why?

2. From the following list of chemicals, select two combinations that would lead to a double replacement reaction. Write the complete, balanced equations for the reactions of the chemicals in solution.

$$KCl \quad HNO_3 \quad AgNO_3 \quad PbCl_2 \quad Na_2SO_4$$

3. When ammonium carbonate $[(NH_4)_2CO_3]$ decomposes, two gases are given off. Name the gases. Suggest a test to identify each one and the observation for the test.

# Chemical properties of consumer products

## Background

Concern for the environment has placed considerable attention on the identification of chemicals that enter our everyday world. Analytical chemistry deals with these concerns in both a quantitative and qualitative sense. In *quantitative analysis*, the concern is for exact amounts of certain chemicals present in a sample; experiments in this manual will deal with this problem (for example, see Experiments 10, 22, 23, 50, and 51). *Qualitative analysis* is limited to establishing the presence or absence of certain chemicals in detectable amounts in a sample. This experiment will focus on the qualitative determination of inorganic chemicals. Later experiments in this manual will deal with organic chemicals.

The simplest approach to the detection of inorganic chemicals is to use tests that will identify the ions that make up the inorganic sample. These ions are cations and anions. *Cations* are ions that carry positive charges; $Na^+$, $NH_4^+$, $Ca^{2+}$, $Cu^{2+}$, and $Al^{3+}$ are representative examples. *Anions* are ions that carry negative charges; $Cl^-$, $HCO_3^-$, $CO_3^{2-}$, $SO_4^{2-}$, and $PO_4^{3-}$ are examples of this type. Since each ion has unique properties, each will give a characteristic reaction or test. By examining an aqueous solution of the chemical, qualitative spot tests often will identify the cation and anion present. The tests used will bring about some chemical change. This change will be seen in the form of a solid precipitate, gas bubbles, or a color change.

This experiment will use chemicals commonly found around the house, so-called consumer chemical products. You may not think of these products as chemicals nor refer to them by their inorganic chemical names. Nevertheless, they are chemicals, and simple qualitative analytical techniques can be used to identify the ions found in their makeup.

**Table salt, NaCl.** Table salt is most commonly used as a flavoring agent. Individuals with high blood pressure (hypertension) are advised to restrict salt intake in order to reduce the amount of sodium ion, $Na^+$, absorbed. When dissolved in water, table salt releases the sodium cation, $Na^+$, and the chloride anion, $Cl^-$. Chloride ion is detected by silver nitrate, $AgNO_3$; a characteristic white precipitate of silver chloride forms.

$$Ag^+(aq) + Cl^-(aq) \longrightarrow AgCl(s) \text{ (White precipitate)}$$

Sodium ions produce a characteristic bright yellow color in a flame.

**Ammonia, $NH_3$.** Ammonia is a gas with a strong irritating odor. The gas dissolves readily in water, giving an aqueous ammonia solution; the solution is commonly referred to as ammonium hydroxide. Aqueous ammonia solutions are used as cleaning agents because of their ability to solubilize grease, oils, and waxes. Ammonia solutions are basic and will change moistened red litmus paper to blue. Ammonium salts (for example, ammonium chloride, $NH_4Cl$) react with strong bases to form ammonia gas.

$$NH_4^+(aq) + OH^-(aq) \longrightarrow NH_3(g) + H_2O(l)$$

**Baking soda, sodium bicarbonate, NaHCO₃.** Baking soda, sodium bicarbonate, $NaHCO_3$, acts as an antacid in some commercial products (e.g., Alka Seltzer) and as a leavening agent, helping to "raise" a cake. When sodium bicarbonate reacts with acids, carbon dioxide, a colorless, odorless gas, is released.

$$HCO_3^-(aq) + H^+(aq) \longrightarrow CO_2(g) + H_2O(1)$$

The presence of $CO_2$ can be confirmed with barium hydroxide solution, $Ba(OH)_2$; a white precipitate of barium carbonate results.

$$CO_2(g) + Ba(OH_2(aq) \longrightarrow H_2O(1) + BaCO_3(s) \text{ (White precipitate)}$$

**Epsom salt, MgSO₄ · 7H₂O.** Epsom salt has several uses; it may be taken internally as a laxative or purgative, or it may be used externally as a solution for soaking one's feet. When dissolved in water, Epsom salt releases magnesium cations, $Mg^{2+}$, and sulfate anions, $SO_4^{2-}$. The magnesium cation may be detected by first treating with a strong base, such as NaOH, and then with the organic dye $p$-nitrobenzene-azoresorcinol. The magnesium hydroxide, $Mg(OH)_2$, which initially forms, combines with the dye to give a blue color. This behavior is specific for the magnesium cation.

$$Mg^{2+}(aq) + 2OH^-(aq) \longrightarrow Mg(OH)_2(s) \xrightarrow{\text{dye}} \text{Blue complex}$$

The sulfate anion, $SO_4^{2-}$, reacts with barium chloride, $BaCl_2$, to form a white precipitate of barium sulfate, $BaSO_4$.

$$Ba^{2+}(aq) + SO_4^{2-}(aq) \longrightarrow BaSO_4(s) \text{ (White precipitate)}$$

**Bleach, sodium hypochlorite, NaOCl.** Bleach sold commercially is a dilute solution of sodium hypochlorite, NaOCl, usually 5% in concentration. The active agent is the hypochlorite anion. In solution, it behaves as if free chlorine, $Cl_2$, were present. Chlorine is an effective oxidizing agent. Thus in the presence of iodide salts, such as potassium iodide, KI, iodide anions are oxidized to iodine, $I_2$; chlorine is reduced to chloride anions, $Cl^-$.

$$Cl_2(aq) + 2I^-(aq) \longrightarrow I_2(aq) + 2CI^-(aq)$$

The iodine gives a reddish-brown color to water. However, since iodine is more soluble in organic solvents, such as hexane, $C_6H_{14}$, the iodine dissolves in the organic solvent. The organic solvent separates from the water, and the iodine colors the organic solvent violet.

**Sodium phosphate, Na₃PO₄.** In some communities that use well water for their water supply, dissolved calcium and magnesium salts make the water "hard." Normal soaps do not work well as a result. In order to increase the efficiency of their products, especially in hard water areas, some commercial soap preparations, or detergents, contain sodium phosphate, $Na_3PO_4$. The phosphate anion is the active ingredient and keeps the calcium and magnesium ions from interfering with the soap's cleaning action. Other products containing phosphate salts are plant fertilizers; here, ammonium phosphate, $(NH_4)_3PO_4$, serves as the source of phosphorus. The presence of the phosphate anion can be detected with ammonium molybdate, $(NH_4)_2MoO_4$. In acid solution, phosphate anions combine with the molybdate reagent to form a bright yellow precipitate.

$$PO_4^{3-}(aq) + 12MoO_4^{2-}(aq) + 3NH_4^+(aq) + 24H^+(aq) \longrightarrow (NH_4)_3PO_4(MoO_3)_{12}(s) + 12H_2O(l)$$
$$\text{(Yellow precipitate)}$$

## Objectives

1. To examine the chemical properties of some common substances found around the house.
2. To use spot tests to learn which inorganic cations and anions are found in these products.

## Procedure

## CAUTION!

Although we are using chemical substances common to our everyday life, conduct this experiment as you would any other. Wear safety glasses; do not taste anything; mix only those substances as directed.

### Analysis of Table Salt, NaCl

1. Place a small amount (covering the tip of a small spatula) of table salt in a test tube (100 × 13 mm). Add 1 mL (approx. 20 drops) of distilled water and mix to dissolve. Add 2 drops of 0.1 M $AgNO_3$. Record your observation (1).

2. Take a small spatula and clean the tip by holding it in a Bunsen burner flame until the yellow color disappears. Allow to cool but do not let the tip touch anything. Place a few crystals of table salt on the clean spatula tip and heat in the flame of the Bunsen burner. Record your observation (2).

### Analysis of Household Ammonia, NH₃, and Ammonium Ions, NH₄⁺

1. Place 1 mL of household ammonia in a test tube (100 × 13 mm). Hold a piece of dry red litmus paper over the mouth of the test tube (be careful not to touch the glass with the paper). Record your observation (3). Moisten the red litmus paper with distilled water and hold it over the mouth of the test tube. Record your observation (4).

2. Place a small amount (covering the tip of a small spatula) of ammonium chloride, $NH_4Cl$, in a test tube (100 × 13 mm). Add 0.5 mL (about 10 drops) of 6 M NaOH to the test tube. Hold a moist piece of red litmus inside the mouth of the test tube (be careful not to touch the glass with the paper). Does the litmus change color? If the litmus paper does not change color, gently warm the test tube (do not boil the solution). Record your observation (5).

3. Place a small amount (covering the tip of a small spatula) of commercial fertilizer in a test tube (100 × 13 mm). Add 0.5 mL (about 10 drops) of 6 M NaOH to the test tube. Test as above with moist red litmus paper. Record your observation and conclusion (6).

## Analysis of Baking Soda, NaHCO₃

1. Place a small amount (covering the tip of a small spatula) of baking soda in a test tube (100 × 13 mm). Dissolve the solid in 1 mL of distilled water. Add 5 drops of 6 M $H_2SO_4$ and tap the test tube to mix. Record your observation (7).

2. Test the escaping gas for $CO_2$. Make a loop in a wire; the loop should be about 5 mm in diameter. Dip the wire loop into 5% barium hydroxide, $Ba(OH)_2$, solution; a drop of solution should cling to the loop. Carefully lower the wire loop down into the mouth of the test tube. Avoid touching the walls. Record what happens to the drop (8).

## Analysis of Epsom Salt, MgSO₄ · 7H₂O

1. Place a small amount (covering the tip of a small spatula) of Epsom salt into a ˜ test tube (100 × 13 mm). Dissolve in 1 mL (about 20 drops) of distilled water. Add 5 drops of 6 M NaOH. Then add 5 drops of the "organic dye" solution (0.01% p-nitrobenzene–azoresorcinol). Record your observation (9).

2. Place a small amount (covering the tip of a small spatula) of Epsom salt into a test tube (100 × 13 mm). Dissolve in 1 mL (about 20 drops) of distilled water. Add 1 drop of 3 M $HNO_3$, followed by 2 drops of 1 M $BaCl_2$ solution. Record your observation (10).

## Analysis of Bleach, NaOCl

Place a small amount (covering the tip of a small spatula) of potassium iodide, KI, in a test tube (100 × 13 mm). Dissolve in 1 mL (about 20 drops) of distilled water. Add 1 mL of bleach to the solution, followed by 10 drops of hexane, $C_6H_{14}$. Cork the test tube and shake vigorously. Set aside and allow the layers to separate. Note the color of the upper organic layer and record your observation (11).

## Analysis of Sodium Phosphate, Na₃PO₄

Label three clean test tubes (100 × 13 mm) no. 1, no. 2, and no. 3. In test tube no. 1, place 2 mL of 1 M $Na_3PO_4$; in test tube no. 2, place a small amount (covering the tip of a small spatula) of a detergent; in test tube no. 3, place a small amount (covering the tip of a small spatula) of a fertilizer. Add 2 mL of distilled water to the solids in test tubes no. 2 and no. 3 and mix. Add 6 M $HNO_3$ dropwise to all three test tubes until the solutions test acid to litmus paper (blue litmus turns red when treated with acid). Mix each solution well and then add 10 drops of the $(NH_4)_2MoO_4$ reagent to each test tube. Warm the test tube in a water bath maintained at 60–70°C. Compare the three solutions and record your observations (12).

## Chemicals and Equipment

1. Bunsen burner
2. Copper wire
3. Litmus paper, blue
4. Litmus paper, red
5. Commercial ammonia solution, $NH_3$
6. Ammonium chloride, $NH_4Cl$
7. Commercial baking soda, $NaHCO_3$
8. Commercial bleach, $NaOCl$
9. Detergent, $Na_3PO_4$
10. Epsom salt, $MgSO_4 \cdot 7H_2O$
11. Garden fertilizer, $(NH_4)_3PO_4$
12. Table salt, $NaCl$
13. Ammonium molybdate reagent, $(NH_4)_2MoO_4$
14. 1 M $BaCl_2$
15. 5% $Ba(OH)_2$
16. 3 M $HNO_3$
17. 6 M $HNO_3$
18. Potassium iodide, KI
19. 0.1 M $AgNO_3$
20. 6 M NaOH
21. 1 M $Na_3PO_4$
22. 6 M $H_2SO_4$
23. 0.01% $p$-nitrobenzene-azoresorcinol ("organic dye" solution)
24. Hexane, $C_6H_{14}$

## Experiment 9

### PRE-LAB QUESTIONS

1. Below are some observations that were made when tests were carried out on solutions for specific ions. Based on the observed result, what is the most likely ion present in the solution?

    **a.** Heating produced a gas with a strong irritating odor; the gas changed moistened red litmus to blue.

    **b.** A white precipitate formed with barium chloride, $BaCl_2$.

    **c.** A solution treated with potassium iodide, KI, turned reddish-brown and produced a violet colored layer with hexane.

    **d.** A solid produced a bright yellow color in a flame.

    **e.** A white precipitate formed with silver nitrate solution [$AgNO_3$(aq)].

2. Below is a list of the materials to be analyzed. Complete the table by providing the name and formula of the salt found in each product, the name and formula of the cation, and the name and formula of the anion.

| Product | Salt | Cation | Anion |
|---|---|---|---|
| **1.** Baking soda | | | |
| **2.** Bleach | | | |
| **3.** Detergent | | | |
| **4.** Epsom salt | | | |
| **5.** Fertilizer | | | |
| **6.** Table salt | | | |

## Experiment 9

## REPORT SHEET

### Analysis of table salt, NaCl

1. $AgNO_3$ + NaCl _____

2. Color of flame _____

### Analysis of household ammonia, $NH_3$, and ammonium ions, $NH_4^+$

3. Color of dry litmus with ammonia fumes _____

4. Color of wet litmus with ammonia fumes _____

5. Color of wet litmus with $NH_4Cl$ + NaOH _____

6. Presence of ammonium ions in fertilizer _____

### Analysis of baking soda, $NaHCO_3$

7. $H_2SO_4$ + $NaHCO_3$ _____

8. Presence of $CO_2$ gas _____

### Analysis of Epsom salt, $MgSO_4 \cdot 7H_2O$

9. Presence of magnesium cation _____

10. Presence of sulfate anion _____

### Analysis of bleach, NaOCl

11. Color of hexane layer _____

### Analysis of sodium phosphate, $Na_3PO_4$

12. Presence of phosphate     no. 1 _____

no. 2 _____

no. 3 _____

## POST-LAB QUESTIONS

1. Club soda is also called "carbonated water." When the cap is removed from a bottle of soda, a gas is released. What is the gas? What test would identify the gas? What would you see when you do the test?

2. Brine is a heavily salted solution used for pickeling. What tests could be used to identify the ions present? What should you see when you do the tests?

3. An aspirin product on the market claims to use magnesium hydroxide, $Mg(OH)_2$, as a buffering agent. How could you verify whether magnesium ions are present?

4. Baking soda, sodium bicarbonate, $NaHCO_3$, is used to produce a "rising action" in the dough for baked goods. When mixed with an acidic material, such as lemon juice, a gas is given off that is trapped in the dough and results in the "rising action." Write an equation that shows what happens to this salt when it is stirred with the lemon juice in the mixing bowl. What brings about the "rising action" in the dough?

5. A company was accused of dumping phosphate rich waste into a local pond. How could the authorities determine the presence of phosphate in the water?

# Water analysis

## Background

The water we drink every day became an important issue in the 1990s. It is easy to forget that just 150 years ago there was no guaranteed drinking water in most parts of the United States. Cholera was a health hazard. In 1854, 1 out of every 20 Chicago residents died of cholera. Around the world huge sections of cities were razed to rebuild contaminated water transport systems. Today's tap water is safe, and the Safe Drinking Water Act administered by the U.S. Environmental Protection Agency mandates frequent monitoring of the drinking water supplies. Filtration and chlorination are the main methods to protect against pathogens. The taste of the water is, however, another matter. Chlorination, if it's overdone, gives a swimming pool flavor to the tap water.

$$Cl_2(g) + H_2O \longrightarrow Cl^-(aq) + HClO\ (aq) \tag{1}$$

The hypochlorous acid, HClO, formed in this reaction is the actual oxidizing agent that kills bacteria. The taste and odor of chlorinated water comes from the reaction products of HClO with organic compounds. Normally few products are formed in the above reaction. Only in the presence of a very strong reducing agent can chlorine be converted to chloride. Thus, the extent of chlorination of the drinking water increases the chloride content only slightly.

It became fashionable to drink bottled water at the dinner table, in restaurants and even during walking and hiking. Spring water, according to the U.S. Food and Drug Administration, must originate underground and flow naturally to the surface. However, most spring water is pumped from a bore hole. This allows a steady flow and it is free of bacterial contamination which may not be the case in water from spring-fed streams or lakes. Artesian or well water has similarities with spring water.

How the water tastes depends on its mineral content. Distilled water, deionized water (from which the ions are removed by an ion-exchange column), and water obtained by reverse osmosis are perfectly tasteless. Sodium chloride adds a salty taste; magnesium and calcium ions, among others, provide good flavor and "mouth feel." In the present experiment you will check the sodium chloride content of water from different sources. Besides adding to the taste, the chloride content of the water may also indicate the origin of water. Sea water obviously has high chloride content; brackish water, which may be the result of tides mixing sea and fresh water, has also undesirably high chloride content. High chloride may signal other pollution effects as well. The limit of chloride in drinking water is set at 250 mg NaCl/L. The desirable limit in drinking water is less than 25 mg NaCl/L.

The analysis of chloride in water will be performed by a modified Volhard method. The principle of this method is the insolubility of AgCl in water. Thus when silver nitrate and sodium chloride are mixed, the net ionic reaction can be written as:

$$Ag^+(aq) + Cl^-(aq) \longrightarrow AgCl(s) \tag{2}$$

To insure that all the chloride ions will be precipitated, excess $AgNO_3$ is added to the solution. The excess $Ag^+$ ion will be measured by back-titration with $SCN^-$, thiocyanate ion. The net ionic equation describing this reaction is

$$Ag^+(aq) + SCN^-(aq) \longrightarrow AgSCN(s) \tag{3}$$

AgSCN is a white precipitate. An indicator is added to the titration mixture in the form of iron(III) ion which will indicate the end point of the titration by the appearance of a reddish brown color:

$$Fe^{3+}(aq) + SCN^-(aq) \longrightarrow Fe(SCN)^{2+}(aq) \tag{4}$$
$$\text{Colorless} \quad \text{Colorless} \qquad \text{Reddish brown}$$

Nitric acid is also added to the solution to facilitate the AgCl precipitation and to preserve the iron in the 3+ oxidation state.

The titration must be performed under vigorous stirring, because the AgSCN precipitate has a tendency to adsorb silver ions. Thus stirring minimizes this adsorption process and all the silver ions will be titrated properly. A criterion of the true end point is that the color will be stable for 1 min.

To calculate the chloride content of a sample, expressed as mg NaCl/L, one subtracts the volume of $NH_4SCN$ used to titrate the sample from that used to titrate the deionized water according to the following equation:

$$\text{mg NaCl/L} = \frac{58.5[\text{mL NH}_4\text{SCN}_{\text{deionized}} - \text{mL NH}_4\text{SCN}_{\text{sample}}]}{\text{mL of sample}} \times 10 \tag{5}$$

## Objectives

1. To practice back-titration technique.
2. To compare the chloride content of drinking water.

## Procedure

1. Rinse a 50-mL buret with 0.01 M $NH_4SCN$ solution. Fill the buret with the 0.01 M $NH_4SCN$. Tilting the filled buret at a 45° angle, turn the stopcock open to allow the solution to fill the tip of the buret. Air bubbles should be completely removed from the tip by this maneuver. Clamp the buret onto a ring stand (Fig. 22.1). By slowly opening the stopcock, allow the bottom of the meniscus to drop to the 0.0-mL mark. Record the reading of the meniscus on your Report Sheet (1a; Trial 1).

2. Using a 50-mL volumetric pipet add exactly 50 mL of deionized water into a 100-mL Erlenmeyer flask. With the aid of a 5-mL graduated pipet add 3 mL 6 M $HNO_3$ solution. Place the Erlenmeyer flask on a magnetic stirrer, add a stir-bar, and set the magnetic stirrer to a steady, not too violent rotation. Add 0.5 mL 40% $Fe(NO_3)_3$ solution using a 1-mL graduated pipet. Finally, using a 5-mL volumetric pipet add exactly 5.0 mL 0.01 M $AgNO_3$ to the Erlenmeyer flask. Make certain that after each delivery you keep the pipets in their respective solutions. **Do not mix them up.**

3. Place the buret above the Erlenmeyer flask so that its tip reaches 1 cm below the rim. Open the stopcock of the buret slightly and allow the *dropwise* addition of the

thiocyanate solution to the flask. A reddish-brown color will appear where the thiocyanate drop hits the solution, but then it disappears upon stirring. Continue the dropwise addition until the faint reddish-brown color persists for 1 min. That is the end point of your titration. Record the meniscus readings from your buret on your Report Sheet (1b; Trial 1).

4. Wash your Erlenmeyer flask thoroughly and rinse it with deionized water. Repeat steps nos. 2 and 3 with a new aliquot of 50 mL of deionized water. Record the meniscus readings of the thiocyanate solution in the buret before and after the titration on your Record Sheet (1a and 1b; Trial 2).

5. Repeat steps nos. 2 and 3 with duplicate samples of (a) spring water and (b) tap water. **Make certain that before taking a new sample in your 50-mL volumetric pipet you rinse the pipet with deionized water.** Repeat steps nos. 2 and 3 also with duplicate samples of (c) brackish water. For brackish water use a 25-mL volumetric pipet. Record the meniscus readings under (2a and 2b for Trials 1 and 2) for spring water, under (3a and 3b for Trials 1 and 2) for tap water, and under (4a and 4b for Trials 1 and 2) for brackish water.

6. Calculate the amount of $NH_4SCN$ used in each titration by taking the differences in meniscus readings before and after the titration. Record them on your Report Sheet under (2c), (3c), and (4c). Average the duplicate Trials and record them on your Report Sheet under (2d), (3d), and (4d).

7. Using the average titration values you just recorded, calculate the NaCl content of each sample employing the formula of equation (5). Keep in mind that for the brackish water you used only a 25-mL sample.

---

### Chemicals and Equipment

1. 50-mL buret
2. Buret clamp
3. 50-mL volumetric pipet
4. 25-mL volumetric pipet
5. 5-mL graduated pipet
6. 5-mL volumetric pipet
7. 1-mL graduated pipet
8. Magnetic stirrer
9. Magnetic stir-bar
10. 6 M $HNO_3$
11. 0.01 M $NH_4SCN$
12. 0.01 M $AgNO_3$
13. 40% $Fe(NO_3)_3$
14. Deionized water
15. Tap water
16. Spring water
17. Brackish water

## Experiment 10

**PRE-LAB QUESTIONS**

1. Chlorine gas is toxic and has a pungent irritable odor. Do we smell chlorine in overchlorinated water?

2. What do we mean by "back-titration"?

3. What is the chemical formula of the complex ion that gives a reddish-brown color?

## Experiment 10

## REPORT SHEET

|  | **TRIALS** | |
| --- | --- | --- |
|  | **T1** | **T2** |

**1.** Titration of deionized water

   **a.** Meniscus reading before titration    _____mL   _____mL

   **b.** Meniscus reading after titration    _____mL   _____mL

   **c.** Titer used: (1b) − (1a)    _____mL   _____mL

   **d.** Average titer used: (T1 + T2)/2    _____mL

**2.** Titration of spring water

   **a.** Meniscus reading before titration    _____mL   _____mL

   **b.** Meniscus reading after titration    _____mL   _____mL

   **c.** Titer used: (2b) − (2a)    _____mL   _____mL

   **d.** Average titer used: (T1 + T2)/2    _____mL

   **e.** Average NaCl content of spring water [equation (5)]    _____mg/L

**3.** Titration of tap water

   **a.** Meniscus reading before titration    _____mL   _____mL

   **b.** Meniscus reading after titration    _____mL   _____mL

   **c.** Titer used: (3b) − (3a)    _____mL   _____mL

   **d.** Average titer used: (T1 + T2)/2    _____mL

   **e.** Average NaCl content of tap water [equation (5)]    _____mg/L

**4.** Titration of brackish water

    **a.** Meniscus reading before titration     _____mL       _____mL

    **b.** Meniscus reading after titration     _____mL       _____mL

    **c.** Titer used: (4b) − (4a)     _____mL       _____mL

    **d.** Average titer used: (T1 + T2)/2           _____mL

    **e.** Average NaCl content of brackish water [equation (5)]     _____mg/L

## POST-LAB QUESTIONS

**1.** Was the NaCl content of your tap water within the desirable limit of drinking water? Explain.

**2.** What would happen to the NaCl content of spring water if you added chlorine to make it safe against pathogens? Explain.

**3.** Somebody gave you a sample of water from the Dead Sea containing about 70 g/L NaCl. Could you measure this amount of NaCl in a 50-mL sample adding the same amount of silver nitrate reagent used in this experiment (5 mL 0.01 N)?

**4.** On the basis of your analysis of the brackish water, could you use it as a drinking water?

# Experiment  11

## Calorimetry: the determination of the specific heat of a metal

### Background

Any chemical or physical change involves a change in energy. Heat is a form of energy that can be observed as a flow of energy. Heat can pass spontaneously from an object at a high temperature to an object at a lower temperature. Two objects in contact at different temperatures, given enough time, will eventually reach the same temperature. The flow of heat energy can also be either into or out of a system under study.

The amount of heat can be measured in a device called a *calorimeter*. A calorimeter is a container with insulated walls. The insulation prevents a rapid heat exchange between the contents of the calorimeter and the surroundings. In the closed environment of the system, there is no loss or gain of heat. Since the change in temperature of the contents of the calorimeter is used to measure the magnitude of the heat flow, a thermometer is included with the calorimeter.

The specific heat of any substance can be determined in a calorimeter. The *specific heat* is an intensive physical property of a substance and is the quantity of heat (in calories) necessary to raise the temperature of one gram of substance by one degree Celsius. The specific heats for some common substances are listed in Table 11.1. Notice that specific heat has the units calories per gram per degree Celsius. From Table 11.1, the specific heat of water is 1.00 cal/g °C; this means that it would take one calorie to raise the temperature of one gram of water by 1°C. In contrast, iron has a specific heat of 0.11 cal/g °C; it would take only 0.11 calorie to raise the temperature of one gram of iron by 1°C. Just by comparing these two substances, you can see that water is a convenient coolant and explains its use in the internal combustion engine of automobiles. A small quantity of water is capable of absorbing a relatively large amount of heat, yet shows only a modest rise in temperature.

In general, when a given mass of a substance undergoes a temperature change, the heat energy required for the change is given by the equation

$$Q = m \times S \times \Delta T$$

where $Q$ is the change in heat energy, $m$ is the mass of the substance in grams, $S$ is the specific heat of the substance, and $\Delta T$ is the change in temperature (the difference between the *final* and *initial* temperatures); thus

$$\text{calories} = g \times (\text{cal/g °C}) \times °C$$

The specific heat of a metal can be found with a water calorimeter. This can be

## Table 11.1 Specific Heat Values for Some Common Substances

| Substance | Specific Heat (cal/g °C) | Substance | Specific Heat (cal/g °C) |
|-----------|--------------------------|-----------|--------------------------|
| Lead (Pb) | 0.038 | Glass | 0.12 |
| Tin (Sn) | 0.052 | Table salt (NaCl) | 0.21 |
| Silver (Ag) | 0.056 | Aluminum (Al) | 0.22 |
| Copper (Cu) | 0.092 | Wood | 0.42 |
| Zinc (Zn) | 0.093 | Ethyl alcohol ($C_2H_6O$) | 0.59 |
| Iron (Fe) | 0.11 | Water ($H_2O$) | 1.00 |

### EXAMPLE 1

If 20 g of water is heated so that its temperature rises from 20° to 25°C, then we know that 100 cal have been absorbed.

$$Q = m \times S \times \Delta T$$

$$20 \text{ g} \times 1.0 \frac{\text{cal}}{\text{g °C}} \times (25 - 20)°C = 100 \text{ cal}$$

conveniently done by using the Principle of Conservation of Energy: *Energy can neither be created nor destroyed in any process, but can be transferred from one part of a system to another.* Experimentally, the amount of heat absorbed by a known mass of water can be measured when a known mass of hot metal is placed in the water. The temperature of the water will rise as the temperature of the metal falls. Using the known heat capacity of water, the amount of heat added to the water can be calculated, just as in Example 1. This is exactly the amount of heat given up by the metal.

Heat(cal) lost by metal = Heat(cal) gained by water

$$Q_{metal} = Q_{water}$$

$$m_m \times S_m \times (\Delta T)_m = m_w \times S_w \times (\Delta T)_w$$

All the terms in the above equation are either known or can be determined experimentally, except for the value $S_m$, the specific heat of the metal. The unknown can then be calculated.

$$S_m = \frac{m_w \times S_w \times (\Delta T)_w}{m_m \times (\Delta T)_m}$$

## EXAMPLE 2

An unknown hot metal at 100.0°C with a mass of 50.03 g was mixed with 40.11 g of water at a temperature of 21.5°C. A final temperature of 30.6°C was reached. The heat gained by the water is calculated by

$$Q_w = (40.11 \text{ g}) \times 1.00 \, \frac{\text{cal}}{\text{g °C}} \times (30.6 - 21.5)°C = 365 \text{ cal}$$

The heat lost by the metal is equal to the heat gained by the water.

$$Q_m = Q_w = 365 \text{ cal}$$

The specific heat of the unknown metal is calculated to be

$$S_m = \frac{365 \text{ cal}}{(50.03 \text{ g}) \times (100.0 - 30.6)°C} = 0.105 \, \frac{\text{cal}}{\text{g °C}}$$

The specific heat of iron is 0.11 cal/g °C; thus, from the value of $S_m$ determined experimentally, the unknown metal is iron.

If the specific heat of the metal is known, an approximate atomic weight can be determined. This can be done using the relationship between the specific heat of solid metallic objects and their atomic weights observed by Pierre Dulong and Alexis Petit in 1819; it is known as the Law of Dulong and Petit.

$$S_m \times \text{Atomic Weight} = 6.3 \text{ cal/mole °C}$$

## EXAMPLE 3

The specific heat from Example 2 is 0.11 cal/g °C (to two significant figures). The approximate atomic weight is calculated to be

$$\text{Atomic Weight} = \frac{6.3 \text{ cal/mole °C}}{0.11 \text{ cal/g °C}} = 57 \text{ g/mole}$$

The atomic weight of iron is 56 g/mole (to two significant figures).

The calculations assume no heat is lost from the calorimeter to the surroundings and that the calorimeter absorbs a negligible amount of heat. However, this is not entirely correct. The calorimeter consists of the container, the stirrer, and the thermometer. All three get heated along with the water. As a result, the calorimeter absorbs heat. Therefore, the heat capacity for the calorimeter will be obtained experimentally, and the value derived applied whenever the calorimeter is used.

$$Q_{\text{calorimeter}} = C_{\text{calorimeter}} \Delta T$$

EXAMPLE 4

The temperature of 50.0 mL of warm water is 36.9°C. The temperature of 50.0 mL of cold water in a calorimeter is 19.9°C. When the two were mixed together in the calorimeter, the temperature after mixing was 28.1°C. The heat capacity of the calorimeter is calculated as follows (assume the density of water is 1.00 g/mL):

The heat lost by the warm water is

$$(28.1 - 36.9)°C \times 50.0 \text{ g} \times 1.00 \text{ cal/g }°C = -440 \text{ cal}$$

The heat gained by the cold water is

$$(28.1 - 19.9)°C \times 50.0 \text{ g} \times 1.00 \text{ cal/g }°C = 410 \text{ cal}$$

The heat lost to the calorimeter is

$$-440 \text{ cal} + 410 \text{ cal} = -30 \text{ cal}$$

The heat capacity of the calorimeter is

$$\frac{30 \text{ cal}}{(28.1 - 19.9)°C} = 3.7 \text{ cal/}°C$$

In this experiment, you also will plot the water temperature in the calorimeter versus time. Since the calorimeter walls and cover are not perfect insulators, some heat will be lost to the surroundings. In fact, when the hot water (or hot metal) is added to the colder water in the calorimeter, some heat will be lost before the maximum temperature is reached. In order to compensate for this loss, the maximum temperature is obtained by extrapolation of the curve as shown in Fig. 11.2. This gives the maximum temperature rise that would have been recorded had there been no heat loss through the calorimeter walls. Once $T_{max}$ is found, $\Delta T$ can be determined.

## Objectives

1. To construct a simple calorimeter.
2. To measure the heat capacity of the calorimeter.
3. To measure the specific heat of a metal.

## Procedure

**Determination of the Heat Capacity of the Calorimeter**

1. Construct a calorimeter as shown in Fig. 11.1. The two dry 8-oz. Styrofoam cups are inserted one into the other, supported in a 250-mL beaker. The plastic lid should fit tightly on the cup. With a suitable-sized cork borer, make two holes in the lid; one hole should be near the center for the thermometer and one hole to the side for the stirring wire. In order to keep the thermometer bulb 2 cm above the bottom of the inner cup, fit

a rubber ring (cut from latex rubber tubing) around the thermometer and adjust the ring by moving it up or down the thermometer.

**Figure 11.1**
The Styrofoam calorimeter.

- Rubber ring
- 250-mL beaker
- 2 cm

2. Since the density of water is nearly 1.00 g/mL over the temperature range for this experiment, the amount of water used in the calorimeter will be measured by volume. With a volumetric pipet, place 50.0 mL of cold water in the calorimeter cup; determine and record the mass (1). Cover the cup with the lid-thermometer-stirrer assembly. Stir the water for 5 min., observing the temperature during the time; record the temperature at 1-min. intervals on the Data Sheet. When the system is at equilibrium, record the temperature to the nearest 0.2°C (3).

3. With a volumetric pipet, place 50.0 mL of water in a clean, dry 150-mL beaker; determine and record the mass (2). Heat the water with a low flame until the temperature of the water is about 70°C. Allow the hot water to stand for a few minutes, stirring occasionally during this time period. Quickly record the temperature to the nearest 0.2°C (4) and pour the water completely into the calorimeter that has been assembled and has reached equilibrium (Fig. 11.1).

4. Replace the cover assembly and stir the contents gently. Observe the temperature for 5 min. and record the temperature on the Data Sheet (p. 118) every 30 sec. during that 5-min. period. Plot the temperature as a function of time, as shown in Figure 11.2. (Use the graph paper on p. 119.) Determine from your curve the maximum temperature by extrapolation and record it (5). Determine the ΔT. From the data, calculate the heat capacity of the calorimeter according to the calculations on the Report Sheet (p. 115).

**Figure 11.2**
Plot of temperature vs. time.

## Determination of the Specific Heat of a Metal

1. Dry the Styrofoam cups used for the calorimeter calibration. Reassemble the apparatus as in Fig. 11.1.

2. With a volumetric pipet, place 50.0 mL of cold water in the calorimeter cup; record the mass (1).

3. Obtain an unknown metal sample from your instructor. Record the number of the unknown on the Report Sheet (p. 116).

4. Weigh a clean, dry 50-mL beaker to the nearest 0.01 g (2). Place about 40 g of your unknown sample in the beaker and reweigh to the nearest 0.01 g (3). Determine the mass of the metal by subtraction (4). Pour the sample into a 25 × 150 mm clean, dry test tube.

5. Place the test tube in the water bath as shown in Fig. 11.3. Be sure that all of the metal in the test tube is below the surface of the water. Heat the water to a gentle boil and keep the test tube in the bath for 10 min. Make certain that water does not splash into the test tube.

**Figure 11.3**
Assembly for heating the metal.

6. While the metal is heating, follow the temperature of the cold water in the calorimeter for 5 min.; record the temperature on the Data Sheet at 1-min. intervals. After 5 min., record the temperature on the Report Sheet of the cold water to the nearest 0.2°C (5).

7. After 10 min. of heating the metal, observe and record the temperature on the Report Sheet of the boiling water in the beaker to the nearest 0.2°C (6). Obtain and use another thermometer for the calorimeter.

8. All steps must be done *quickly* and *carefully* at this point. Remove the test tube from the boiling water; dry the outside glass with a paper towel; remove the lid on the calorimeter; add the hot metal to the calorimeter. Be careful no water is added to or lost from the calorimeter on the transfer.

9. Record the calorimeter temperature on the Data Sheet as soon as the apparatus has been reassembled. Note the time when the temperature is determined. Stir the water. Continue to follow the temperature, recording the temperature on the Data Sheet every 30 sec. for the next 5 min.

10. Plot the temperature as a function of time, as shown in Fig. 11.2. (Use the graph paper on page 120.) Determine from your curve the maximum temperature; record the temperature on the Report Sheet (7). Determine the $\Delta T$. From the data, determine the specific heat and the atomic mass of the metal.

---

### Chemicals and Equipment

1. Metal pellets
2. Styrofoam cups (2)
3. Lid for Styrofoam cups
4. Metal stirring loop
5. Thermometers, 110°C (2)
6. Latex rubber ring
7. Volumetric pipet, 50-mL
8. Thermometer clamp
9. Test tube, 25 × 150 mm

---

## Experiment 11

### PRE-LAB QUESTIONS

1. Explain why heat exchange can be studied in a calorimeter.

2. When a hot iron rod is thrust into a container of cold water, the rod cools and the water warms. Is this observation consistent with the Principle of Conservation of Energy? Explain.

3. What is the advantage in using water as a coolant rather than ethyl alcohol in an internal combustion engine?

4. What are the units of specific heat?

## Experiment 11

### REPORT SHEET

*Determination of the heat capacity of the calorimeter*

1. Mass of the cold water
   50.0 mL × 1.00 g/mL _____ g

2. Mass of the warm water
   50.0 mL × 1.00 g/mL _____ g

3. Temperature of the equilibrated system:
   cold water and calorimeter _____ °C

4. Temperature of the warm water _____ °C

5. Maximum temperature from the graph _____ °C

6. $\Delta T$ of cold water and calorimeter
   (5) − (3) _____ °C

7. $\Delta T$ of warm water
   (5) − (4) _____ °C

8. Heat lost by warm water
   (2) × 1.00 cal/g °C × (7) _____ cal

9. Heat gained by cold water and
   the calorimeter: −(8) _____ cal

10. Heat gained by cold water
    (1) × 1.00 cal/g °C × (6) _____ cal

11. Heat gained by the calorimeter
    (9) − (10) _____ cal

12. Heat capacity of calorimeter, $C_{cal}$
    (11)/(6) _____ cal/°C

## Determination of the specific heat of a metal

1. Mass of cold water
   50.0 mL × 1.00 g/mL          _____ g

2. Mass of 50-mL beaker          _____ g

3. Mass of the beaker plus metal          _____ g

4. Mass of metal: (3) − (2)          _____ g

5. Temperature of the equilibrated system          _____ °C

6. Temperature of hot metal
   (Temperature of boiling water)          _____ °C

7. Maximum temperature from the graph          _____ °C

8. $\Delta T$ of cold water and calorimeter
   (7) − (5)          _____ °C

9. Heat gained by the calorimeter and water
   $[(1) \times 1.00 \text{ cal/g °C} \times (8)] + [C_{cal} \times (8)]$          _____ cal

10. $\Delta T$ of the metal
    (7) − (6)          _____ °C

11. Heat lost by the metal: − (9)          _____ cal

12. Specific heat of the metal
    $$\frac{(11)}{(4) \times (10)}$$          _____ cal/g °C

13. Atomic mass
    $$\frac{6.3 \text{ cal/mole °C}}{(12)}$$          _____ g/mole

    Unknown number _____      Metal unknown _____

## POST-LAB QUESTIONS

1. In this experiment you plot temperature as a function of time and from your curve, you determine the maximum temperature by extrapolation. Instead of doing this, suppose you just used the maximum temperature from the data recorded on p. 118. Explain how this will affect the heat capacity, Q, of the metal.

2. Suppose you used a glass beaker for the calorimeter rather than the styrofoam cup assembly for this experiment. What errors would be introduced into the experiment by this substitution?

3. The specific heats (cal/g $°C$) of some material are as follows: wood is 0.42, cement is 0.2, and iron is 0.11. Which material would be best for a walk that would lie in the hot sun for prolonged periods of the day? Explain your choice.

4. Table 11.1 has the following values for specific heat (cal/g $°C$): tin, 0.052; copper, 0.092, glass, 0.12. Each is used to make a small cup weighing 100 g.

   a. Which material requires the least amount of heat to raise the temperature of the cup by 10°C?

   b. How much heat is needed to raise the glass cup by 10°C?

   c. If the same amount of heat from **b**, above, were used to heat the tin cup, then the temperature of the tin cup would rise by how many degrees?

## Temperature Data Sheet

| Calorimeter Calibration | | Specific Heat Determination | |
|---|---|---|---|
| *Time (min.)* | *Temp (°C)* | *Time (min.)* | *Temp (°C)* |
| 0 | _____ | 0 | _____ |
| 1.0 | _____ | 1.0 | _____ |
| 2.0 | _____ | 2.0 | _____ |
| 3.0 | _____ | 3.0 | _____ |
| 4.0 | _____ | 4.0 | _____ |
| 5.0 | _____ | 5.0 | _____ |
| 5.5 | _____ | 5.5 | _____ |
| 6.0 | _____ | 6.0 | _____ |
| 6.5 | _____ | 6.5 | _____ |
| 7.0 | _____ | 7.0 | _____ |
| 7.5 | _____ | 7.5 | _____ |
| 8.0 | _____ | 8.0 | _____ |
| 8.5 | _____ | 8.5 | _____ |
| 9.0 | _____ | 9.0 | _____ |
| 9.5 | _____ | 9.5 | _____ |
| 10.0 | _____ | 10.0 | _____ |

**Time, min.**

**Temperature, °C**

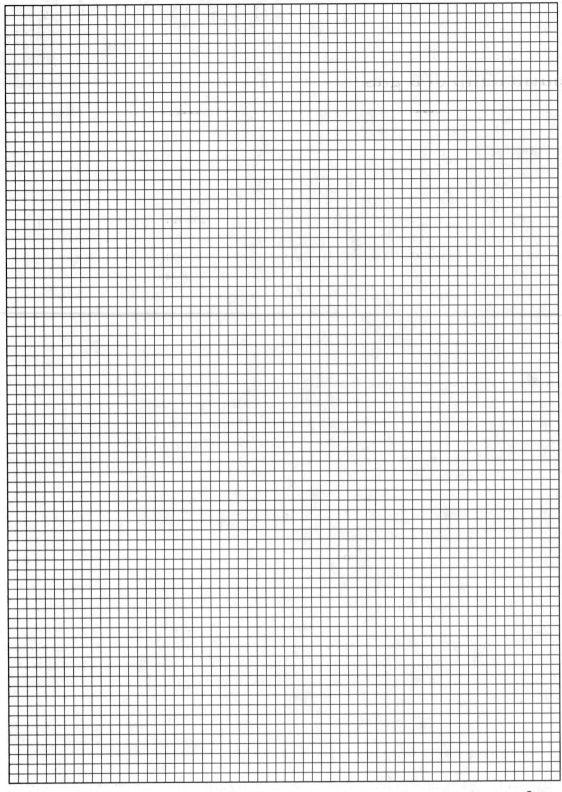

**Temperature, °C**

**Time, min.**

# Experiment  12

## Boyle's Law: the pressure–volume relationship of a gas

### Background

The British scientist Robert Boyle made many contributions in the fields of medicine, astronomy, physics, and chemistry. However, he is best known for his work on the behavior of gases. In 1662, Boyle found that when the temperature is held constant, the pressure of a trapped amount of gas (any gas) is inversely proportional to its volume. That is, when the pressure of the gas increases, the volume of the gas decreases; when the pressure of the gas decreases, the volume of the gas increases. Boyle's Law can be written mathematically as follows:

$$V = k \times \frac{1}{P} \quad \text{or} \quad V = \frac{k}{P} \quad \text{or} \quad PV = k$$

where $V$ is the volume of the gas, $P$ is the pressure of the gas, and $k$ is a constant that depends on the temperature and amount of the gas. By looking at these equations, it is easy to see the inverse relationship. For example, if pressure on a sample of trapped gas is *doubled*, the volume of the sample will be reduced by *half* of the value it had been before the increase in pressure. On the other hand, if the pressure is reduced by *half*, the volume will become *doubled*.

If there is a pressure change, for example from $P_1$ to $P_2$, then the volume also changes from $V_1$ to $V_2$. The relationship between the initial pressure and volume to the new pressure and volume can be expressed as follows.

$$P_1V_1 = P_2V_2 \quad \text{or} \quad \frac{P_1}{P_2} = \frac{V_2}{V_1}$$

Volume may be expressed in a variety of units—liter (L), milliliter (mL), cubic meter (m³), cubic centimeter (cc or cm³). Similarly, pressure can be expressed in a variety of units, but the standard unit of pressure is the atmosphere (atm); one atmosphere (1 atm) is defined as the pressure needed to support a column of mercury 760 mm in height at 0°C at sea level. In honor of Evangelista Torricelli, the Italian inventor of the barometer, the unit **torr** is used and is equal to 1 mm Hg. Thus

$$1 \text{ mm Hg} = 1 \text{ torr}$$

$$1 \text{ atm} = 760 \text{ mm Hg} = 760 \text{ torr}$$

Breathing is a good example of how Boyle's Law works. We breathe as a result of the movements that take place in the diaphragm and the rib cage. When the diaphragm contracts (moves downward) and the rib cage is raised (expands), the volume of the chest cavity increases. This action decreases the pressure in the lungs and this pressure is lower than the outside pressure. The result: air flows from the outside, higher-pressure area into the lungs and they expand—we inhale. When we breathe out, the process is reversed: the diaphragm is relaxed (moves upward) and the rib cage is lowered (contracts). This decreases the volume of the chest cavity and increases the pressure inside the lungs. With pressure greater inside the lungs than the outside, air flows out and the lungs contract—we exhale.

In this experiment, a volume of air is trapped in a capillary tube by a column of mercury. The mercury acts as a movable piston. Depending on how the capillary tube is tilted, the mercury column moves and thus causes the volume of trapped air to change. The pressure of the trapped air supports not only the pressure exerted by the atmosphere but also the pressure of the mercury column. The pressure exerted by the mercury column varies depending on the angle of the tilt. If $\theta$ is the angle of the tilt, the pressure of the mercury column can be calculated by the following equation:

$$P_{Hg} = (\text{length of Hg column}) \times \sin \theta$$

The total pressure of the trapped gas is the sum of the atmospheric pressure and the pressure due to the mercury column.

You need only measure the length of the column of air, $L_{air}$, since the length is directly related to the volume. The column of air is geometrically a regular cylinder. The radius of the cylinder, in this case the capillary tube, remains the same. The volume of a regular cylinder is a constant ($\pi r^2$) times the height of the cylinder ($L_{air}$); the only quantity that varies in this experiment is the value $L_{air}$.

## Objectives

1. To show the validity of Boyle's Law.
2. To measure the volume of a fixed quantity of air as the pressure changes at constant temperature.

## Procedure*

### CAUTION!

Mercury can be spilled easily. While mercury has a low vapor pressure, its vapor is extremely toxic. Mercury can also be absorbed through the skin. If any mercury is spilled, notify the instructor immediately for proper clean-up.

1. Obtain a Boyle's Law apparatus and a 30°-60°-90° plastic triangle. The Boyle's Law apparatus consists of a piece of glass tubing that contains a column of mercury and is attached to a ruler by means of rubber bands (Fig. 12.1).

2. Record the temperature on the Report Sheet (1).

3. Record the barometric pressure, $P_{at}$, in mm Hg on the Report Sheet (2).

* Adapted from R. A. Hermens, *J. Chem. Educ.* **60** (1983), 764.

**4.** Measure the length of the column of mercury, $L_{Hg}$, to the nearest 0.5 mm; record this length on the Report Sheet (3).

**Figure 12.1**
The Boyle's Law apparatus.

5. The length of the column of trapped air is to be measured when the tube is at various angles to the bench top, as outlined in the following table.

| Angle of Tube | Position of Open End of Tube |
|---|---|
| 0° | Horizontal |
| 90° | Up |
| 90° | Down |
| 60° | Up |
| 60° | Down |
| 30° | Up |
| 30° | Down |

The column length of the trapped air, $L_{air}$, is measured (from the glass seal to the mercury) to the nearest 0.5 mm and is recorded in the table on the Report Sheet (4). The correct angle can be obtained with the aid of the 30°-60°-90° triangle by placing the Boyle's Law apparatus along the appropriate edge.

---

**CAUTION!**

Do not touch the glass tube during the measurements, to avoid any temperature changes. Do not jar the tube at any time. This will avoid separation or displacement of the mercury column when measurements have begun.

---

6. Calculate the reciprocal, $\dfrac{1}{L_{air}}$, and enter on the Report Sheet (5).

7. Using the appropriate formula from the table on the Report Sheet (6), calculate the pressure, P, of the column of air (7).

8. Plot the data on graph paper as follows: y-axis, the calculated pressure, P; x-axis, the reciprocal of the length of the trapped air, $\dfrac{1}{L_{air}}$.

9. Replot the data with P (y-axis) versus $L_{air}$ (x-axis).

| Chemicals and Equipment |
| --- |
| 1. Boyle's Law apparatus |
| 2. 30°-60°-90° plastic triangle |

## Experiment 12

### PRE-LAB QUESTIONS

1. Complete the following statements about Boyle's Law (assume temperature is constant).

   **a.** As the pressure increases, the volume _____.

   **b.** As the pressure _____, the volume increases.

2. On the radio you hear the weather report announce the daily barometric pressure. What is the cause of this pressure?

3. Would the barometric pressure be greater at sea level or at the top of Mt. Everest (elev. 29,035 ft.)? Explain.

4. Make the following conversions.

   1.80 atm   = _____ torr

   720 torr   = _____ atm

   1.50 atm   = _____ mm Hg

   740 mm Hg = _____ atm

## Experiment 12

### REPORT SHEET

1. Temperature                                _____°C

2. Barometric pressure, $P_{at}$           _____ mm Hg

3. Length of mercury column, $L_{Hg}$     _____ mm

| Boyle's Law Data | | | | |
|---|---|---|---|---|
| Angle of Tube (opening*) | (4) Length of Trapped Air, $L_{air}$, mm | (5) $\dfrac{1}{L_{air}}$ | (6) Pressure (calculation) | (7) Pressure P |
| 0° | | | $P_{at}$ | |
| 90° (U) | | | $P_{at} + L_{Hg}$ | |
| 90° (D) | | | $P_{at} - L_{Hg}$ | |
| 60° (U) | | | $P_{at} + (L_{Hg} \sin 60°)$ | |
| 60° (D) | | | $P_{at} - (L_{Hg} \sin 60°)$ | |
| 30° (U) | | | $P_{at} + (L_{Hg} \sin 30°)$ | |
| 30° (D) | | | $P_{at} - (L_{Hg} \sin 30°)$ | |

*U = up; D = down

### POST-LAB QUESTIONS

1. For this experiment you are cautioned not to touch the glass tube during measurements. Why take this precaution?

**2.** In this experiment, the column length of the trapped air was measured rather than the actual volume. Why could this be done as a valid determination of Boyle's Law?

**3.** At a fixed temperature, what happens to the pressure if the volume expands by a factor of three?

**4.** A hospital dispensary has a 10-L bottle of compressed oxygen gas at a pressure of 50.0 atm. The valve was opened and the oxygen gas escaped at room temperature and 1 atm. What is the final volume occupied by the oxygen gas? Show your calculations.

**5.** Ten grams of oxygen gas fills a 10-L container at 750 mm Hg. If the temperature is held constant, what will its volume be at 770 mm Hg? Show your calculations.

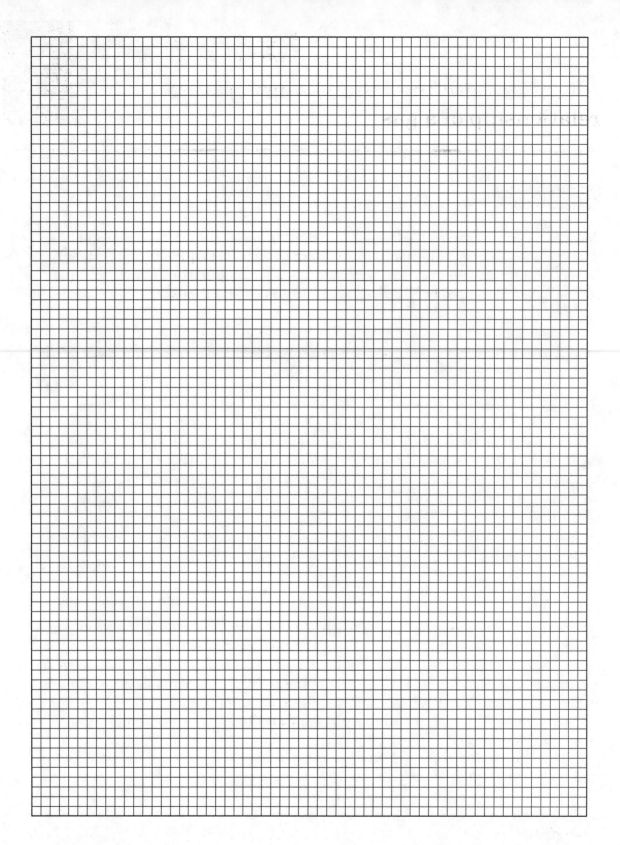

# Experiment 13

## Charles's Law: the volume–temperature relationship of a gas

---

### Background

Jacques Charles observed that for a fixed quantity of gas, the volume at constant pressure changes when temperature changes: the volume increases (V ↑) when the temperature increases (T ↑); the volume decreases (V ↓) when the temperature decreases (T ↓). Although first described by Charles in 1787, it was not until 1802 that Joseph Gay-Lussac expressed the relationship mathematically.

Charles's Law states that when the pressure is held constant, the volume of a fixed mass of ideal gas is in direct proportion to the temperature in degrees Kelvin. Charles's Law can be written mathematically as follows:

$$\mathbf{V} = \mathbf{k} \times \mathbf{T} \quad \text{or} \quad \frac{\mathbf{V}}{\mathbf{T}} = \mathbf{k} \tag{1}$$

where $\mathbf{V}$ is the volume of the gas, $\mathbf{T}$ is the temperature in degrees Kelvin, and $\mathbf{k}$ is a constant that depends on the pressure and amount of gas. The direct relationship is clear by looking at the equations. If a sample of gas at a fixed pressure has its temperature *doubled*, the volume in turn is *doubled*. Conversely, decreasing the temperature by *one-half* brings about a decrease in volume by *one-half*.

The law applies, for a given pressure and quantity of gas, at all sets of conditions. Thus for two sets of $\mathbf{T}$ and $\mathbf{V}$, the following can be written:

$$\frac{\mathbf{V_1}}{\mathbf{T_1}} = \frac{\mathbf{V_2}}{\mathbf{T_2}} \quad \text{or} \quad \mathbf{V_1 T_2} = \mathbf{V_2 T_1} \quad \text{or} \quad \frac{\mathbf{V_1 T_2}}{\mathbf{V_2 T_1}} = 1 \tag{2}$$

where at constant pressure, $\mathbf{V_1}$ and $\mathbf{T_1}$ refer to the set of conditions at the beginning of the experiment, and $\mathbf{V_2}$ and $\mathbf{T_2}$ refer to the set of conditions at the end of the experiment.

Charles's Law can be illustrated by a hot-air balloon. The material that the balloon is made from is stretchable, so the pressure of the air inside is constant. As the air inside is heated ($\mathbf{T}$ ↑), the volume of the air increases (expands; $\mathbf{V}$ ↑) and the balloon fills out. With the mass the same but the volume larger, the density decreases (see Experiment 3). Since the air inside is less dense than the air outside, the balloon rises.

This experiment determines the volume of a sample of air when measured at two different temperatures with the pressure held constant.

### Objectives

1. To measure the volume of a fixed quantity of air as the temperature changes at constant pressure.
2. To verify Charles's Law.

1. Use a clean and dry 250-mL Erlenmeyer flask (Flask no. 1). Fit the flask with a prepared stopper assembly, consisting of a no. 6 one-hole rubber stopper which has a 5-cm to 8-cm length of glass tubing inserted through the hole. If an assembly needs to be constructed, use the following procedure:

   a. Select a sharpened brass cork borer with a diameter that just allows the glass tubing to pass through it easily.

   b. Lubricate the outside of the cork borer with glycerine and push it through the rubber stopper from the bottom.

   c. Once the cork borer is through the stopper, pass the glass tubing through the cork borer so that the tubing is flush with the bottom of the stopper.

   d. Grasp the tubing and the stopper with one hand to hold these two pieces stationary; with the other hand carefully remove the borer. The glass tubing stays in the stopper. (Check to be certain that the end of the glass tubing is flush with the bottom of the rubber stopper.)

2. Mark the position of the bottom of the rubber stopper on Flask no. 1 with a marking pencil. Connect a 2-ft. piece of latex rubber tubing to the glass tubing.

3. Place 300 mL of water and three (3) boiling stones in an 800-mL beaker. Support the beaker on a ring stand using a ring support and wire gauze, and heat the water with a Bunsen burner to boiling (Fig. 13.1) (or place the beaker on a hot plate and heat to boiling). Keep the water at a gentle boil. Record the temperature of the boiling water on the Report Sheet (1).

**Figure 13.1**
Equipment to study
Charles's Law.

200 mL of water

4. Prepare an ice-water bath using a second 800-mL beaker half-filled with a mixture of ice and water. Record the temperature of the bath on the Report Sheet (3). Set aside for use in step no. 8.

5. Put about 200 mL of water into a second 250-mL Erlenmeyer flask (Flask no. 2) and place the end of the rubber tubing into the water. Make sure that the end of the rubber tubing reaches to the bottom of the flask and stays submerged at all times. (You may wish to hold it in place with a clamp attached to a ring stand.)

6. With a clamp holding the neck of Erlenmeyer Flask no. 1, lower the flask as far as it will go into the boiling water. Secure onto the ring stand (Fig. 13.1). Adjust the water level in the beaker to cover as much of the Erlenmeyer flask as possible.

7. Boil gently for 5 min. Air bubbles should emerge from the rubber tubing submerged in Flask no. 2. Add water to the beaker if boiling causes the water level to go down.

8. When bubbles no longer emerge from the end of the submerged tubing (after 5 min.), carefully lift Flask no. 1 from the boiling water bath and quickly place it into the ice-water bath. Record what you observe happening as Flask no. 1 cools (2). **Be sure to keep the end of the rubber tubing always submerged in the water in Flask no. 2.**

---

### CAUTION!

The water, the glassware, and the ironware are hot.

---

9. When no more water is drawn into Flask no. 1, raise the flask until the level of water inside the flask is at the same height as the water in the ice-water bath. Then remove the stopper from the Flask no. 1.

10. Take a graduated cylinder and measure the water in Flask no. 1. Record the volume to the nearest 0.1 mL on the Report Sheet (4).

11. Determine the volume of Erlenmeyer Flask no. 1 as follows:

   a. First, fill it with water to the level marked by the marking pencil. Insert the stopper with the glass tubing into the flask to be sure the bottom of the stopper touches the water with no air space present. Adjust the water level if necessary.

   b. Remove the stopper and measure the volume of the water in the flask by pouring it into a graduated cylinder. If a 100-mL graduated cylinder is used, it will be necessary to empty and refill it until all the water from Flask no. 1 has been measured.

   c. The *total* volume of water should be measured to the nearest 0.1 mL. Record this value on the Report Sheet (5).

12. Do the calculations to verify Charles's Law.

## Chemicals and Equipment

1. Boiling stones
2. Bunsen burner (or hot plate)
3. 250-mL Erlenmeyer flasks (2)
4. 800-mL beakers (2)
5. Clamps
6. Glass tubing (6- to 8-cm length; 7-mm OD)
7. Marking pencil
8. One-hole rubber stopper (size no. 6)
9. Ring stand
10. Ring support
11. Rubber tubing (2-ft. length)
12. Thermometer, 110°C
13. Wire gauze

## Experiment 13

**PRE-LAB QUESTIONS**

1. Complete the following equation for Charles's Law relating temperature and volume.

$$T =$$

2. Temperature in Charles's Law is expressed in units of _____.

3. Indicate whether the following statements about Charles's Law are true (T) or false (F):

   **a.** As temperature doubles from 10°C to 20°C, the volume doubles.    _____

   **b.** As the volume decreases by one-half, the temperature decreases by one-half.    _____

   **c.** Volume is inversely related to the temperature at constant pressure.    _____

   **d.** $V \times T = k$    _____

4. Balloons are filled to capacity outdoors where the temperature is 25°F. They are brought indoors where the temperature is 70°F. Explain what will happen to the balloons as they warm up indoors.

## Experiment 13

### REPORT SHEET

1. Temperature, boiling water ($T_2$)  _____ °C  _____ K

2. Observation as Flask no. 1 cools:

3. Temperature, ice water ($T_1$)  _____ °C  _____ K

4. Volume of water sucked into Flask no. 1 ($V_w$)  _____ mL

5. Volume of Flask no. 1  _____ mL

6. Volume of air at the temperature of boiling water (5)  _____ mL
   ($V_2$)

7. Volume of the air at the temperature of ice water ($V_1$)  _____ mL
   ($V_1 = V_2 - V_w$)

8. Verify Charles's Law  _____

   $$\frac{V_2 \times T_1}{V_1 \times T_2} =$$

9. Percent deviation from Charles's Law  _____ %

   $$\% = \frac{1.00 - (8)}{1.00} \times 100 =$$

## POST-LAB QUESTIONS

1. A student assumed the volume of the 250-mL Erlenmeyer flask to be 250 mL without actually measuring the volume. How would this assumption affect the results?

2. Another student allowed all of the water in the beaker to boil away. What effect does this have on the temperature of the gas in the flask?

3. In step no. 9 of the **Procedure,** Flask no. 1 is raised until the level of water inside the flask is at the same height as the water in the ice-water bath. Why was this done?

4. A gas in a balloon has a volume of 25.0 L at a temperature of 20.0°C. What temperature must be reached in order for the balloon to expand to a volume of 50.0 L, at constant pressure? Show all your work.

5. A student carried out the procedure to verify Charles's Law and obtained the following experimental data:

   a. Temperature of boiling water, T = 100.0°C;

   b. Room temperature, T = 20.0°C;

   c. Volume of water drawn into the flask, V = 75.5 mL;

   d. Total volume of the flask, V = 263.5 mL.

   Verify Charles's Law according to equation (2) (see **Background**), and determine any percent deviation. Show all your work.

# Experiment  14

## Properties of gases: determination of the molecular weight of a volatile liquid

In the world in which we live, the gases with which we are familiar (for example $O_2$, $N_2$, $CO_2$, $H_2$) possess a molecular volume and show interactions between molecules. When cooled or compressed, these real gases eventually condense into the liquid phase. In the hypothetical world, there are hypothetical gases which we refer to as *ideal* gases. The molecules of these gases are presumed to possess negligible volume. There are no attractions between molecules of ideal gases, and, as a result, molecules do not stick together upon collision. The gases possess the volume of the container. Volumes of all gases decrease linearly with decreasing temperature. When we extrapolate these lines, they all converge to one temperature, $-273.15°C$, where the volume of the gas is presumable zero. Obviously this is an imaginary situation because gases cannot exist at such low temperatures; they condense and become liquids and then solids. But $-273.15°C$ is the lowest possible temperature, and although we can never reach it, scientists have come pretty close, to one millionth of a degree Celsius above it. For that reason it is rational to start a temperature scale at the lowest possible temperature and call that zero. This is the absolute or Kelvin scale. On the Kelvin scale, this low temperature is termed *absolute zero* and is $-273.15$ degrees below zero on the Celsius scale.

The relationship that unites pressure, $P$, volume, $V$, and temperature, $T$, for a given quantity of gas, $n$, in the sample is the ideal gas equation:

$$P \times V = n \times R \times T$$

This equation expresses pressure in atmospheres, volume in liters, temperature in degrees Kelvin, and the quantity of gas in moles. These four quantities are related exactly through the use of the ideal gas constant, $R$; the value for R is 0.0821 L atm/K mole. (Notice that the units for R are in terms of V, P, T, and n.)

A container of fixed volume at a given temperature and pressure holds only one possible quantity of gas. This quantity can be calculated by using the ideal gas equation:

$$n = \frac{P \times V}{R \times T}$$

The number of moles, $n$, can be determined from the mass of the gas sample, $m$, and the molar mass of the gas, $M$:

$$n = \frac{m}{M}$$

Substituting this expression into the ideal gas equation gives

$$P \times V = \frac{m}{M} \times R \times T$$

Solving for the molar mass, **M**:

$$M = \frac{m \times R \times T}{P \times V}$$

This equation gives us the means for determining the molar mass of a gas sample from the measurements of mass, temperature, pressure, and volume. Most real gases at low pressures (1 atm or less) and high temperatures (300 K or more) behave as an ideal gas and thus, under such conditions, the ideal gas law is applicable for real gases as well.

    If a volatile liquid (one of low boiling point and low molar mass) is used, the method of Dumas (John Dumas, 1800 - 1884) can give a reasonably accurate value of the molar mass. In this experiment, a small quantity of a volatile liquid will be vaporized in a pre-weighed flask of known volume at a known temperature and barometric pressure. Since the boiling point of the liquid will be below that of boiling water when the flask is submerged in a boiling water bath, the liquid will vaporize completely. If the density of the sample gas is greater than the density of the air, the gas will drive out the air and fill the flask with the gaseous sample. The gas pressure in the flask is in equilibrium with atmospheric pressure and, therefore, can be determined from a barometer. The temperature of the gas is the same as the temperature of the boiling water. Cooling the flask condenses the vapor. The weight of the vapor can be measured by weighing the liquid, and M, the molar mass of the gas, can be calculated.

---

### EXAMPLE

A sample of an unknown liquid is vaporized in an Erlenmeyer flask with a volume of 250 mL. At 100°C the vapor exerts a pressure of 0.975 atm. The condensed vapor weighs 0.685 g. Determine the molar mass of the unknown liquid.

    Using the equation

$$M = \frac{m \times R \times T}{P \times V}$$

we can substitute all the known values:

$$M = \frac{(0.685 \text{ g}) (0.0821 \text{ L atm/mole K}) (373 \text{ K})}{(0.975 \text{ atm}) (0.250 \text{ L})} = 86.1 \text{ g/mole}$$

    Hexane, with the molecular formula $C_6H_{14}$, has a molar mass of 86.17 g/mole.

---

### Objectives

1. Experimentally determine the mass of the vapor of a volatile liquid.
2. Calculate the molar mass of the liquid by applying the ideal gas equation to the vapor.

## Procedure

1. Obtain a sample of unknown liquid from your instructor. Record the code number of the unknown liquid on the Report Sheet (1). The unknown liquid will be one of the liquids found in Table 14.1.

| Table 14.1 | Volatile Liquid Unknowns | | |
|---|---|---|---|
| **Liquid** | **Formula** | **Molar Mass** | **Boiling Point (°C at 1 atm)** |
| Pentane | $C_5H_{12}$ | 72.2 | 36.2 |
| Acetone | $C_3H_6O$ | 58.1 | 56.5 |
| Methanol | $CH_4O$ | 32.0 | 64.7 |
| Hexane | $C_6H_{14}$ | 86.2 | 69.0 |
| Ethanol | $C_2H_6O$ | 46.1 | 78.5 |
| 2-Propanol | $C_3H_8O$ | 60.1 | 82.3 |

2. Weigh together a clean, dry 125-mL Erlenmeyer flask, a 2.5- by 2.5-in. square of aluminum foil, and a 4-in. piece of copper wire. Record the total weight to the nearest 0.001 g (2).

3. Pour approximately 3 mL of the liquid into the flask. Cover the mouth of the flask with the aluminum foil square and crimp the edges tightly over the neck of the flask. Secure the foil by wrapping the wire around the neck and twisting the ends by hand.

4. With a larger square of aluminum foil (3 × 3 in.), secure a second cover on the mouth of the flask with a rubber band.

5. Carefully punch a *single, small* hole in the foil covers with a needle or pin. The assembly is now prepared for heating (Fig. 14.1).

**Figure 14.1**
Assembly for vaporization
of the liquid.

Pinhole

Liquid

6. Take a 1000-mL beaker and add 300 mL of water and a few boiling chips. Heat the water to boiling using a hot plate. Regulate the heat so the boiling does not cause the water to splash.

7. Immerse the flask containing the volatile unknown liquid in the boiling water so that most of the flask is beneath the hot water as shown in Fig. 14.2. (You may need to weigh down the flask with a test tube clamp or a lead sinker.)

**Figure 14.2**
Assembly for the determination of molar mass.

8. Observe the unknown liquid. There is more liquid than is required to fill the flask with vapor. As the liquid evaporates, the level will decrease and excess vapor will escape through the pin hole. When all the liquid appears to be gone, continue to heat for an additional 5 min.

9. Record the temperature of the boiling water (5). Record the temperature of the vapor in the flask (6).

10. Using tongs, carefully remove the flask from the water and set it on the laboratory bench.

11. While the flask cools to room temperature, record the barometric pressure in the laboratory (7).

12. When the flask has cooled to room temperature, wipe dry the outside of the flask with a paper towel. Carefully remove only the second foil cover; blot dry the first foil cover with a paper towel. Take a look inside the flask. Are droplets of liquid present?

13. Weigh the flask, foil cover, wire, and condensed liquid. Record to the nearest 0.001 g on the Report Sheet (3). Determine the weight of the condensed liquid (4).

14. Remove the cover from the flask; save the aluminum foil and wire. Discard the unknown liquid that is in the flask in a container for waste organic liquids. Then rinse the flask with water and refill with water to the rim; dry the outside of the flask. Carefully and without spilling, weigh the flask with water, foil, and wire to the nearest 0.01 g (8).

15. Determine the weight of the water (9). Determine the temperature of the water in the flask (10). Look up the density of water at that temperature in the *Handbook of Chemistry and Physics,* and, using this value, calculate the volume of water in the flask (11). (You may check the volume in the flask by measuring the water with a graduated cylinder.)

16. Calculate the molar mass of the unknown (12). Identify the unknown (13).

---

### Chemicals and Equipment

1. Aluminum foil
2. Tongs
3. Hot plate
4. Copper wire
5. Rubber bands
6. Boiling chips
7. Absorbent paper towels
8. Unknown liquid chosen from Table 14.1
9. Lead sinkers

---

## Experiment 14

**PRE-LAB QUESTIONS**

**1.** What is an *ideal gas*?

**2.** What happens to an ideal gas at absolute zero if it would not condense and solidify? How does a real gas differ from an ideal gas?

**3.** What is the coldest possible temperature in degrees Fahrenheit, in degrees Celsius, and in degrees Kelvin?

**4.** A gas weighing 17.10 g occupies a 5.00-L flask at 30°C and 750 torr of pressure. What is the molar mass of the gas? (Remember to use proper units: V in liters, T in Kelvin, P in atmospheres.) Show your work.

## Experiment 14

**REPORT SHEET**

1. Code number of unknown liquid _____

2. Weight of 125-mL Erlenmeyer flask, aluminum foil, and copper wire _____ g

3. Weight of cooled flask, foil, wire, and condensed liquid _____ g

4. Weight of condensed liquid: (3) − (2) _____ g

5. Temperature of boiling water _____ °C _____ K

6. Temperature of vapor in flask _____ °C _____ K

7. Barometric pressure _____ atm

8. Weight of 125-mL Erlenmeyer flask, aluminum foil, copper wire, and water _____ g

9. Weight of water: (8) − (2) _____ g

10. Temperature of water _____ °C

11. Volume of the flask: (9)/(density of water) _____ mL _____ L

12. Molar mass of the unknown:

$$M = \frac{(4) \times (0.0821 \text{ L atm/K mole}) \times (6)}{(7) \times (11)}$$ _____ g/mole

13. The unknown is _____

## POST-LAB QUESTIONS

1. When do real gases approximate the behavior of an ideal gas? Do the conditions for this experiment meet these criteria? Explain.

2. Instead of measuring the volume of the Erlenmeyer flask as directed in step no. 15, a student recorded the volume written on the flask, in this case 125 mL. Recalculate the molar mass using this value for the volume and determine the percent error that results.

3. How will the calculated molar mass be affected if all of the liquid in the flask did not completely vaporize?

4. A student took the atmospheric pressure to be 1 atm instead of recording the actual barometric pressure (which in the student's case was 750 mm Hg). Will the calculated molar mass be too high, too low, or show no change?

5. A student had an unknown liquid and wished to determine its molecular weight. The following data was collected using the procedure of this experiment:

   a. Weight of the 125-mL Erlenmeyer flask and cover (aluminum foil and copper wire), 79.621 g;

   b. Weight of cooled flask and cover containing condensed liquid, 79.983 g;

   c. Temperature of boiling water, 100.0°C;

   d. Volume of the 125-mL Erlenmeyer flask, 128.0 mL;

   e. Atmospheric pressure, 760.0 mm Hg.

   Using this data, calculate the liquid's molar mass and identify the unknown from Table 14.1. Show all of your work.

# Experiment  15

## Physical properties of chemicals: melting point, sublimation, and boiling point

If you were asked to describe a friend, most likely you would start by identifying particular physical characteristics. You might begin by giving your friend's height, weight, hair color, eye color, or facial features. These characteristics would allow you to single out the individual from a group.

Chemicals also possess distinguishing physical properties which enable their identification. In many circumstances, a thorough determination of the physical properties of a given chemical can be used for its identification. If faced with an unknown sample, a chemist may compare the physical properties of the unknown to properties of known substances that are tabulated in the chemical literature; if a match can be made, an identification can be assumed (unless chemical evidence suggests otherwise).

The physical properties most commonly listed in handbooks of chemical data are color, crystal form (if a solid), refractive index (if a liquid), density (discussed in Experiment 3), solubility in various solvents (discussed in Experiment 16), melting point, sublimation characteristics, and boiling point. When a new compound is isolated or synthesized, these properties almost always accompany the report in the literature.

The transition of a substance from a solid to a liquid to a gas, and the reversal, represent physical changes. In a physical change there is only a change in the form or state of the substance. No chemical bonds break; no alteration in the chemical composition occurs. Water undergoes state changes from ice to liquid water to steam; however, the composition of molecules in all three states remains $H_2O$.

$$H_2O(s) \rightleftarrows H_2O(l) \rightleftarrows H_2O(g)$$
$$\text{Ice} \qquad\qquad \text{Liquid} \qquad\qquad \text{Steam}$$

The *melting* or *freezing point* of a substance refers to the temperature at which the solid and liquid states are in equilibrium. The terms are interchangeable and correspond to the same temperature; how the terms are applied depends upon the state the substance is in originally. The melting point is the temperature at equilibrium when starting in the solid state and going to the liquid state. The freezing point is the temperature at equilibrium when starting in the liquid state and going to the solid state.

Melting points of pure substances occur over a very narrow range and are usually quite sharp. The criteria for purity of a solid is the narrowness of the melting point range and the correspondence to the value found in the literature. Impurities will lower the melting point and cause a broadening of the range. For example, pure benzoic acid has a reported melting point of 122.13°C; benzoic acid with a melting point range of 121–122°C is considered to be quite pure.

The *boiling point* or *condensation* point of a liquid refers to the temperature when its vapor pressure is equal to the external pressure. If a beaker of liquid is brought to a boil in your laboratory, bubbles of vapor form throughout the liquid. These bubbles rise rapidly to the surface, burst, and release vapor to the space above the liquid. In this case, the liquid is in contact with the atmosphere; the normal boiling point of the liquid will be the temperature when the pressure of the vapor is equal to the atmospheric pressure (1 atm or 760 mm Hg). Should the external pressure vary, so will the boiling point. A liquid will boil at a higher temperature when the external pressure is higher and will boil at a lower temperature when the external pressure is reduced. The change in state from a gas to a liquid represents condensation and is the reverse of boiling. The temperature for this change of state is the same as the boiling temperature but is concerned with the approach from the gas phase.

Just as a solid has a characteristic melting point, a liquid has a characteristic boiling point. At one atmosphere, pure water boils at 100°C, pure ethanol (ethyl alcohol) boils at 78.5°C, and pure diethyl ether boils at 34.6°C. The vapor pressure curves shown in Fig. 15.1 illustrate the variation of the vapor pressure of these liquids with temperature. One can use these curves to predict the boiling point at a reduced pressure. For example, diethyl ether has a vapor pressure of 422 mm Hg at 20°C. If the external pressure were reduced to 422 mm Hg, diethyl ether would boil at 20°C.

*Sublimation* is a process that involves the direct conversion of a solid to a gas without passing through the liquid state. Relatively few solids do this at atmospheric pressure. Some examples are the solid compounds naphthalene (mothballs), caffeine, iodine, and solid carbon dioxide (commercial Dry Ice). Water, on the other hand, sublimes at −10°C and at 0.001 atm. Sublimation temperatures are not as easily obtained as melting points or boiling points.

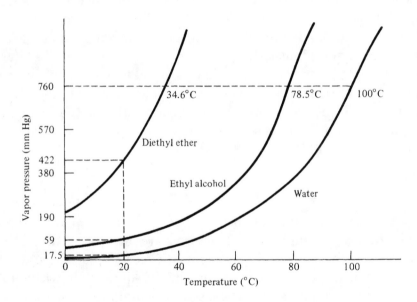

**Figure 15.1** • Diethyl ether, ethyl alcohol (ethanol), and water vapor pressure curves.

---

**Objectives**

1. To use melting points and boiling points in identifying substances.
2. To use sublimation as a means of purification.

## Melting Point Determination

1. Unknowns are provided by the instructor. Obtain approximately 0.1 g of unknown solid and place it on a small watch glass. Record the code number of the unknown on the Report Sheet (1). (The instructor will weigh out a 0.1-g sample as a demonstration; take approximately that amount with your spatula.) Carefully crush the solid on a watch glass into a powder with the flat portion of a spatula.

2. Obtain a melting point capillary tube. One end of the tube will be sealed. The tube is packed with solid in the following way:

   **Step A** Press the open end of the capillary tube vertically into the solid sample (Fig. 15.2 A). A small amount of sample will be forced into the open end of the capillary tube.

   **Step B** Invert the capillary tube so that the closed end is pointing toward the bench top. Gently tap the end of the tube against the lab bench top (Fig. 15.2 B). Continue tapping until the solid is forced down to the closed end. A sample depth of 5–10 mm is sufficient.

**Figure 15.2**
Packing a capillary tube.

A. Forcing solid into the capillary tube.

Step A

B. Tapping to force down solid.

Step B

C. Alternative method for bringing the solid down.

Step C

   **Step C** An alternative method for bringing the solid sample to the closed end uses a piece of glass tubing of approximately 20 to 30 cm. Hold the capillary tube, closed end down, at the top of the glass tubing, held vertically; let the capillary tube drop through the tubing so that it hits the lab bench top. The capillary tube will bounce and bring the solid down. Repeat if necessary (Fig. 15.2 C).

**3.** The melting point may be determined using either a commercial melting point apparatus or a Thiele tube.

   **a.** A commercial melting point apparatus will be demonstrated by your instructor.

   **b.** The use of the Thiele tube is as follows:

   - Attach the melting point capillary tube to the thermometer by means of a rubber ring. Align the mercury bulb of the thermometer so that the tip of the melting point capillary containing the solid is next to it (Fig. 15.3).

**Figure 15.3**
Proper alignment of the capillary tube and the mercury bulb.

Aligned

   - Use an extension clamp to support the Thiele tube on a ring stand. Add mineral oil or silicone oil to the Thiele tube, and fill to a level above the top of the side arm. Use a thermometer clamp to support the thermometer with the attached melting point capillary tube in the oil. The bulb and capillary tube should be immersed in the oil; keep the rubber ring and open end of the capillary tube out of the oil (Fig. 15.4).

   - Heat the arm of the Thiele tube very slowly with a Bunsen burner flame. Use a small flame and gently move the burner along the arm of the Thiele tube.

   - You should position yourself so that you can follow the rise of the mercury in the thermometer as well as observe the solid in the capillary tube. Record the temperature when the solid begins to liquefy (2) (the solid will appear to shrink). Record the temperature when the solid is completely a liquid (3). Express these readings as a melting point range (4).

   - Identify the solid by comparing the melting point with those listed in Table 15.1 for different solids (5).

**Figure 15.4**
Thiele tube apparatus.

Rubber band

Oil level

| Table 15.1 | Melting Points of Selected Solids | |
|---|---|---|
| **Solid** | **Melting Point (°C)** | **Use** |
| Acetamide | 82 | Plasticizer; stabilizer |
| Acetanilide | 114 | Manufacture of other medicinals and dyes |
| Adipic acid | 152 | Manufacture of nylon |
| Benzophenone | 48 | Manufacture of antihistamines, hypnotics, insecticides |
| Benzoic acid | 122 | Preserving foods; antifungal agent |
| *p*-Dichlorobenzene | 54 | Moth repellent; insecticidal fumigant |
| Naphthalene | 80 | Moth repellent; insecticide |
| Stearic acid | 70 | Suppositories; ointments |

4. Do as many melting point determinations as your instructor may require. Just remember to use a new melting point capillary tube for each melting point determination.

5. Dispose of the solids as directed by your instructor.

## Purification of Naphthalene by Sublimation

1. Place approximately 0.5 g of impure naphthalene into a 100-mL beaker. (Your instructor will weigh out 0.5 g of sample as a demonstration; with a spatula take an amount which approximates this quantity.)

2. Into the 100-mL beaker place a smaller 50-mL beaker. Fill the smaller one halfway with ice cubes or ice chips. Place the assembled beakers on a wire gauze supported by a ring clamp (Fig. 15.5).

**Figure 15.5**
Setup for sublimation
of naphthalene.

Ice cubes

Pure naphthalene

Crude naphthalene

3. Using a small Bunsen burner flame, gently heat the bottom of the 100-mL beaker by passing the flame back and forth beneath the beaker.

4. You will see solid flakes of naphthalene collect on the bottom of the 50-mL beaker. When a sufficient amount of solid has been collected, turn off the burner.

5. Pour off the ice water from the 50-mL beaker and carefully scrape the flakes of naphthalene onto a piece of filter paper with a spatula.

6. Take the melting point of the pure naphthalene and compare it to the value listed in Table 15.1 (6).

7. Dispose of the crude and pure naphthalene as directed by your instructor.

## Boiling Point Determination

---

### CAUTION!

The chemicals used for boiling point determinations are flammable. Be sure all Bunsen burner flames are extinguished before starting this part of the experiment.

---

1. Obtain an unknown liquid from your instructor and record its code number on the Report Sheet (7).

2. Clamp a clean, dry test tube (100 × 13 mm) onto a ring stand. Add to the test tube approximately 3 mL of the unknown liquid and two small boiling chips. Lower the test tube into a 250-mL beaker which contains 100 mL of water and two boiling chips. Adjust the depth of the test tube so that the unknown liquid is below the water level of the water bath (Fig. 15.6).

**Figure 15.6**
Setup for determining the boiling point.

3. With a thermometer clamp, securely clamp a thermometer and lower it into the test tube through a neoprene adapter. Adjust the thermometer so that it is approximately 1 cm above the surface of the unknown liquid.

4. Use a piece of aluminum foil to cover the mouth of the test tube. (Be certain that the test tube mouth has an opening; the system should not be closed.)

5. Gradually heat the water in the beaker with a hot plate and watch for changes in temperature. As the liquid begins to boil, the temperature above the liquid will rise. When the temperature no longer rises but remains constant, record the temperature to the nearest 0.1°C (8). This is the observed boiling point. From the list in Table 15.2, identify your unknown liquid by matching your observed boiling point with the compound whose boiling point best corresponds (9).

| Table 15.2 | Boiling Points of Selected Liquids | |
|---|---|---|
| **Liquid** | **Boiling Point (°C at 1 atm)** | **Use** |
| Acetone | 56 | Solvent; paint remover |
| Cyclohexane | 81 | Solvent for lacquers and resins |
| Ethyl acetate | 77 | Solvent for airplane dopes; artificial fruit essence |
| Hexane | 69 | Liquid in thermometers with blue or red dye |
| Methanol (methyl alcohol) | 65 | Solvent; radiator antifreeze |
| 1-Propanol | 97 | Solvent |
| 2-Propanol (isopropyl alcohol) | 83 | Solvent for shellac; essential oils; body rubs |

6. Do as many boiling point determinations as required by your instructor.

7. Dispose of the liquid as directed by your instructor.

## Chemicals and Equipment

1. Aluminum foil
2. Boiling chips
3. Bunsen burner
4. Hot plate
5. Commercial melting point apparatus (if available)
6. Melting point capillary tubes
7. Rubber rings
8. Thiele tube melting point apparatus
9. Thermometer clamp
10. Glass tubing
11. Acetamide
12. Acetanilide
13. Acetone
14. Adipic acid
15. Benzophenone
16. Benzoic acid
17. Cyclohexane
18. *p*-Dichlorobenzene
19. Ethyl acetate
20. Hexane
21. Methanol (methyl alcohol)
22. Naphthalene (pure)
23. Naphthalene (impure)
24. 1-Propanol
25. 2-Propanol (isopropyl alcohol)
26. Stearic acid

## Experiment 15

### PRE-LAB QUESTIONS

**1.** What is meant by the term *physical change*?

**2.** When water boils and passes from the liquid phase to the gaseous phase, has there been any change in the chemical composition of the water?

**3.** How could you get water to boil at 60°C?

**4.** Solid naphthalene is heated hot enough to become a gas at 1 atm. Describe all the phase changes that naphthalene passes through. What is the name of the process that naphthalene undergoes?

**5.** Frozen ethanol is heated and eventually becomes a gas at 1 atm. Describe all the phase changes that ethanol passes through and the names for these phase changes.

**5.** How can you determine whether a solid substance is pure or impure?

## Experiment 15

### REPORT SHEET

#### *Melting point determination*

|  | Trial No. 1 | Trial No. 2 |
|---|---|---|
| **1.** Code number of unknown | _____ | _____ |
| **2.** Temperature melting begins | _____ °C | _____ °C |
| **3.** Temperature melting ends | _____ °C | _____ °C |
| **4.** Melting point range | _____ °C | _____ °C |
| **5.** Identification of unknown | _____ | _____ |

#### *Purification of naphthalene by sublimation*

| | | |
|---|---|---|
| **6.** Melting point range | _____ °C | _____ °C |

#### *Boiling point determination*

| | | |
|---|---|---|
| **7.** Unknown number | _____ | _____ |
| **8.** Observed boiling point | _____ °C | _____ °C |
| **9.** Identification of unknown | _____ | _____ |

## POST-LAB QUESTIONS

1. A student did a melting point determination for a sample of benzoic acid and found a melting point of 108-118°C. What conclusion can the student make about the sample?

2. Woolen clothing was placed in a closed chest along with some *p*-dichlorobenzene, moth balls. Four months later when the chest was opened, the moth balls were gone, but the chest had a strong odor from the moth balls. What happened to the moth balls?

3. A student in Denver, Colorado (nicknamed the "Mile-High City") did this laboratory at her school. However, none of the boiling points from Table 15.2 could be obtained; all were noticeably lower. Assuming everything was done as directed with a properly calibrated thermometer, what would account for the observations?

4. Aspirin is a white solid easily synthesized in the laboratory (see Experiment 33). A pure sample melts at 135°C. What would a synthetic chemist do to quickly determine whether the sample just made is pure or impure? What would the chemist look for?

# Solubility and solution

## Background

Most materials encountered in every day life are *mixtures*. This means that more than one component is found together in a system. Think back to your morning breakfast beverage; orange juice, coffee, tea, and milk are examples of mixtures.

Some mixtures have special characteristics. A mixture that is uniform throughout, with no phase boundaries, is called a *homogeneous mixture*. If you were to sample any part of the system, the same components in the same proportions would be found in each sample. The most familiar of these homogeneous mixtures is the liquid *solution*; here a *solute* (either a solid or a liquid) is thoroughly and uniformly dispersed into a *solvent* (a liquid). If the solution were allowed to remain standing, the components would not separate, no matter how much time was allowed to pass.

There are limits as to how much solute may be dispersed or dissolved in a given amount of solvent. This limit is the *solubility* and is defined as the *maximum weight of solute that dissolves in 100 g of a given solvent at a given temperature*. For example, sucrose (or table sugar) is soluble to the extent of 203.9 g per 100 g of water at 20°C. This means that if you have 100 g of water, you can dissolve up to 203.9 g of table sugar, but no more, in that quantity of water at 20°C. If more is added, the extra amount sinks to the bottom undissolved. A solution in this state is referred to as *saturated*. A solution with less than the maximum at the same temperature is called *unsaturated*. Solubility also varies with temperature (Fig. 16.1).

Liquids dissolved in liquids similarly may form homogeneous solutions. Some liquids have limited solubility in water. Diethyl ether, $CH_3CH_2OCH_2CH_3$ (an organic liquid), is soluble to the extent of 4 g per 100 g of water at 25°C; an excess of the diethyl ether will result in a separation of phases with the less dense organic liquid floating on the water. Some liquids mix in all proportions; these liquids are completely *miscible*. The mixture of commercial antifreeze, ethylene glycol, $HOCH_2CH_2OH$, and water, used as a coolant in automobile radiators, is such a solution.

The solubility of a given solute in a particular solvent depends on a number of factors. One generalization which can be used for determining solubility is "like dissolves like." This means that the more similar the polarity of a solute is to the polarity of the solvent, the more likely the two will form a homogeneous solution. A polar solvent, such as water, will dissolve a polar compound: an ionic salt like common table salt, NaCl, will dissolve in water; a polar covalent solid like table sugar, sucrose, will dissolve in water. Nonpolar solvents such as naphtha or turpentine will dissolve nonpolar material, such as grease or oil. On the other hand, oil and water do not mix because of their different polar characteristics.

**Figure 16.1** • The effect of temperature on the solubility of some solutes in water.

When ionic salts dissolve in water, the individual ions separate. These positively and negatively charged particles in the water medium are mobile and can move from one part of a solution to another. Because of this movement, solutions of ions can conduct electricity. *Electrolytes* are substances which can form ions when dissolved in water and can conduct an electric current. These substances are also capable of conducting an electric current in the molten state. *Nonelectrolytes* are substances which do not conduct an electric current. Electrolytes may be further characterized as either strong or weak. A strong electrolyte dissociates almost completely when in a water solution; it is a good conductor of electricity. A weak electrolyte has only a small fraction of its particles dissociated into ions in water; it is a poor conductor of electricity. Table 16.1 lists examples of compounds behaving as electrolytes or nonelectrolytes in a water solution.

| Table 16.1 Selected Electrolytes and Nonelectrolytes | | |
|---|---|---|
| **Strong Electrolytes** | **Weak Electrolytes** | **Nonelectrolytes** |
| Sodium chloride, $NaCl$ | Acetic acid, $CH_3CO_2H$ | Methanol, $CH_3OH$ |
| Sulfuric acid, $H_2SO_4$ | Carbonic acid, $H_2CO_3$ | Benzene, $C_6H_6$ |
| Hydrochloric acid, $HCl$ | Ammonia, $NH_3$ | Acetone, $(CH_3)_2CO$ |
| Sodium hydroxide, $NaOH$ | | Sucrose, $C_{12}H_{22}O_{11}$ |

## Objectives

1. To show how temperature affects solubility.
2. To demonstrate the difference between electrolytes and nonelectrolytes.
3. To show how the nature of the solute and the solvent affects solubility.

## Procedure

### Saturated Solutions

1. Place 10 mL of distilled water into a 50-mL beaker; record the temperature of the water (1).

2. While stirring with a glass rod, add sucrose to the water in 2-g portions; keep adding until no more sucrose dissolves. The solution should be saturated. Record the mass of sucrose added (2).

3. Heat the solution on a hot plate to 50°C; maintain this temperature. Again add to the solution, while stirring, sucrose in 2-g portions until no more sucrose dissolves. Record the mass of sucrose added (3).

4. Heat the solution above 50°C until all of the solid dissolves. With beaker tongs, remove the beaker from the hot plate and set it on the bench top, out of the way. Place an applicator stick, or suspend a string, into the solution and allow the solution to cool. Continue with the next part of this experiment and return to this part after the solution cools to room temperature.

5. Observe what happened to the solution when it cooled to room temperature. Offer an explanation for what has taken place (5). (If no crystals have formed, drop into the solution a single crystal of sucrose or stir the solution with a stirring rod.)

### Electrical Conductivity

This part of the experiment can be done in pairs. Obtain and set up a conductivity apparatus (Fig. 16.2). It consists of two terminals connected to a light bulb and a plug for connection to a 110-volt electrical wall outlet.

**Figure 16.2**
Conductivity apparatus.

Nonelectrolyte           Electrolyte

**CAUTION!**

To avoid a shock, do not touch the terminals when the apparatus is plugged in. Be sure to unplug the apparatus between tests and while rinsing and drying. Do not let the terminals touch each other.

The following solutions are to be tested with the conductivity apparatus:

**a.** distilled water

**b.** tap water

**c.** 1 M NaCl

**d.** 0.1 M NaCl

**e.** 1 M sucrose, $C_{12}H_{22}O_{11}$

**f.** 0.1 M sucrose, $C_{12}H_{22}O_{11}$

**g.** 1 M HCl

**h.** 0.1 M HCl

**i.** glacial acetic acid, $CH_3CO_2H$

**j.** 0.1 M acetic acid, $CH_3CO_2H$

**1.** For each solution follow steps 2, 3, 4, and 5.

**2.** Place about 20 mL of the solution to be tested into a 30-mL beaker that has been rinsed with distilled water. A convenient way to rinse the beaker is with a squeezable plastic wash bottle. Direct a stream of water from the wash bottle into the beaker, swirl the water about, and discard the water into the sink.

**3.** Lower the terminals into the beaker so that the solution covers the terminals. For each test solution, try to keep the same distance between the terminals and the terminals submerged to the same depth.

**4.** Plug the apparatus into the wall socket. Observe the effect on the light bulb. A solution containing an electrolyte conducts electricity—the circuit is completed and the bulb will light. Strong electrolytes will give a bright light; weak electrolytes will give a dim light; nonelectrolytes will give no light. Note the effect of concentration. Record your observations on the Report Sheet.

**5.** Between each test, disconnect the conductivity apparatus from the wall socket, raise the terminals from the solution, and rinse the terminals with distilled water from the wash bottle.

## Solubility: Solute and Solvent Characteristics

**1.** Clean and dry 16 test tubes (100 × 13 mm).

**2.** Place approximately 0.1 g of the following solids into test tubes numbered as indicated (your instructor will weigh exactly 0.1 g of solid as a demonstration; use your spatula to estimate the 0.1-g sample):

  **a.** No. 1: table salt, NaCl

  **b.** No. 2: table sugar, sucrose, $C_{12}H_{22}O_{11}$

  **c.** No. 3: naphthalene, $C_{10}H_8$

  **d.** No. 4: iodine, $I_2$

**3.** Add 4 mL of distilled water to each test tube and shake the mixture (sharp tapping of the test tube with your fingers will agitate the contents enough).

**4.** Record on the Report Sheet whether the solid dissolved completely (soluble), partially (slightly soluble), or not at all (insoluble).

**5.** With new sets of labeled test tubes containing the solids listed above, repeat the solubility tests using the solvents ethanol (ethyl alcohol), $C_2H_5OH$, acetone, $(CH_3)_2CO$, and petroleum ether in place of the water. Record your observations.

**6.** Discard your solutions in waste containers provided. *Do not discard into the sink.*

---

### Chemicals and Equipment

  **1.** Sucrose (solid and solutions)
  **2.** NaCl (solid and solutions)
  **3.** Naphthalene
  **4.** Iodine
  **5.** HCl solutions
  **6.** Acetic acid (glacial and solutions)
  **7.** Ethanol (ethyl alcohol)
  **8.** Acetone
  **9.** Petroleum ether
  **10.** Conductivity apparatus
  **11.** Hot plate
  **12.** Wash bottle
  **13.** Beakers, 30-mL

## Experiment 16

**PRE-LAB QUESTIONS**

**1.** What are the two components of a homogeneous liquid solution?

**2.** When is a liquid solution saturated?

**3.** Refer to Fig. 16.1. Look at the solubility curves for potassium nitrate, $KNO_3$, and sodium nitrate, $NaNO_3$. Which salt is more soluble: (a) at 20°C; (b) at 80°C? At what temperature are the salts equally soluble?

**4.** Account for the following: oil and terpentine mix, but oil and water do not mix.

**5.** What does this statement mean: sodium chloride, $NaCl$, is a strong electrolyte.

## Experiment 16

### REPORT SHEET

#### Saturated solution

1. Temperature of distilled water          _____ °C

2. Mass of sucrose                         _____ g/10 mL

3. Mass of additional sucrose              _____ g

4. Total mass of sucrose: (2) + (3)        _____ g/10 mL at 50°C

5. Observations and explanation

#### Electrical conductivity

Rate the brightness of the light bulb on a scale from 0 to 5: 0 for no light to 5 for very bright light.

| Substance | Observation |
| --- | --- |
| Distilled water | _____ |
| Tap water | _____ |
| 1 M NaCl | _____ |
| 0.1 M NaCl | _____ |
| 1 M sucrose | _____ |
| 0.1 M sucrose | _____ |
| 1 M HCl | _____ |
| 0.1 M HCl | _____ |
| Glacial acetic acid | _____ |
| 0.1 M acetic acid | _____ |

## Solubility: solute and solvent characteristics

Record the solubility as soluble (s), slightly soluble (ss), or insoluble (i).

| Solute | Solvent | | | |
|---|---|---|---|---|
| | Water | Ethanol | Acetone | Petroleum Ether |
| Table salt, NaCl | | | | |
| Table sugar, sucrose | | | | |
| Naphthalene | | | | |
| Iodine | | | | |

## POST-LAB QUESTIONS

1. On the basis of your observations, are HCl solutions electrolytes or nonelectrolytes? Explain your answer.

2. Both acetic acid and HCl solutions react as acids. However, only the HCl solutions caused the light bulb to shine brightly while the acetic acid solutions caused the light bulb to shine dimly. How do you account for these observations?

3. Why does naphthalene dissolve in petroleum ether (an organic solvent similar to gasoline) but not in water?

4. Distilled water shows no conductivity since the light bulb does not go on and thus, is a nonelectrolyte. However, tap water does show some conductivity since the light goes on faintly. Explain this result.

# Water of hydration

Some compounds do not melt when heated but undergo decomposition. In decomposing, the compound can break down irreversibly or reversibly into two or more substances. If it is reversible, recombination leads to reformation of the original material. Hydrates are examples of compounds which do not melt but which decompose upon heating. The decomposition products are an anhydrous salt and water. The original hydrates can be regenerated by addition of water to the anhydrous salt.

Hydrates contain water as an integral part of the crystalline structure of the salt. When a hydrate crystallizes from an aqueous solution, water molecules are bound to the metal ion. A characteristic of the metal is the number of water molecules that bind to the metal ion. These water molecules are called *waters of hydration* and are in a definite proportion. Thus when copper sulfate crystallizes from water, the blue salt copper(II) sulfate pentahydrate, $CuSO_4 \cdot 5H_2O$, forms. As indicated by the formula, 5 waters of hydration are bound to the copper(II) ion in copper sulfate. Notice how the formula is written—the waters of hydration are separated from the formula of the salt by a *dot*.

Heat can transform a hydrate into an anhydrous salt. The water can often be seen escaping as steam. For example, the blue crystals of copper(II) sulfate pentahydrate can be changed into a white powder, the anhydrous salt, by heating to approximately 250°C.

$$CuSO_4 \cdot 5H_2O(s) \xrightarrow{\phantom{xx} 250°C \phantom{xx}} CuSO_4(s) + 5H_2O(g)$$
$$\text{Blue} \qquad\qquad\quad \text{White}$$

This process is reversible; adding water to the white anhydrous copper sulfate salt will rehydrate the salt and regenerate the blue pentahydrate. Not all hydrates behave in this way. For example, the waters of hydration that are bound to the iron (III) ion, $Fe^{3+}$ in the salt iron(III) chloride hexahydrate, $FeCl_3 \cdot 6H_2O$, are held very strongly; intense heat will not drive off the water.

Some anhydrous salts are capable of becoming hydrated upon exposure to the moisture in their surroundings. These salts are called *hygroscopic salts* and can be used as chemical drying agents or *desiccants*. Some salts are such excellent desiccants and are able to absorb so much moisture from their surroundings that they can eventually dissolve themselves! Calcium chloride, $CaCl_2$, is such a salt and is said to be *deliquescent*.

There are commercial areas where these salts and their hydrates are used. Containers holding pharmaceutical pills often have small packets of desiccant to control moisture so the pills last longer. Fertilizers will become wet and sticky as they absorb moisture from the air; some will even "turn to liquid" after some time as they absorb so much water that they dissolve. Some humidity indicators use cobalt or copper salts and vary in color as the moisture in the air varies.

Since many hydrates contain water in a stoichiometric quantity, it is possible to determine the molar ratio of water to salt. First, you would determine the weight of the water lost from the hydrate by heating a weighed sample. From the weight of the water lost, you then can calculate the percent of water in the hydrate. From the weight of the water lost you can also determine the number of water molecules in the hydrate salt and thus the molar ratio.

---

**EXAMPLE**

A sample of Epsom salt, the hydrate of magnesium sulfate, 5.320 g, lost water on heating; the anhydrous salt, which remained, weighed 2.598 g.

**a.** The weight of the water lost:

| Weight of hydrate sample (g) | 5.320 g |
|---|---|
| − Weight of the anhydrous salt (g) | − 2.598 g |
| Weight of the water lost (g) | 2.722 g |

**b.** The percent by mass of water:

$$\frac{\text{Weight of water lost (g)}}{\text{Weight of hydrate sample (g)}} \times 100 = \frac{2.722 \text{ g}}{5.320 \text{ g}} \times 100 = 51.17\%$$

**c.** The number of moles of water lost:

$$\frac{\text{Weight of water lost (g)}}{\text{MW of water (g/mole)}} = \frac{2.722 \text{ g}}{18.02 \text{ g/mole}} = 0.1511 \text{ mole}$$

**d.** The number of moles of $MgSO_4$:

$$\frac{\text{Weight of } MgSO_4 \text{ anhydrous (g)}}{\text{MW of } MgSO_4 \text{ (g/mole)}} = \frac{2.598 \text{ g}}{120.4 \text{ g/mole}} = 0.02158 \text{ mole}$$

**e.** The mole ratio of $H_2O$ to anhydrous $MgSO_4$:

$$\frac{\text{Moles of } H_2O}{\text{Moles of } MgSO_4} = \frac{0.1511}{0.02158} = 7$$

Therefore, the formula of the hydrate of magnesium sulfate is $MgSO_4 \cdot 7H_2O$.

---

**Objectives**

1. To learn some properties and characteristics of hydrates.
2. To verify the percent of water in the hydrate of copper sulfate.
3. To verify that the mole ratio of water to salt in the hydrate of copper sulfate is fixed.

---

**Procedure**

**Properties of Anhydrous CaCl₂**

1. Take a small spatula full of anhydrous $CaCl_2$ and place it on a watch glass.

2. Set the watch glass to the side, out of the way, and continue the rest of the experiment. From time to time during the period, examine the solid and record your observations.

3. What happened to the solid $CaCl_2$ by the end of the period?

## Composition of a Hydrate

1. Obtain from your instructor a porcelain crucible and cover. Clean with soap and water and dry thoroughly with paper towels.

2. Place the crucible and cover in a clay triangle supported by a metal ring on a ring stand (Fig. 17.1). Heat the crucible with a Bunsen burner to red heat for 5 min. Using crucible tongs, place the crucible and cover on a wire gauze and allow it to cool to room temperature.

**Figure 17.1**
a) Heating the crucible.
b) Moving the crucible with crucible tongs.

3. Weigh the crucible and cover to the nearest 0.001 g (1).

4. Repeat this procedure (heating, cooling, weighing) until two successive weights of the covered crucible agree to within 0.005 g or less (2).

---

### CAUTION!

Handle the crucible and cover with the crucible tongs from this point on. This will avoid possible burns and will avoid transfer of moisture and oils from your fingers to the porcelain.

---

5. Add between 3 and 4 g of the hydrate of copper sulfate to the crucible. Weigh the covered crucible to the nearest 0.001 g (3). Determine the exact weight of the hydrate (4) by subtraction.

6. Place the covered crucible and contents in the clay triangle. Move the cover so that it is slightly ajar (Fig. 17.1a). Begin heating the crucible with a small flame for 5 min. (If any spattering of the solid occurs, remove the heat and completely cover the crucible.) Gradually increase the flame until the blue inner cone touches the bottom of the crucible. Heat to red hot for an additional 5 min.

7. Remove the heat. Using crucible tongs, place the covered crucible on a wire gauze. Have the crucible completely covered by the lid. Allow to cool to room temperature. Weigh to the nearest 0.001 g (5).

8. Repeat the procedure (heating, cooling, weighing) until two successive weights of the covered crucible and contents agree to within 0.005 g or less (6).

9. Determine the weight of the anhydrous copper sulfate (7) and the weight of the water lost (8).

10. Carry out the calculations indicated on the Report Sheet.

11. If time permits, repeat the procedure again for a Trial 2.

12. Before you discard the white anhydrous salt (as directed by the instructor), add a few drops of water to the salt. What happens (14)?

---

### Chemicals and Equipment

1. Crucible and cover
2. Ring stand
3. Clay triangle
4. Crucible tongs
5. Calcium chloride, $CaCl_2$
6. Copper(II) sulfate pentahydrate, $CuSO_4 \cdot 5H_2O$

## Experiment 17

### PRE-LAB QUESTIONS

**1.** What happens to most hydrates when they are heated? Do they melt?

**2.** What happens to calcium chloride in a moist environment?

**3.** Does the Law of Constant Composition apply to most hydrates? Explain.

**4.** Plaster of Paris is a hydrate of calcium sulfate, $(CaSO_4)_2 \cdot H_2O$.

  **a.** How many total atoms of all kinds are present?

  **b.** On heating the hydrate, how many moles of water should be driven off per mole of hydrate?

  **c.** Calculate the percent water in the hydrate. Show your work.

  **d.** If you heat 15.00 g of the hydrate and drive off the water, what is the weight of the anhydrous salt remaining? Show your work.

## Experiment 17

**REPORT SHEET**

*Observations on the properties of anhydrous CaCl₂*

| Composition of a hydrate | Trial 1 | Trial 2 |
|---|---|---|
| 1. Weight of crucible and cover, 1st heating | _____ g | _____ g |
| 2. Weight of crucible and cover, 2nd heating | _____ g | _____ g |
| 3. Weight of covered crucible plus sample | _____ g | _____ g |
| 4. Weight of sample (hydrate): (3) − (2) | _____ g | _____ g |
| 5. Weight of covered crucible plus sample, after 1st heating | _____ g | _____ g |
| 6. Weight of covered crucible plus sample, after 2nd heating | _____ g | _____ g |
| 7. Weight of anhydrous salt: (6) − (2) | _____ g | _____ g |
| 8. Weight of water lost: (4) − (7) | _____ g | _____ g |
| 9. Percent of water in hydrate: % = [(8)/(4)] × 100 | _____ % | _____ % |
| 10. Moles of water lost: (8)/18.00 g/mole | _____ mole | _____ mole |
| 11. Moles of anhydrous CuSO₄: (7)/159.6 g/mole | _____ mole | _____ mole |

**12.** Moles of water per mole of $CuSO_4$:
(10)/(11)

_____     _____

**13.** The formula for the hydrated
copper(II) sulfate

_____     _____

**14.** Observation: water added to the anhydrous copper(II) sulfate:

## POST-LAB QUESTIONS

**1.** In the procedure you are cautioned to handle the crucible and cover with tongs. Why the precaution? How will the results be affected if you used your fingers instead of tongs?

**2.** In addition to the water of hydration, your sample contained a volatile impurity. Would the presence of this contaminant cause your calculations for percent (9) to be too high or too low? Explain.

**3.** Gypsum is the hydrate of calcium sulfate, $CaSO_4 \cdot 2H_2O$. Calculate the percent water in the hydrate. Show your work.

**4.** Given a 10.00-g sample of washing soda, $Na_2CO_3 \cdot 10H_2O$, what weight of anhydrous sodium carbonate can be obtained after driving off the water? Show your work.

**5.** A student found the percent water in $CaCl_2 \cdot 6H_2O$ to be 46.45%. Determine the experimental error. Show your work.

# Experiment 18

## Colligative properties: freezing point depression and osmotic pressure

### Background

Certain properties of solutions depend only on the number of solute particles dissolved in a given amount of solvent and not on the nature of these particles. Such properties are called *colligative properties*. For example, one such property is the *freezing point depression*. One mole of any solute dissolved in 1000 g of water lowers the freezing point of the water by 1.86°C. We call this value, 1.86 degree/mole/1000 g water, the *freezing point depression constant* of water, **$K_f$**. Each solvent has a characteristic freezing point depression constant that is related to its heat of fusion. The nature of the solute does not matter.

This principle can be used in a number of practical ways. One application is the use of antifreeze in car radiators. Since water expands on freezing, the water in a car's cooling system can crack the engine block of a parked car when the outside temperature falls below 0°C. The addition of a common antifreeze, ethylene glycol, prevents this because the freezing point is depressed and the water–ethylene glycol mixture freezes at a much lower temperature.

The freezing point depression, $\Delta T$, is proportional to the number of particles of the solute (moles) in 1000 g of solvent and the proportionality constant is the freezing point depression constant, $K_f$.

$$\Delta T = K_f \times \text{mole solute/1000 g solvent} \tag{1}$$

For example, if we add 275 g of ethylene glycol (molecular weight 62.0) per 1000 g of water in a car radiator, what will the freezing point of the solution be?

$$\Delta T = \frac{1.86°C}{\text{mole/1000 g}} \times \frac{275 \text{ g}}{62.0 \text{ g}} \times \frac{1 \text{ mole}}{1000 \text{ g}} = 8.26°C$$

The freezing point of water will be lowered from 0°C to −8.26°C.

If a solute is ionic, then each mole of solute dissociates. For NaCl we get two moles of ions, and for $Na_2SO_4$ three moles of ions for each mole of the solute. For convenience in calculation, we define a new term, *osmole*, as moles multiplied by the number of particles produced by one molecule of solute in solution.

In the present experiment, we will obtain the freezing point depression constant, $K_f$, for lauric acid, $CH_3(CH_2)_{10}COOH$, which will serve as a solvent. We will use benzoic acid, $C_6H_5COOH$, as a solute. In order to obtain the $K_f$, you will measure the freezing points of lauric acid and a mixture of the lauric acid–benzoic acid system. In actuality, you will measure the melting points of the solids. Freezing point or melting point is the temperature of transition between solid and liquid. Melting points (going from solid to liquid) can be measured more accurately than freezing points (going from liquid to solid). This is so because in freezing point measurements supercooling may occur which would yield a lower than correct freezing (melting) point.

In addition to freezing point depression, there are several other colligative properties, among which osmotic pressure is the most important biologically. Osmotic pressure develops whenever a semipermeable membrane separates a solution from a solvent. A *semipermeable membrane* is a material that contains tiny holes that are big enough to allow small solvent molecules to pass through but not big enough to allow large solute molecules to pass (Fig. 18.1). The passage of solvent molecules from the solvent side (right compartment) to the solution side (left compartment) of the semipermeable membrane generates the *osmotic pressure* that can be measured by the difference in the heights of the two columns.

Living cells, among them the red blood cells, are surrounded by semipermeable membranes. The osmolarity of most cells is 0.30 osmol. For example, a 0.89% w/v NaCl solution, normally referred to as physiological saline solution, has an osmolarity of 0.30. Thus when a cell is put in physiological saline solution, the osmolarity on both sides of the membrane is the same and therefore no osmotic pressure is generated across the membrane. Such a solution is called *isotonic*. On the other hand, if a cell is put in water (pure solvent) or in a solution which has lower osmolarity than the cell, there will be a net flow of water into the cell driven by the osmotic pressure. Such a solution is called *hypotonic*. A cell placed in a hypotonic solution will swell and eventually may burst. If that happens to a red blood cell, the process is called *hemolysis*. In contrast, a solution with higher osmolarity than the cell is called a *hypertonic* solution. A cell suspended in a hypertonic solution will shrivel; there is a net flow of water from the cell into the surroundings. When that happens to a red blood cell, the process is called *crenation*.

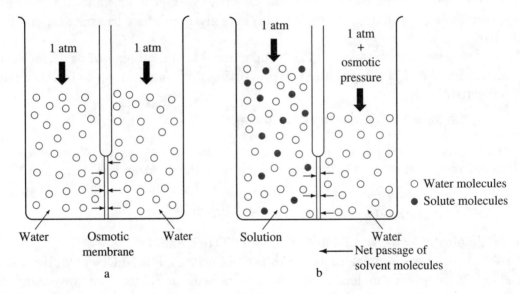

**Figure 18.1** • Osmotic pressure. (a) Two compartments separated by an osmotic semipermeable membrane both contain only solvent molecules that can pass through the membrane. (b) The compartment on the right contains only solvent, the one on the left both solute and solvent. Solute molecules cannot pass through the membrane. The solvent molecules move to the left compartment in an effort to dilute the solution, raising the liquid level on that side.

1. To demonstrate freezing point depression and obtain the freezing point depression constant.
2. To show the effect of tonicity on cells.

## Procedure

### Effect of the Tonicity of Solutions on Cells

1. Take five clean test tubes. Label them: a, b, c, d, e.

2. Add 2 mL of the following solutions to the labelled test tubes:

   **a.** Distilled water

   **b.** 0.1 M glucose

   **c.** 0.5 M glucose

   **d.** 0.89% NaCl

   **e.** 3% NaCl

3. To each test tube add thin (about 0.5 mm thick) slices of freshly cut carrot, scallion, and celery sections.

4. Put the test tubes in a test tube rack and wait until you have finished all the other experiments.

5. Observe the appearance of the sections with the naked eye and also under a microscope.

6. Repeat step no. 1 and step no. 2 using a new set of five clean test tubes.

7. Using an eyedropper, add five drops of fresh whole bovine blood to each test tube. Tap the bottom of the test tubes to ensure proper mixing.

8. Observe the color and the appearance of the solutions after 20 min. both by the naked eye and also under a microscope. *For the proper handling and disposal of blood samples, read the instructions in Appendix 3, Exp. 18.*

### Freezing Point Depression Measurements

1. Assemble a simple freezing point (melting point) apparatus. A beaker will serve as a water bath. A hot plate or Bunsen burner will provide the source of heat. A test tube will serve as a secondary water bath in which a thermometer is suspended (Fig. 18.2).

**Figure 18.2**
Melting point apparatus.

2. Benzoic acid–lauric acid mixtures can be prepared in front of the class as follows (or as an alternative the instructor can prepare this in advance):

Weigh out 3 g of lauric acid and place it in a 25-mL beaker. Weigh out 0.6 g of benzoic acid. Heat the lauric acid gently on a hot plate until it melts (50°C). Add the benzoic acid to the beaker. Mix it thoroughly until a uniform solution is obtained. Cool the beaker in cold water to obtain a solid sample. Grind the sample to a fine powder in a mortar with a pestle.

3. Each student will pack four capillary melting tubes with samples: (a) lauric acid (b) three tubes with the 17% benzoic acid solution.

4. Pack the melting tubes as follows:

   a. Scoop up a very small amount of sample into the melting point capillary tube by pressing the open end of the tube vertically into the sample (Fig. 18.3).

   **Figure 18.3**
   Sampling.

   b. Invert the capillary tube. Stroking the capillary with a file, allow the solid to pack at the bottom of the capillary. You only need a 1–5 mm long packed sample in the capillary tube. (See also Experiment 15, Fig. 15.2, p. 151.)

**5.** Attach the capillary tube to the thermometer, using a narrow rubber band near the top of the tube. Be certain to align the tip of the thermometer with the tip of the capillary tube (Fig. 18.4).

**Figure 18.4**
Positioning the capillary.

Aligned

**6.** Measure the melting point of each sample as follows:

Clamp the thermometer with the capillary tube attached and immerse it in the secondary thermostat filled with water. Lower the secondary thermostat into the beaker filled with water and start the heating process. Observe the melting point of each sample and record it. Melting occurs when you observe the first shrinkage in your sample or the appearance of tiny bubbles. **(Do not wait until the whole sample in the capillary becomes translucent!)** After taking the melting point of the first sample, allow the thermostat to cool to room temperature by adding some cold water. You should start the heating process to observe the melting point of the second sample only after the water in both the primary and secondary thermostat is below the melting point of the sample or at room temperature.

## Chemicals and Equipment

1. Capillary tubes
2. Test tubes
3. Thermometer
4. Rubber band
5. Beakers
6. Clamp
7. Hot plate (or Bunsen burner)
8. Microscope
9. Razor blade or dissecting knife
10. Lauric acid
11. Benzoic acid
12. Fresh whole bovine blood
13. 0.1 M and 0.5 M glucose solutions
14. 0.89% and 3% NaCl solutions
15. Fresh carrot, scallion, and celery (cut to 0.5 mm sections)
16. Thermometer clamp

## Experiment 18

### PRE-LAB QUESTIONS

1. Write the structure and calculate the molecular weight of (a) lauric acid, $C_{12}H_{24}O_2$, and (b) benzoic acid, $C_7H_6O_2$.

2. What is the expected melting point of lauric acid? (Obtain this information from your textbook: Table 20.1)

3. A 0.89% w/V NaCl solution is isotonic with red blood cells. What would be the tonicity of a 0.65 M KCl solution? Isotonic, hypotonic, or hypertonic?

4. What is the osmolarity of (a) 5% glucose and (b) 5% sodium benzoate, $(C_7H_5O_2)^- Na^+$?

5. What kind of information can we obtain if we know the freezing point depression constant, $K_f$, of a solvent and the freezing point depression of a 10% solution of an unknown substance in the solvent referenced above?

## Experiment 18

**REPORT SHEET**

*Tonicity of solutions*

| | Observations | | | |
|---|---|---|---|---|
| | **Appearance of Plant Cells** | | **Appearance of Red Blood Cells** | |
| **Solutions** | **Naked Eye** | **Microscope** | **Naked Eye** | **Microscope** |
| Distilled water | | | | |
| 0.1 M glucose | | | | |
| 0.5 M glucose | | | | |
| 0.89% NaCl | | | | |
| 3.0% NaCl | | | | |

*Freezing point depression*

1. Melting point of lauric acid _____

2. Melting point of 17% w/w benzoic acid                (a) _____

                                                        (b) _____

                                                        (c) _____

3. Freezing point depression of 17% benzoic acid: (1) – (2a)   (a) _____

                                        (1) – (2b)   (b) _____

                                        (1) – (2c)   (c) _____

4. Average freezing point depression of 17% benzoic acid      _____

5. Mole benzoic acid in 1000 g lauric acid in 17% w/w sample  _____

6. $K_f$ calculated from equation (1) for 17% benzoic acid: (4)/(5)   _____

## POST-LAB QUESTIONS

1. What was the maximum deviation from the average (the difference between the average and a measurement) in your freezing point depression (a) in degrees Celsius and (b) in percent?

2. Using the average freezing point depression constant obtained for lauric acid in your experiment, calculate what would be the freezing point of a 10.0% w/w benzoic acid solution.

3. Which would give you a lower freezing point: (a) 5.0% aqueous NaCl solution, or (b) 10.0% aqueous ethylene glycol (HO–CH$_2$–CH$_2$–OH), a non dissociating antifreeze solution?

4. Which of your test solutions was (a) isotonic with red blood cells (tonicity = 0.30 osmolar)? (b) hypotonic? (c) hypertonic?

5. Did you observe any difference in the behavior of plant cells versus red blood cells in hypotonic and hypertonic solutions? What were those differences? Red blood cells have only semipermeable plasma membranes, while plant cells have an additional cell wall made of polysaccharides. Would that explain your observations? How?

6. You used 0.5 M glucose and 3% NaCl aqueous solutions in your tonicity experiments. What would be the freezing points of each of these solutions, considering that the freezing point depression constant of water is 1.86°C/osmol?

# Factors affecting rate of reactions

## Background

Some chemical reactions take place rapidly; others are very slow. For example, antacid neutralizes stomach acid (HCl) rapidly but hydrogen and oxygen react with each other to form water very slowly. A tank containing a mixture of $H_2$ and $O_2$ shows no measurable change even after many years. The study of rates of reactions is called *chemical kinetics*. The *rate of reaction* is the change in concentration of a reactant (or product) per unit time. For example, in the reaction

$$2HCl(aq) + CaCO_3(s) \rightleftharpoons CaCl_2(aq) + H_2O(l) + CO_2(g)$$

we monitor the evolution of $CO_2$, and we find that 4.4 g of carbon dioxide gas was produced in 10 min. Since 4.4 g corresponds to 0.1 moles of $CO_2$, the rate of the reaction is 0.01 moles $CO_2$/min. On the other hand, if we monitor the HCl concentration, we may find that at the beginning we had 0.6 M HCl and after 10 min. the concentration of HCl was 0.4 M. This means that we used up 0.2 M HCl in 10 min. Thus the rate of reaction is 0.02 moles HCl/L-min. From the above we can see that when describing the rate of reaction (it is not sufficient to give a number), we have to specify the units and also the reactant (or product) we monitored.

   In order that a reaction should take place, molecules or ions must first collide. Not every collision yields a reaction. In many collisions, the molecules simply bounce apart without reacting. A collision that results in a reaction is called an *effective collision*. The minimum energy necessary for the reaction to happen is called the *activation energy* (Fig. 19.1). In this energy diagram, we see that the rate of reaction depends on this activation energy.

**Figure 19.1** • Energy diagram for a typical reaction.

The lower the activation energy the faster the rate of reaction; the higher the activation energy the slower the reaction. This is true for both exothermic and endothermic reactions.

A number of factors affect the rates of reactions. In our experiments we will see how these affect the rates of reactions.

1. **Nature of reactants.** Some compounds are more reactive than others. In general, reactions that take place between ions in aqueous solutions are rapid. Reactions between covalent molecules are much slower.

2. **Concentration.** In most reactions, the rate increases when the concentration of either or both reactants is increased. This is understandable on the basis of the collision theory. If we double the concentration of one reactant, it will collide in each second twice as many times with the second reactant as before. Since the rate of reaction depends on the number of effective collisions per second, the rate is doubled (Fig. 19.2).

**Figure 19.2**
Concentration affecting the rate of reaction.

Few collisions

More collisions

Concentration increased

Lower concentration

Higher concentration

3. **Surface area.** If one of the reactants is a solid, the molecules of the second reactant can collide only with the surface of the solid. Thus the surface area of the solid is in effect its concentration. An increase in the surface area of the solid (by grinding to a powder in a mortar) will increase the rate of reaction.

4. **Temperature**. Increasing the temperature makes the reactants more energetic than before. This means that more molecules will have energy equal to or greater than the activation energy. Thus one expects an increase in the rate of reaction with increasing temperature. As a rule of thumb, every time the temperature goes up by 10°C, the rate of reaction doubles. This rule is far from exact, but it applies to many reactions.

5. **Catalyst.** Any substance that increases the rate of reaction without itself being used up in the process is called a *catalyst*. A catalyst increases the rate of reaction by lowering the activation energy (Fig. 19.3). Thus many more molecules can cross the energy barrier (activation energy) in the presence of a catalyst than in its absence. Almost all the chemical reactions in our bodies are catalyzed by specific catalysts called enzymes.

**Figure 19.3** • Energy diagrams of reactions with and without a catalyst.

## Objectives

1. To investigate the relationship between the rate and the nature of reactants.
2. To measure the rate of reaction as a function of concentration.
3. To demonstrate the effect of temperature on the rate of reaction.
4. To investigate the effect of surface area and the effect of a catalyst on the rate of reaction.

## Procedure

1. **Nature of reactants.** Label five (10 × 75 mm) test tubes 1 through 5. Place in each test tube one 1-cm polished strip of magnesium ribbon. Add 1 mL of acid to each test tube as follows: no. 1) 3 M $H_2SO_4$; no. 2) 6 M HCl; no. 3) 6 M $HNO_3$; no. 4) 2 M $H_3PO_4$; and no. 5) 6 M $CH_3COOH$. The reaction will convert the magnesium ribbon to the corresponding salts with the liberation of hydrogen gas. You can assess the rate of reaction qualitatively, by observing the speed with which the gas is liberated (bubbling) and/or by noticing the time of disappearance of the magnesium ribbon. Do all of the reactions in the five test tubes at the same time; assess the rates of reaction; then list, in decreasing order, the rates of reaction of magnesium with the various acids on your Report Sheet (1).

2. Place 1 mL of 6 M HCl in each of three labeled test tubes. Add a 1-cm polished strip of magnesium to the first, zinc to the second, and copper to the third. Do all of the reactions in the three test tubes at the same time; assess the rates of reaction of the three metals by the speed of evolution of $H_2$ gas; then list, in decreasing order, the rates of reaction of the metals with the acid on your Report Sheet (2).

3. **Concentration**. The *iodine clock reaction* is a convenient reaction for observing concentration effects. The reaction is between potassium iodate, $KIO_3$, and sodium bisulfite, $NaHSO_3$; the net ionic reaction is given by the following equation.

$$IO_3^-(aq) + 3HSO_3^-(aq) \rightleftharpoons I^-(aq) + 3SO_4^{2-}(aq) + 3H^+(aq)$$

We can monitor the rate of reaction by the disappearance of the bisulfite. We do so by adding more $IO_3^-$ than $HSO_3^-$ at the start of the reaction. When we have used up all the bisulfite, there is still some iodate left. This will then react with the product iodide, $I^-$, and results in the formation of $I_2$.

$$IO_3^-(aq) + 5I^-(aq) + 6H^+(aq) \rightleftharpoons 3I_2(aq) + 3H_2O(l)$$

We can detect the appearance of iodine with the aid of starch indicator; this reagent forms a blue complex with iodine. The time it takes for the blue color to suddenly appear indicates when all the bisulfite was used up in the first reaction. That's why the name: iodine clock. Thus you should measure the time (with a stopwatch, if available) elapsed between mixing the two solutions and the appearance of the blue color. Place the reactants in two separate 150-mL beakers according to the outline in Table 19.1.

| Table 19.1 | Reactant Concentration and Rate of Reaction | | | | | |
|---|---|---|---|---|---|---|
| | **Beaker A** | | | | **Beaker B** | |
| **Trial** | **0.1 M KIO$_3$** | **Starch** | **Water** | **0.01 M NaHSO$_3$** | **Water** |
| 1 | 2.0 mL | 2 mL | 46 mL | 5 mL | 45 mL |
| 2 | 4.0 mL | 2 mL | 44 mL | 5 mL | 45 mL |
| 3 | 6.0 mL | 2 mL | 42 mL | 5 mL | 45 mL |

Use a graduated pipet to measure each reactant and a graduated cylinder to measure the water. Simultaneously pour the two reactants into a third beaker and time the appearance of the blue color. Repeat the experiment with the other two trial concentrations. Record your data on the Report Sheet (3).

4. **Surface area.** Using a large mortar and pestle, crush and pulverize about 0.5 g of marble chips. Place the crushed marble chips into one large test tube and 0.5 g of uncrushed marble chips into another. Add 2 mL of 6 M HCl to each test tube and note the speed of bubbling of the $CO_2$ gas. Record your data on the Report Sheet (4).

5. **Temperature.** Add 5 mL of 6 M HCl to three clean test tubes. Place the first test tube in an ice bath, the second in a beaker containing warm water (50°C), and the third in a beaker with tap water (20°C). Wait 5 min. To each test tube add a piece of zinc ribbon (1 cm × 0.5 cm × 0.5 mm). Note the time you added the zinc metal. Finally, note the time when the bubbling of gas stops in each test tube and the zinc metal disappears. Record the time of reaction (time of the disappearance of the zinc—the time of the start of the reaction) on your Report Sheet (5).

6. **Catalyst.** Add 2 mL of 3% $H_2O_2$ solution to two clean test tubes. The evolution of oxygen bubbles will indicate if hydrogen peroxide decomposed. Note if anything happens. Add a few grains of $MnO_2$ to one of the test tubes. Note the evolution of oxygen, if any. Record your data on the Report Sheet (6).

## Chemicals and Equipment

1. Mortar and pestle
2. 10-mL graduated pipet
3. 5-mL volumetric pipet
4. Magnesium ribbon
5. Zinc ribbon
6. Copper ribbon
7. 3 M $H_2SO_4$
8. 6 M HCl
9. 6 M $HNO_3$
10. 2 M $H_3PO_4$
11. 6 M $CH_3COOH$
12. 0.1 M $KIO_3$
13. 0.01 M $NaHSO_3$
14. Starch indicator
15. Marble chips
16. 3% hydrogen peroxide
17. Manganese dioxide

## Experiment 19

**PRE-LAB QUESTIONS**

1. In the reaction $2H_2O_2(l) \longrightarrow 2H_2O(l) + O_2(g)$
   the following accumulation of oxygen was observed:

| Time | $O_2$ in g/L |
|------|------|
| 8:00 | 0 |
| 9:20 | 0.016 |
| 10:40 | 0.032 |
| 12:00 | 0.048 |

What is the rate of reaction?

2. In the same reaction catalyzed by $MnO_2$, you collected 0.30 g/L oxygen in a 15 min. period. What is the rate of reaction in the presence of $MnO_2$ catalyst?

   **a.** in g/L min.

   **b.** in moles/L min.

How much faster is the catalyzed reaction?

3. Which reaction will be faster? Explain.

   **a.** $Pb^{2+}(aq) + S^{2-}(aq) \longrightarrow PbS (s)$

   **b.** $CH_3-CH = CH_2 + Br_2 \longrightarrow CH_3-CHBr-CH_2Br$

4. Some antacids come in the form of a compressed pill, while others are in the form of loose powder. Which form would give faster relief for heartburn considering that they contained the same ingredients and were taken in equal amounts?

## Experiment 19

## REPORT SHEET

1. **Nature of reactants**     **Name of the acid**

   Fastest reaction     _____

   _____

   _____

   _____

   Slowest reaction     _____

2. **Nature of reactants**     **Name of the metal**

   Fastest reaction     _____

   _____

   Slowest reaction     _____

3. **Effect of concentration**

   Trial no.     Time

   1     _____

   2     _____

   3     _____

4. **Surface area**

   Fast reaction     _____

   Slow reaction     _____

5. **Effect of temperature**

   Trial at          4°C          20°C          50°C

   Reaction time     _____  _____  _____

6. **Catalyst**     **Observation**

   No catalyst     _____

   $MnO_2$     _____

## POST-LAB QUESTIONS

1. If, in the reaction between magnesium and HCl, we would have used 3 M HCl rather than 6 M HCl, would the rate of reaction increase, decrease, or remain the same?

2. If in the reaction between 6 M HCl and Mg had we used globular chunks of magnesium instead of ribbon (both having the same weight), would the rate of reaction increase, decrease, or stay the same? Explain.

3. Assume that the zinc ribbon you added to the test tubes in Section 5 was 0.5 g. Calculate the rate of reaction as moles of Zn per min. for each temperature.

4. In general, we expect that doubling the concentration of a reactant will approximately double the rate of reaction. Was this expectation justified in the iodine clock reaction?

5. Assume that we do a reaction of zinc with 6 M HCl at room temperature (20°C). How much faster will these two chemicals react at 40°C (see Section 5 of the **Procedure**)?

# Experiment 20

## Law of chemical equilibrium and Le Chatelier's principle

### Background

Two important questions are asked about every chemical reaction: (a) How much product is produced and (b) How fast is it produced? The first question involves chemical equilibrium and the second question belongs to the domain of chemical kinetics. (We dealt with kinetics in Experiment 19.) Some reactions are irreversible and they go to completion (100% yield). When you ignite methane gas in your gas burner in the presence of air (oxygen), methane burns completely and forms carbon dioxide and water.

$$CH_4(g) + 2O_2(g) \longrightarrow CO_2(g) + 2H_2O(g)$$

Other reactions do not go to completion. They are reversible. In such cases, the reaction can go in either direction: forward or backward. For example, the reaction

$$Fe^{3+}(aq) + SCN^-(aq) \rightleftharpoons FeSCN^{2+}(aq)$$

is often used to illustrate reversible reactions. This is so because it is easy to observe the progress of the reaction visually. The yellow $Fe^{3+}$ ion reacts with thiocyanate ion to form a deep red complex ion, $FeSCN^{2+}$. This is the forward reaction. At the same time, the complex red ion also decomposes and forms the yellow iron(III) ion and thiocyanate ion. This is the backward (reverse) reaction. At the beginning when we mix iron(III) ion and thiocyanate ion, the rate of the forward reaction is at a maximum. As time goes on, this rate decreases because we have less and less iron(III) and thiocyanate to react. On the other hand, the rate of the reverse reaction (which began at zero) gradually increases. Eventually the two rates become equal. When this point is reached, we call the process a *dynamic equilibrium*, or just *equilibrium*. When in equilibrium at a particular temperature, a reaction mixture obeys the *Law of Chemical Equilibrium*. This Law imposes a condition on the concentration of reactants and products expressed in the equilibrium constant (K). For the above reaction between iron(lII) and thiocyanate ions, the equilibrium constant can be written as

$$K = [FeSCN^{2+}]/[Fe^{3+}][SCN^-]$$

or in general

$$K = [products]/[reactants]$$

The brackets, [ ], indicate concentration, in moles/L, at equilibrium. As the name implies, the *equilibrium constant* is a constant at a set temperature for a particular reaction. Its magnitude tells if a reaction goes to completion or if it is far from completion (reversible reaction). A number much smaller than 1 for K indicates that at equilibrium only a few molecules of products are formed, meaning the mixture consists mainly of reactants. We say that the equilibrium lies far to the left. On the other hand, a completion of a reaction (100% yield) would have a very large number (infinite?) for the equilibrium constant. In this case, obviously the equilibrium lies far to the right. The above reaction between iron(III) and thiocyanate has an equilibrium constant of 207, indicating that the equilibrium lies to the right but does not go to completion. Thus at equilibrium, both reactants and product are present, albeit the products far outnumber the reactants.

The Law of Chemical Equilibrium is based on the constancy of the equilibrium constant. This means that if one disturbs the equilibrium, for example by adding more reactant molecules, there will be an increase in the number of product molecules in order to maintain the product/reactant ratio unchanged and thus preserving the numerical value of the equilibrium constant. The *Le Chatelier Principle* expresses this as follows: *If an external stress is applied to a system in equilibrium, the system reacts in such a way as to partially relieve the stress.* In our present experiment, we demonstrate the Le Chatelier Principle in two manners: (a) disturbing the equilibrium by changing the concentration of a product or reactant; (b) changing the temperature.

**(a1)** In the first experiment, we add ammonia to a pale blue copper(II) sulfate solution. The ionic reaction is

$$Cu(H_2O)_4{}^{2+}(aq) + 4NH_3(aq) \rightleftharpoons Cu(NH_3)_4{}^{2+}(aq) + 4H_2O(l)$$
$$\text{pale blue} \qquad \text{colorless} \qquad \text{(color?)}$$

A change in the color indicates the copper-ammonia complex formation. Adding a strong acid, HCl, to this equilibrium causes the ammonia, $NH_3$, to react with the acid:

$$NH_3(aq) + H^+(aq) \rightleftharpoons NH_4{}^+(aq)$$

Thus we removed some reactant molecules from the equilibrium mixture. As a result we expect the equilibrium to shift to the left, reforming hydrated copper(II) ions with the reappearance of pale blue color.

**(a2)** In the second reaction, we demonstrate the common ion effect. When we have a mixture of $H_2PO_4{}^-/HPO_4{}^{2-}$ solution, the following equilibrium exists:

$$H_2PO_4{}^-(aq) + H_2O(l) \rightleftharpoons H_3O^+(aq) + HPO_4{}^{2-}(aq)$$

If we add a few drops of aqueous HCl to the solution, we will have added a common ion, $H_3O^+$, that already was present in the equilibrium mixture. We expect, on the basis of the Le Chatelier Principle, that the equilibrium will shift to the left. Thus the solution will not become acidic.

**(a3)** In the iron(III)–thiocyanate reaction

$$Fe^{3+}(aq) + 3Cl^-(aq) + K^+(aq) + SCN^-(aq) \rightleftharpoons Fe(SCN)^{2+}(aq) + 3Cl^-(aq) + K^+(aq)$$
$$\text{yellow} \qquad\qquad\qquad \text{colorless} \qquad\qquad \text{red} \qquad\qquad \text{colorless}$$

the chloride and potassium ions are spectator ions. Nevertheless, their concentration may also influence the equilibrium. For example, when the chloride ions are in excess, the yellow color of the $Fe^{3+}$ will disappear with the formation of a colorless $FeCl_4{}^-$ complex

$$Fe^{3+}(aq) + 4Cl^-(aq) \rightleftharpoons FeCl_4{}^-(aq)$$
$$\text{yellow} \qquad\qquad\qquad \text{colorless}$$

**(b1)** Most reactions are accompanied by some energy changes. Frequently, the energy is in the form of heat. We talk of endothermic reactions if heat is consumed during the reaction. In endothermic reactions, we can consider heat as one of the reactants. Conversely, heat is evolved in an exothermic reaction, and we can consider heat as one of the products. Therefore, if we heat an equilibrium mixture of an endothermic reaction, it will behave as if we added one of its reactants (heat) and the equilibrium will shift to the right. Heating the equilibrium mixture of an exothermic reaction, the equilibrium will shift to the left. We will demonstrate the effect of temperature on the reaction:

$$Co(H_2O)_6^{2+}(aq) + 4Cl^-(aq) \rightleftharpoons CoCl_4^{2-}(aq) + 6H_2O(l)$$

You will observe a change in the color depending on whether the equilibrium was established at room temperature or at 100°C (in boiling water). From the color change, you should be able to tell whether the reaction was endothermic or exothermic.

## Objectives

1. To study chemical equilibria.
2. To investigate the effects of (a) changing concentrations and (b) changing temperature in equilibrium reactions.

## Procedure

**a1.** Place 20 drops (about 1 mL) of 0.1 M $CuSO_4$ solution into a clean and dry test tube. Add (dropwise) 1 M $NH_3$ solution, mixing the contents after each drop. Continue to add until the color changes. Note the new color and the number of drops of 1 M ammonia added and record it on your Report Sheet (1). To the equilibrium mixture thus obtained, add (dropwise, counting the number of drops added) 1 M HCl solution until the color changes back to pale blue. Record your observations on your Report Sheet (2).

**a2.** Place 2 mL of $H_2PO_4^-/HPO_4^{2-}$ solution into a clean and dry test tube. Using red and blue litmus papers, test if the solution is acidic or basic. Record your findings on the Report Sheet (3). Add a drop of 1 M HCl to the litmus papers. Record your observations on the Report Sheet (4). Add one drop of 1 M HCl solution to the test tube. Mix it and test it with litmus papers. Record your observation on the Report Sheet (5).

**a3.** Prepare a stock solution by adding 1 mL of 0.1 M iron(III) chloride, $FeCl_3$, and 1 mL of 0.1 M potassium thiocyanate, KSCN, to 50 mL of distilled water in a beaker. Set up four clean and dry test tubes and label them. To each test tube, add about 2 mL of the stock equilibrium mixture you just prepared. Use the first test tube as a standard to which you can compare the color of the other solutions. To the second test tube add 10 drops of 0.1 M iron(III) chloride solution; to the third add 10 drops of 0.1 M KSCN solution. To the fourth add five drops of saturated NaCl solution. Observe the color in each test tube and record your observations on the Report Sheet (6) and (7).

**b1.** Place 5 drops of 1 M $CoCl_2$ in a dry and clean test tube. Add concentrated HCl dropwise until a color change occurs. Record your observations on the Report Sheet (8). Place 1 mL $CoCl_2$ in a clean and dry test tube. Note the color. Immerse the test tube into a boiling water bath. Record your observations on the Report Sheet (9).

---

**CAUTION!**

Concentrated HCl is toxic and can cause skin burns. Do not allow skin contact and do not inhale the HCl vapors.

---

**Chemicals and Equipment**

1. 0.1 M $CuSO_4$
2. 1 M $NH_3$
3. 1 M HCl
4. Saturated NaCl
5. Concentrated HCl
6. 0.1 M KSCN
7. 0.1 M $FeCl_3$
8. 1 M $CoCl_2$
9. $H_2PO_4^-$/$HPO_4^{2-}$ solution
10. Litmus paper

## Experiment 20

### PRE-LAB QUESTIONS

1. For the reaction $NH_3 + H^+ \rightleftharpoons NH_4^+$ at 20°C, the equilibrium concentrations were as follows: $NH_3 = 2.0 \times 10^{-4}$ M, $H^+ = 2.0 \times 10^{-4}$ M, and $NH_4^+ = 18.0$ M. Calculate the equilibrium constant for the reaction.

2. The reaction above is exothermic and was run at 40°C, would the equilibrium concentration of $NH_3$ be greater, the same, or smaller than $2.0 \times 10^{-4}$ M?

3. If the reaction between iron(III) and thiocyanate ions yielded an equilibrium concentration of 0.2 M for each of these ions, what would be the equilibrium concentration of the red iron–thiocyanate complex? (*Hint:* The equilibrium constant can be found in the **Background**.)

## Experiment 20

**REPORT SHEET**

1. What is the color of the copper–ammonia complex? _____

   How many drops of 1 M ammonia did you add
   to cause a change in color? _____

2. How many drops of 1 M HCl did you add to cause
   a change in color back to pale blue? _____

3. Testing the phosphate solution, what was the color
   of the red litmus paper? _____

   What was the color of the blue litmus paper? _____

4. Testing the 1 M HCl solution, what was the color
   of the red litmus paper? _____

   What was the color of the blue litmus paper? _____

5. After adding one drop of 1 M HCl to the phosphate
   solution and testing it with litmus paper, what was
   the color of the red litmus paper? _____

   What was the color of the blue litmus paper? _____

   Was your phosphate solution acidic, basic, or neutral
   **a.** before the addition of HCl? _____

   **b.** after the addition of HCl? _____

   Was your HCl solution acidic, basic, or neutral? _____

6. Compare the colors in each of the test tubes containing
   the iron(III) chloride–thiocyanate mixtures:

   no. 1 _____

   no. 2 _____

   no. 3 _____

   no. 4 _____

7. In which direction did the equilibrium shift in test tube

    no. 2 _____

    no. 3 _____

    no. 4 _____

8. What is the color of the $CoCl_2$ solution

    **a.** before the addition of HCl? _____

    **b.** after the addition of HCl? _____

9. What is the color of the $CoCl_2$ solution

    **a.** at room temperature? _____

    **b.** at boiling water temperature? _____

10. In which direction did the equilibrium shift
    upon heating? _____

11. From the above shift, determine if the reaction
    was exothermic or endothermic. _____

## POST-LAB QUESTIONS

1. In the first experiment with the copper–ammonia complex, you added ammonia to change the color and, later, equal strength HCl to change it back to blue. Did you require more, less, or an equal number of drops from each to accomplish the color change? On the basis of stoichiometry, what was your expectation?

2. Adding HCl to the $H_2PO_4^-$/$HPO_4^{2-}$ mixture is called a common ion effect. Explain why.

3. Silver ion reacts with chloride ion in solution to form the precipitate AgCl. What would happen to the color of a dilute solution containing $FeCl_4^-$ if we added to it a solution containing silver ion?

4. The manufacture of ammonia from nitrogen and hydrogen is an exothermic reaction. Which temperature would give a greater yield of ammonia: room temperature or 100°C? Explain.

# pH and buffer solutions

We frequently encounter acids and bases in our daily life. Fruits, such as oranges, apples, etc., contain acids. Household ammonia, a cleaning agent, and Liquid Plumber are bases. *Acids* are compounds that can donate a proton (hydrogen ion). *Bases* are compounds that can accept a proton. This classification system was proposed simultaneously by Johannes Brønsted and Thomas Lowry in 1923, and it is known as the Brønsted-Lowry theory. Thus any proton donor is an acid, and a proton acceptor is a base.

When HCl reacts with water

$$HCl + H_2O \rightleftharpoons H_3O^+ + Cl^-$$

HCl is an **acid** and $H_2O$ is a **base** because HCl **donated a proton** thereby becoming $Cl^-$, and water **accepted a proton** thereby becoming $H_3O^+$.

In the reverse reaction (from right to left) the $H_3O^+$ is an acid and $Cl^-$ is a base. As the arrow indicates, the equilibrium in this reaction lies far to the right. That is, out of every 1000 HCl molecules dissolved in water, 990 are converted to $Cl^-$ and only 10 remain in the form of HCl at equilibrium. But $H_3O^+$ (hydronium ion) is also an acid and can donate a proton to the base, $Cl^-$. Why do hydronium ions not give up protons to $Cl^-$ with equal ease and form more HCl? This is because different acids and bases have different strengths. HCl is a stronger acid than hydronium ion, and water is a stronger base than $Cl^-$.

In the Brønsted-Lowry theory, every acid–base reaction creates its *conjugate acid–base pair*. In the above reaction HCl is an acid which, after giving up a proton, becomes a conjugate base, $Cl^-$. Similarly, water is a base which, after accepting a proton, becomes a conjugate acid, the hydronium ion.

conjugate base–acid pair

$$HCl + H_2O \rightleftharpoons H_3O^+ + Cl^-$$

conjugate acid–base pair

Some acids can give up only one proton. These are *monoprotic* acids. Examples are ⒽCl, ⒽNO₃, HCOOⒽ, and CH₃COOⒽ. The hydrogens circled are the ones donated. Other acids yield two or three protons. These are called *diprotic* or *triprotic* acids. Examples are $H_2SO_4$, $H_2CO_3$, and $H_3PO_4$. However, in the Brønsted-Lowry theory, each acid is considered monoprotic, and a diprotic acid (such as carbonic acid) donates its protons in two distinct steps:

1. $H_2CO_3 + H_2O \rightleftharpoons H_3O^+ + HCO_3^-$

2. $HCO_3^- + H_2O \rightleftharpoons H_3O^+ + CO_3^{2-}$

Thus the compound $HCO_3^-$ is a conjugate base in the first reaction and an acid in the second reaction. A compound that can act either as an acid or a base is called *amphiprotic*.

In the self-ionization reaction

$$H_2O + H_2O \rightleftharpoons H_3O^+ + OH^-$$

one water acts as an acid (proton donor) and the other as a base (proton acceptor). In pure water, the equilibrium lies far to the left, that is, only very few hydronium and hydroxyl ions are formed. In fact, only $1 \times 10^{-7}$ moles of hydronium ion and $1 \times 10^{-7}$ moles of hydroxide ion are found in one liter of water. The dissociation constant for the self-ionization of water is

$$K_d = \frac{[H_3O^+][OH^-]}{[H_2O]^2}$$

This can be rewritten as

$$K_w = K_d [H_2O]^2 = [H_3O^+][OH^-]$$

$K_w$, the **ion product of water**, is still a constant because very few water molecules reacted to yield hydronium and hydroxide ions; hence the concentration of water essentially remained constant. At room temperature, the $K_w$ has the value of

$$K_w = 1 \times 10^{-14} = [1 \times 10^{-7}] \times [1 \times 10^{-7}]$$

This value of the ion product of water applies not only to pure water but to any aqueous (water) solution. This is very convenient because if we know the concentration of the hydronium ion, we automatically know the concentration of the hydroxide ion and vice versa. For example, if in a 0.01 M HCl solution HCl dissociates completely, the hydronium ion concentration is $[H_3O^+] = 1 \times 10^{-2}$ M. This means that the $[OH^-]$ is

$$[OH^-] = K_w/[H_3O^+] = 1 \times 10^{-14}/1 \times 10^{-2} = 1 \times 10^{-12} \text{ M}$$

To measure the strength of an aqueous acidic or basic solution, P. L. Sorensen introduced the pH scale.

$$pH = -\log[H_3O^+]$$

In pure water, we have seen that the hydronium ion concentration is $1 \times 10^{-7}$ M. The logarithm of this is $-7$ and, thus, the pH of pure water is 7. Since water is an amphiprotic compound, pH 7 means a neutral solution. On the other hand, in a 0.01 M HCl solution (dissociating completely), we have $[H_3O^+] = 1 \times 10^{-2}$ M. Thus its pH is 2. The pH scale shows that acidic solutions have a pH less than 7 and basic solutions have a pH greater than 7.

pH  0  1  2  3  4  5  6  7  8  9  10  11  12  13  14
     acidic        neutral      basic

The pH of a solution can be measured conveniently by special instruments called pH meters. All that must be done is to insert the electrodes of the pH meter into the solution to be measured and read the pH from a scale. pH of a solution can also be obtained, although less precisely, by using a pH indicator paper. The paper is impregnated with organic compounds that change their color at different pH values. The color shown by the paper is then compared with a color chart provided by the manufacturer.

There are certain solutions that resist a change in the pH even when we add to them acids or bases. Such systems are called *buffers*. A mixture of a weak acid and its conjugate base usually forms a good buffer system. An example is carbonic acid, which is the most important buffer in our blood and maintains it close to pH 7.4. Buffers resist large changes in pH because of the Le Chatelier principle governing equilibrium conditions. In the carbonic acid–bicarbonate (weak acid–conjugate base) buffer system,

$$H_2CO_3 + H_2O \rightleftharpoons HCO_3^- + H_3O^+$$

any addition of an acid, $H_3O^+$, will shift the equilibrium to the left. Thus this reduces the hydronium ion concentration, returning it to the initial value so that it stays constant; hence the change in pH is small. If a base, $OH^-$, is added to such a buffer system, it will react with the $H_3O^+$ of the buffer. But the equilibrium then shifts to the right, replacing the reacted hydronium ions, hence again, the change in pH is small.

Buffers stabilize a solution at a certain pH. This depends on the nature of the buffer and its concentration. For example, the carbonic acid–bicarbonate system has a pH of 6.37 when the two ingredients are at equimolar concentration. A change in the concentration of the carbonic acid relative to its conjugate base can shift the pH of the buffer. The Henderson-Hasselbalch equation below gives the relationship between pH and concentration.

$$pH = pK_a + \log \frac{[A^-]}{[HA]}$$

In this equation the $pK_a$ is the $-\log K_a$, where $K_a$ is the dissociation constant of carbonic acid

$$K_a = \frac{[HCO_3^-][H_3O^+]}{[H_2CO_3]}$$

[HA] is the concentration of the acid and $[A^-]$ is the concentration of the conjugate base. The $pK_a$ of the carbonic acid–bicarbonate system is 6.37. When equimolar conditions exist, then $[HA] = [A^-]$. In this case, the second term in the Henderson-Hasselbalch equation is zero. This is so because $[A^-]/[HA] = 1$, and the log 1 = 0. Thus at equimolar concentration of the acid–conjugate base, the pH of the buffer equals the $pK_a$; in the carbonic acid–bicarbonate system this is 6.37. If, however, we have ten times more bicarbonate than carbonic acid, $[A^-]/[HA] = 10$, then log 10 = 1 and the pH of the buffer will be

$$pH = pK_a + \log [A^-]/[HA] = 6.37 + 1.0 = 7.37$$

This is what happens in our blood—the bicarbonate concentration is ten times that of the carbonic acid and this keeps our blood at a pH of 7.4. Any large change in the pH of our blood may be fatal (acidosis or alkalosis). Other buffer systems work the same way. For example, the second buffer system in our blood is

$$H_2PO_4^- + H_2O \rightleftharpoons HPO_4^{2-} + H_3O^+$$

The $pK_a$ of this buffer system is 7.21. It requires a 1.6 to 1.0 molar ratio of $HPO_4^{2-}$ to $H_2PO_4^-$ to maintain our blood at pH 7.4.

### Objectives

1. To learn how to measure pH of a solution.
2. To understand the operation of buffer systems.

## Procedure

### Measurement of pH

1. Add one drop of 0.1 M HCl to the first depression of a spot plate. Dip a 2-cm long universal pH paper into the solution. Remove the excess liquid from the paper by touching the plate. Compare the color of the paper to the color chart provided (Fig. 21.1). Record the pH on your Report Sheet (1).

2. Repeat the same procedure with 0.1 M acetic acid, 0.1 M sodium acetate, 0.1 M carbonic acid (or club soda or seltzer), 0.1 M sodium bicarbonate, 0.1 M ammonia, and 0.1 M NaOH. For each solution, use a different depression of the spot plate. Record your results on the Report Sheet (1).

**Figure 21.1**
pH paper dispenser.

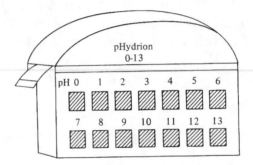

3. Depending on the availability of the number of pH meters this may be a class exercise (demonstration), or 6–8 students may use one pH meter. Add 5 mL of 0.1 M acetic acid to a dry and clean 10-mL beaker. Wash the electrode over a 200-mL beaker with distilled or deionized water contained in a wash bottle. The 200-mL beaker serves to collect the wash water. Gently wipe the electrode with Kimwipes (or other soft tissues) to dryness. Insert the dry electrode into the acetic acid solution. Your pH meter has been calibrated by your instructor. Switch "on" the pH meter and read the pH from the position of the needle on your scale. Alternatively, if you have a digital pH meter, a number corresponding to the pH will appear (Fig. 21.2).

---

**CAUTION!**

Make sure the electrode is immersed into the solution but does not touch the walls or the bottom of the beaker. Electrodes are made of thin glass, and they break easily if you don't handle them gently.

---

**Figure 21.2**

pH meter.

4. Repeat the same procedures with 0.1 M sodium acetate, 0.1 M carbonic acid, 0.1 M sodium bicarbonate, and 0.1 M ammonia. Make certain that for each solution you use a dry and clean beaker, and before each measurement wash the electrode with distilled water and dry with Kimwipes. Record your data on the Report Sheet (2).

5. Prepare four buffer systems in four separate, labeled, dry, and clean 50-mL beakers, as follows:

   a. 5 mL 0.1 M acetic acid + 5 mL 0.1 M sodium acetate

   b. 1 mL 0.1 M acetic acid + 10 mL 0.1 M sodium acetate

   c. 5 mL 0.1 M carbonic acid + 5 mL 0.1 M sodium bicarbonate

   d. 1 mL 0.1 M carbonic acid + 10 mL 0.1 M sodium bicarbonate

   Measure the pH of each buffer system with the aid of universal pH paper. Record your data on the Report Sheet (3), (6), (9), and (12).

6. Divide each of your buffers (a–d) into two halves (5 mL each) and place them into clean and dry 10-mL beakers. To the first sample of buffer (a), add 0.5 mL 0.1 M HCl. Mix and measure the pH with the aid of universal pH paper. Record your data on the Report Sheet (4). To the second sample of buffer (a), add 0.5 mL 0.1 M NaOH. Mix and measure the pH with pH paper. Record your data on the Report Sheet (5).

7. Repeat the same measurements with buffers (b), (c), and (d). Record your data on the Report Sheet for the appropriate buffer system under (7), (8), (10), (11), (13), and (14).

8. Place 5 mL of distilled water in each of two 10-mL beakers. Measure the pH of distilled water with the aid of universal pH paper. Record the data on the Report Sheet (15). To the first sample of distilled water add 0.5 mL of 0.1 M HCl. Mix and measure the pH with the aid of universal pH paper and record it on the Report Sheet (16). To the second sample of distilled water add 0.5 mL of 0.1 M NaOH. Mix and measure the pH as before and record it on the Report Sheet (17).

1. pH meter
2. pH paper
3. Kimwipes
4. Wash bottle
5. 0.1 M HCl
6. 0.1 M acetic acid ($CH_3COOH$)
7. 0.1 M sodium acetate ($CH_3COO^-Na^+$)
8. 0.1 M carbonic acid or club soda or seltzer
9. 0.1 M $NaHCO_3$
10. 0.1 M $NH_3(aq)$ (aqueous ammonia)
11. 0.1 M NaOH
12. Spot plate
13. 10-mL beakers

## Experiment 21

### PRE-LAB QUESTIONS

1. Phosphoric acid, $H_3PO_4$, is a triprotic acid. Show the formula of the conjugate base after two protons have been donated.

2. The $pK_a$ of $H_2PO_4^-$ is 7.21. What is the pH of a solution in which the $[HPO_4^{2-}]/[H_2PO_4^-]$ ratio is (a) 10 and (b) 0.1?

3. The pH of normal blood is 7.4. A diabetic patient's blood gave a pH reading of 6.4. How much more hydronium ion ($H_3O^+$) is in the blood of the diabetic patient?

## Experiment 21

**REPORT SHEET**

| *pH of solutions* | 1. by pH paper | 2. by pH meter |
|---|---|---|
| 0.1 M HCl | _____ | _not done_ |
| 0.1 M acetic acid | _____ | _____ |
| 0.1 M sodium acetate | _____ | _____ |
| 0.1 M carbonic acid | _____ | _____ |
| 0.1 M sodium bicarbonate | _____ | _____ |
| 0.1 M ammonia | _____ | _____ |
| 0.1 M NaOH | _____ | _not done_ |

| *Buffer systems* | pH |
|---|---|
| 3.  5 mL 0.1 M $CH_3COOH$ + 5 mL 0.1 M $CH_3COO^-Na^+$ (**a**) | _____ |
| 4.  after addition of 0.5 mL 0.1 M HCl | _____ |
| 5.  after addition of 0.5 mL 0.1 M NaOH | _____ |
| 6.  1 mL 0.1 M $CH_3COOH$ + 10 mL 0.1 M $CH_3COO^-Na^+$ (**b**) | _____ |
| 7.  after addition of 0.5 mL 0.1 M HCl | _____ |
| 8.  after addition of 0.5 mL 0.1 M NaOH | _____ |
| 9.  5 mL 0.1 M $H_2CO_3$ + 5 mL 0.1 M $NaHCO_3$ (**c**) | _____ |
| 10. after addition of 0.5 mL 0.1 M HCl | _____ |
| 11. after addition of 0.5 mL 0.1 M NaOH | _____ |
| 12. 1 mL 0.1 M $H_2CO_3$ + 10 mL 0.1 M $NaHCO_3$ (**d**) | _____ |
| 13. after addition of 0.5 mL 0.1 M HCl | _____ |
| 14. after addition of 0.5 mL 0.1 M NaOH | _____ |
| 15. distilled water | _____ |
| 16. after addition of 0.5 mL 0.1 M HCl | _____ |
| 17. after addition of 0.5 mL 0.1 M NaOH | _____ |

## POST-LAB QUESTIONS

1. Calculate the expected pH values of the buffer systems from the experiment (a–d), using the Henderson-Hasselbalch equation and the $pK_a$ values:

   carbonic acid/bicarbonate = 6.37; acetic acid/acetate = 4.75

   **a.**

   **b.**

   **c.**

   **d.**

   Are they in agreement with your measured pH values?

2. Compared to each of the four buffer systems, how many more units of pH change did you observe in the distilled water upon addition of 0.5 mL 0.1 M HCl? What can you conclude from these results?

   **a.**

   **b.**

   **c.**

   **d.**

3. Which of the four buffers you prepared (a–d) is the best buffer?

4. Looking at the pH values you obtained in your experiments with a pH meter, estimate the useful range of your pH meter.

# Experiment 22

## Analysis of vinegar by titration

### Background

In order to measure how much acid or base is present in a solution we often use a method called *titration*. If a solution is acidic, titration consists of adding base to it until all the acid is neutralized. To do this, we need two things: (1) a means of measuring how much base is added and (2) a means of telling just when the acid is completely neutralized.

How much base is added requires the knowledge of the number of moles of the base. The number of moles can be calculated if we know the volume and the molarity of the base.

$$\text{moles} = \mathbf{V} \times \mathbf{M}$$

When we deal with monoprotic acids and bases that produce one $OH^-$ per molecule, the titration is completed when the number of moles of acid equals the number of moles of base.

$$\text{Moles}_{acid} = \text{Moles}_{base}$$

or

$$\mathbf{V}_{acid}\mathbf{M}_{acid} = \mathbf{V}_{base}\mathbf{M}_{base}$$

This is called the *titration equation*.

We use an *indicator* to tell us when the titration is completed. Indicators are organic compounds that change color when there is a change in the pH of the solution. The *end point* of the titration is when a sudden change in the pH of the solution occurs. Therefore, we can tell the completion of the titration when we observe a change in the color of our solution to which a few drops of indicator have been added.

Commercial vinegar contains 5–6% acetic acid. Acetic acid, $CH_3COOH$, is a monoprotic acid. It is a weak acid and when titrated with a strong base such as NaOH, upon completion of the titration, there is a sudden change in the pH in the range from 6.0 to 9.0. The best way to monitor such a change is to use the indicator phenolphthalein, which changes from colorless to a pink hue at pH 8.0–9.0.

With the aid of the titration equation, we can calculate the concentration of acetic acid in the vinegar. To do so, we must know the volume of the acid (5 mL), the molarity of the base (0.2 M), and the volume of the base used to reach the end point of the titration. This will be read from the *buret*, which is filled with the 0.2 M NaOH solution at the beginning of the titration to its maximum capacity. The base is then slowly added (dropwise) from the buret to the vinegar in an Erlenmeyer flask. Continuous swirling ensures proper mixing. The titration is stopped when the indicator shows a permanent pink coloration. The buret is read again. The volume of the base added is the difference between the initial volume (25 mL) and the volume left in the buret at the end of titration.

1. To learn the techniques of titration.
2. To determine the concentration of acetic acid in vinegar.

## Procedure

1. Rinse a 25-mL buret (or 50-mL buret) with about 5 mL of 0.2 M NaOH solution. (Be sure to record the exact concentration of the base.) After rinsing, fill the buret with 0.2 M NaOH solution about 2 mL above the 0.0-mL mark. Use a clean and dry funnel for filling. Tilting the buret at a 45° angle, slowly turn the stopcock to allow the solution to fill the tip. Collect the excess solution dripping from the tip into a beaker to be discarded later. The air bubbles must be completely removed from the tip. If you do not succeed the first time, repeat it until the liquid in the buret forms one continuous column from top to bottom. Clamp the buret onto a ring stand (Fig. 22.1a). By slowly opening the stopcock, allow the bottom of the meniscus to drop to the 0.0-mL mark. Collect the excess solution dripping from the tip into a beaker to be discarded later. Read the meniscus carefully to the nearest 0.1 mL (Fig. 22.1b).

a. Titration setup.

b. Reading the meniscus.

17.58 mL—incorrect
17 mL—incorrect
17.5 mL—correct

**Figure 22.1** • Titration setup.

2. With the aid of a 5-mL volumetric pipet, add 5 mL vinegar to a 100-mL Erlenmeyer flask. Allow the vinegar to drain completely from the pipet by holding the pipet in such a manner that its tip touches the wall of the flask. Record the volume of the vinegar for trial 1 on your Report Sheet (1). Record also the molarity of the base (2) and the initial reading of the base in the buret on your Report Sheet (3). Add a few drops of phenolphthalein indicator to the flask and about 10 mL of distilled water. The distilled water is added to dilute the natural color that some commercial vinegars have. In this way, the natural color will not interfere with the color change of the indicator.

3. While holding the neck of the Erlenmeyer flask in your left hand and swirling it, open the stopcock of the buret slightly with your right hand and allow the *dropwise* addition of the base to the flask. At the point where the base hits the vinegar solution the color may *temporarily* turn pink, but this color will disappear upon mixing the solution by swirling. Continue the titration until a faint *permanent* pink coloration appears. Stop the titration. Record the readings of the base in your buret on your Report Sheet (4). Read the meniscus to the nearest 0.1 mL (Fig. 22.1b).

### CAUTION!

Be careful not to add too much base. It is an error called *overtitration*. If the indicator in your flask turns deep pink or purple, you have overtitrated and will need to repeat the entire titration with a new sample of vinegar.

4. Repeat the procedures in steps 1–3 with a new 5-mL vinegar sample for trial 2. Record these results on your Report Sheet.

5. With the aid of the titration equation, calculate the molarity of the vinegar (5) for trials 1 and 2.

6. Average the two molarities. Using the molar mass of 60 g/mole for acetic acid, calculate the percent concentration of acetic acid in vinegar.

### Chemicals and Equipment

1. 25-mL buret (or 50-mL buret)
2. Buret clamp
3. 5-mL pipet
4. Small funnel
5. 0.2 M standardized NaOH
6. Phenolphthalein indicator

## Experiment 22

**PRE-LAB QUESTIONS**

1. In a monoprotic acid, such as acetic acid, $CH_3COOH$, only one hydrogen is donated. In the above structure there are four hydrogens. Which one is donated?

2. What is the molarity of 1.44% $HCOOH$?

3. How many moles of acetic acid are in 1 L vinegar containing 5% acetic acid?

4. What is the molarity of an unknown acid if 35 mL of the acid can be titrated to an end point by 15 mL of 0.25 M NaOH?

## Experiment 22

### REPORT SHEET

| *Titration* | Trial 1 | Trial 2 |
|---|---|---|
| 1. Volume of vinegar sample | _____ | _____ |
| 2. Molarity of NaOH solution | _____ | _____ |
| 3. Initial reading of NaOH in buret | _____ mL | _____ mL |
| 4. Final reading of NaOH in buret | _____ mL | _____ mL |
| 5. Volume of NaOH used in titration: (4) − (3) | _____ mL | _____ mL |
| 6. Molarity of acetic acid in vinegar: (2) × [(5)/(1)] | _____ | _____ |

7. Average molarity of acetic acid        _____

8. Percent (w/v) of acetic acid in vinegar:
   % = (7) × 60 × 0.1        _____ %

   [% w/v = g/100 mL = (7) ~~mol.~~/1000 ~~mL~~ × 60 g/~~mol.~~ × 0.1 × 1000 ~~mL~~/100 mL]

### POST-LAB QUESTIONS

1. Assume that your vinegar contained a small amount of citric acid (a triprotic acid). Using the same experimental data, would you expect the molarity of this sample to be the same as or different than a sample which contained only pure acetic acid?

2. Assume that the tip of your buret was not properly filled with NaOH solution. It contained an air bubble which was eliminated during the titration. Would the calculated molarity of the vinegar be smaller, larger, or the same as the true molarity? Explain.

3. We added about 10 mL of distilled water to the vinegar, but we did not use this volume in the calculation of molarity. Why do you think we can ignore this volume in the calculation?

4. According to your results, how many grams of acetic acid are in a 250-mL bottle of vinegar?

5. You have a 10.0-mL sample of formic acid, HCOOH, to be titrated with 0.200 M NaOH. The initial reading of the volume of NaOH in the buret was 23.1 mL. The reading of the buret at the end point of the titration was 41.8 mL NaOH. What is the % w/v of formic acid in your sample? Show your work.

6. You have a 5% formic acid solution and a 5% acetic acid solution (vinegar). You titrate 5 mL of each with 0.1868 M NaOH. (a) Which will need more NaOH to reach the end point? (b) How much more? Show your calculation.

# Experiment 23

## Analysis of antacid tablets

### Background

The natural environment of our stomach is quite acidic. Gastric juice, which is mostly hydrochloric acid, has a pH of 1.0. Such a strong acidic environment denatures proteins and helps with their digestion by enzymes such as pepsin. Not only is the denatured protein more easily digested by enzymes than the native protein, but the acidic environment helps to activate pepsin. The inactive form of pepsin, pepsinogen, is converted to the active form, pepsin, by removing a chunk of its chain, 42 amino acid units. This can only occur in an acidic environment, and pepsin molecules catalyze this reaction (autocatalysis). But too much acid in the stomach is not good either. In the absence of food, the strong acid, HCl, denatures the proteins in the stomach wall itself. If this goes on unchecked, it may cause stomach or duodenal ulcers.

We feel the excess acidity in our stomach. Such sensations are called "heartburn" or "sour stomach." To relieve "heartburn," we take antacids in tablet or liquid form. Antacid is a medical term. It implies a substance that neutralizes acid. Drugstore antacids contain a number of different active ingredients. Almost all of them are weak bases (hydroxides and/or carbonates). Table 23.1 lists the active ingredients of some commercial antacids.

| Table 23.1 | Active Ingredients of Some Drugstore Antacids |
|---|---|
| Alka-Seltzer: sodium bicarbonate and citrate | |
| Bromo-Seltzer: sodium bicarbonate and citrate | |
| Chooz, Tums: calcium carbonate | |
| Di-gel, Gelusil, Maalox: aluminum hydroxide and magnesium hydroxide | |
| Gaviscon, Remegel: aluminum hydroxide and magnesium carbonate | |
| Rolaids: aluminum sodium dihydroxy carbonate | |

HCl in the gastric juice is neutralized by these active ingredients in the following reactions:

$$NaHCO_3 + HCl \longrightarrow NaCl + H_2O + CO_2$$
$$CaCO_3 + 2HCl \longrightarrow CaCl_2 + H_2O + CO_2$$
$$Al(OH)_3 + 3HCl \longrightarrow AlCl_3 + 3H_2O$$
$$Mg(OH)_2 + 2HCl \longrightarrow MgCl_2 + 2H_2O$$
$$AlNa(OH)_2CO_3 + 4HCl \longrightarrow AlCl_3 + NaCl + 3H_2O + CO_2$$

Besides the active ingredients, antacid tablets also contain inactive ingredients, such as starch, which act as a binder or filler. The efficacy of an antacid tablet is its ability to neutralize HCl. The more HCl that is neutralized, the more effective the antacid pill. (You must have heard the competing advertisement claims of different commercial antacids: "Tums neutralizes one-third more stomach acid than Rolaids.")

Antacids are not completely harmless. The HCl production in the stomach is regulated by the stomach pH. If too much antacid is taken, the pH becomes too high; the result will be the so-called "acid rebound." This means that ultimately, more HCl will be produced than was present before taking the antacid pill.

In the present experiment, you will determine the amount of HCl neutralized by two different commercial antacid tablets. To do so we use a technique called *back-titration*. We add an **excess** amount of 0.2 M HCl to the antacid tablet. The excess acid (more than is needed for neutralization) helps to dissolve the tablet. Then the active ingredients in the antacid tablet will neutralize part of the added acid. The remaining HCl is determined by titration with NaOH. A standardized NaOH solution of known concentration (0.2 M) is used and added slowly until all the HCl is neutralized. We observe this end point of the titration when the added indicator, thymol blue, changes its color from red to yellow. Since HCl is monoprotic and NaOH yields one OH$^-$ for each molecule, we can use the titration equation to obtain the volume of the excess 0.2 M HCl (the volume not neutralized by the antacid).

$$V_{acid} = (V_{base} \times M_{base})/M_{acid}$$

Once this is known, the amount of HCl neutralized by the antacid pill is obtained as the difference between the initially added volume and the back-titrated volume:

$$V_{HCl\ neutralized\ by\ the\ pill} = V_{HCl\ initially\ added} - V_{HCl\ backtitrated}$$

In this way we can compare the effectiveness of different drugstore antacids.

## Objectives

1. To learn the technique of back-titration.
2. To compare the efficacies of drugstore antacid tablets.

## Procedure

1. Rinse a 25-mL buret (or 50-mL buret) with about 5 mL 0.2 M NaOH. After rinsing, fill the buret with 0.2 M NaOH solution about 2 mL above the top mark. Use a clean and dry funnel for filling. Tilting the filled buret at a 45° angle, turn the stopcock open to allow the solution to fill the tip of the buret. Collect the excess solution dripping from the tip into a beaker to be discarded later. The air bubbles should be completely removed from the tip by this maneuver. If you do not succeed the first time, repeat it until the liquid in the buret forms one continuous column from top to bottom. Clamp the buret onto a ring stand (Fig. 22.1a). By slowly opening the stopcock, allow the bottom of the meniscus to drop to the 0.0-mL mark. Collect the excess solution dripping from the tip into a beaker to be discarded later. (Carefully read the meniscus here and, in all other readings, to the nearest 0.1 mL; see Experiment 22, Fig. 22.1b.)

2. Repeat the above procedure with a 100-mL buret and fill it to the 0.0-mL mark with 0.2 M HCl. Clamp this, too, onto a ring stand.

3. Obtain two different antacid tablets from your instructor. Note the name of the tablets on your Report Sheet (1). Weigh each tablet on a balance to the nearest 0.001 g. Report the weight on your Report Sheet (2). Place each tablet in separate 250-mL Erlenmeyer flasks. Label the flasks. Add about 10 mL water to each flask. With the help of stirring rods (one for each flask), break up the tablets.

4. Add exactly 50 mL 0.2 M HCl to each Erlenmeyer flask from the buret. Also, add a few drops of thymol blue indicator. Gently stir with the stirring rods to disperse the tablets. (Some of the inactive ingredients may not go into solution and will settle as a fine powder on the bottom of the flask.) At this point the solution should be red (the color of thymol blue at acidic pH). If either of your solutions does not have red coloration, add 10 mL 0.2 M HCl from the refilled buret and make certain that the red color will persist for more than 30 sec. Record the total volume of 0.2 M HCl added to each flask on your Report Sheet (3).

5. Place the Erlenmeyer flask under the buret containing the 0.2 M NaOH. Record the level of the meniscus of the NaOH solution in the buret before you start the titration (4). While holding and swirling the neck of the Erlenmeyer flask with your left hand, titrate the contents of your solution by adding (dropwise) 0.2 M NaOH by opening the stopcock of the buret with your right hand. Continue to add NaOH until the color changes to yellow and stays yellow for 30 sec. after the last drop. Record the level of the NaOH solution in the buret by reading the meniscus at the end of titration (5).

6. Refill the buret with 0.2 M NaOH as before and repeat the titration for the second antacid.

7. Calculate the volume of the acid obtained in the back-titration and record it on your Report Sheet (6). Calculate the volume of the 0.2 M HCl neutralized by the antacid tablets (7). Calculate the grams of HCl neutralized by 1 g antacid tablet. Record it on your Report Sheet (8).

---

### Chemicals and Equipment

1. 25-mL buret (or 50-mL buret)
2. 100-mL buret
3. Buret clamp
4. Ring stand
5. Balance
6. Antacid tablets
7. 0.2 M NaOH
8. 0.2 M HCl
9. Thymol blue indicator

## Experiment 23

### PRE-LAB QUESTIONS

1. Some antacid pills generate $CO_2$ gas when they neutralize gastric juice. Name at least two such antacid pills.

2. What does the term *back-titration* mean?

3. Maalox contains magnesium hydroxide as an antacid. Looking at the reaction between HCl and $Mg(OH)_2$, how many moles of HCl do you need to neutralize (a) 0.25 (b) 0.142 (c) 3.2 moles of $Mg(OH)_2$?

4. Write a generalized titration equation when the acid is monoprotic (HCl) and when the base yields two $(OH)^-$ per molecule as in the case of $Mg(OH)_2$.

**5.** Write a generalized titration equation when the acid is monoprotic and the base yields three (OH)$^-$ ions per molecule as in the case of Al(OH)$_3$.

**6.** Some patients with elevated blood pressure must restrict the sodium intake in their diet. Which antacid would you recommned for such a patient?

## Experiment 23

## REPORT SHEET

|  | (a) | (b) |
|---|---|---|

1. Name of the antacid tablet                     _____    _____

2. Weight of the antacid tablet                   _____ g  _____ g

3. Total volume of 0.2 M HCl added
   to the antacid before titration                _____ mL _____ mL

4. Reading of 0.2 M NaOH in the buret
   before titration                               _____ mL _____ mL

5. Reading of 0.2 M NaOH in the buret
   after titration                                _____ mL _____ mL

6. Volume of 0.2 M HCl obtained in
   back-titration: (5) − (4)                      _____ mL _____ mL

7. Volume of 0.2 M HCl neutralized
   by one antacid tablet: (3) − (6)               _____ mL _____ mL

8. Grams of HCl neutralized by 1 g antacid
   tablet: $[(7)/(2)] \times 0.2 \times 36.5$     _____    _____

## POST-LAB QUESTIONS

1. Which antacid tablet neutralized more stomach acid (a) per tablet and (b) per gram
   tablet?

2. Certain TV commercials claim that their antacid product neutralizes 47 times its own weight in stomach acid. On the basis of your results, is such a claim justified? (In your calculation, you should keep in mind that stomach acid is 0.1 M HCl. In other words, there are 3.65 g HCl in 1 L of stomach acid, or 1 g HCl equals 274 g stomach acid.)

3. A pill of Rolaids contain 250 mg active ingredient. How many mL of 0.1M HCl (gastric juice) will it neutralize?

4. You want to manufacture an antacid from either $Mg(OH)_2$ or $Al(OH)_3$. You can buy either of these for $2.00 per pound. Which of these will provide, if any, better relief for the same amount of money?

# Experiment 24

## Structure in organic compounds: use of molecular models. I

The study of organic chemistry usually involves those molecules which contain carbon. Thus a convenient definition of *organic chemistry* is the chemistry of carbon compounds.

There are several characteristics of organic compounds that make their study interesting:

a. Carbon forms strong bonds to itself as well as to other elements; the most common elements found in organic compounds, other than carbon, are hydrogen, oxygen, and nitrogen.

b. Carbon atoms are generally tetravalent. This means that carbon atoms in most organic compounds are bound by four covalent bonds to adjacent atoms.

c. Organic molecules are three-dimensional and occupy space. The covalent bonds which carbon makes to adjacent atoms are at discrete angles to each other. Depending on the type of organic compound, the angle may be 180°, 120°, or 109.5°. These angles correspond to compounds which have triple bonds (1), double bonds (2), and single bonds (3), respectively.

$$-C\equiv C- \qquad\qquad \underset{}{>}C=C\underset{}{<} \qquad\qquad -\overset{|}{\underset{|}{C}}-\overset{|}{\underset{|}{C}}-$$

$$\text{(1)} \qquad\qquad\qquad \text{(2)} \qquad\qquad\qquad \text{(3)}$$

d. Organic compounds can have a limitless variety in composition, shape, and structure.

Thus, while a molecular formula tells the number and type of atoms present in a compound, it tells nothing about the structure. The structural formula is a two-dimensional representation of a molecule and shows the sequence in which the atoms are connected and the bond type. For example, the molecular formula, $C_4H_{10}$, can be represented by two different structures: butane (4) and 2-methylpropane (isobutane) (5).

$$\begin{array}{ccccc} & H & H & H & H \\ & | & | & | & | \\ H- & C- & C- & C- & C-H \\ & | & | & | & | \\ & H & H & H & H \end{array}$$

Butane (4)

$$\begin{array}{ccc} H & H & H \\ | & | & | \\ H-C- & C- & C-H \\ | & | & | \\ H & & H \\ & H-\overset{|}{\underset{|}{C}}-H \\ & H \end{array}$$

2-Methylpropane (5)

(Isobutane)

Consider also the molecular formula, $C_2H_6O$. There are two structures which correspond to this formula: dimethyl ether (6) and ethanol (ethyl alcohol) (7).

$$
\begin{array}{c}
\quad H \quad\quad H \\
\quad | \quad\quad | \\
H-C-O-C-H \\
\quad | \quad\quad | \\
\quad H \quad\quad H
\end{array}
$$

Dimethyl ether (6)

$$
\begin{array}{c}
\quad H \ H \\
\quad | \ | \\
H-C-C-O-H \\
\quad | \ | \\
\quad H \ H
\end{array}
$$

Ethanol (7)
(Ethyl alcohol)

In the pairs above, each structural formula represents a different compound. Each compound has its own unique set of physical and chemical properties. Compounds with the same molecular formula but with different structural formulas are called *isomers*.

The three-dimensional character of molecules is expressed by its stereochemistry. By looking at the *stereochemistry* of a molecule, the spatial relationships between atoms on one carbon and the atoms on an adjacent carbon can be examined. Since rotation can occur around carbon-carbon single bonds in open chain molecules, the atoms on adjacent carbons can assume different spatial relationships with respect to each other. The different arrangements that atoms can assume as a result of a rotation about a single bond are called *conformations*. A specific conformation is called a *conformer*. While individual isomers can be isolated, conformers cannot since interconversion, by rotation, is too rapid.

Conformers may be represented by projections through the use of two conventions, as shown in Fig. 24.1. These projections attempt to show on a flat surface how three-dimensional objects, in this case organic molecules, might look in three-dimensional space.

**a)** Sawhorse projection
of ethane

**b)** Newman projection
of ethane

**c)** Ball and stick model
of ethane

**Figure 24.1** • Molecular representations.

The *sawhorse projection* views the carbon-carbon bond at an angle and, by showing all the bonds and atoms, shows their spatial arrangements. The *Newman projection* provides a view along a carbon-carbon bond by sighting directly along the carbon-carbon bond. The near carbon is represented by a circle, and bonds attached to it are represented by lines going to the center of the circle. The carbon behind is not visible (since it is blocked by the near carbon), but the bonds attached to it are partially visible and are represented by lines going to the edge of the circle. With Newman projections, rotations show the spatial relationships of atoms on adjacent carbons easily. Two conformers that represent extremes are shown in Fig. 24.2.

**a)** Eclipsed conformation
of ethane

**b)** Staggered conformation
of ethane

Rotate rear
carbon 60°

**Figure 24.2** • Two conformers of ethane.

The *eclipsed* conformation has the bonds (and the atoms) on the adjacent carbons as close as possible. The *staggered* conformation has the bonds (and the atoms) on adjacent carbons as far as possible. One conformation can interconvert into the other by rotation around the carbon-carbon bond axis.

The three-dimensional character of molecular structure is shown through molecular model building. With molecular models, the number and types of bonds between atoms and the spatial arrangements of the atoms can be visualized for the molecules. This allows comparison of isomers and of conformers for a given set of compounds. The models also will let you see what is meant by *chemical equivalence*. Here *equivalence* relates to those positions or to those hydrogens on carbon(s) in an organic molecule that are equal in terms of chemical reactivity. In the case of hydrogen, replacement of any one of the equivalent hydrogens in a molecule by a substituent (any atom or group of atoms, for example, Cl or OH, respectively) leads to the identical substituted molecule.

## Objectives

1. To use models to visualize structure in organic molecules.
2. To build and compare isomers having a given molecular formula.
3. To explore the three-dimensional character of organic molecules.
4. To demonstrate equivalence of hydrogens in organic molecules.

## Procedure

Obtain a set of ball-and-stick molecular models from the laboratory instructor. The set contains the following parts (other colored spheres may be substituted as available):

- 2 Black spheres representing *Carbon*; this tetracovalent element has four holes;
- 6 Yellow spheres representing *Hydrogen*; this monovalent element has one hole;
- 2 Colored spheres representing the *halogen Chlorine*; this monovalent element has one hole;
- 1 Blue sphere representing *Oxygen*; this divalent element has two holes;
- 8 Sticks to represent bonds.

1. With your models, construct the molecule methane. Methane is a simple hydrocarbon consisting of one carbon and four hydrogens. After you put the model together, answer the questions below in the appropriate space on the Report Sheet.

**a.** With the model resting so that three hydrogens are on the desk, examine the structure. Move the structure so that a different set of three hydrogens are on the desk each time. Is there any difference between the way that the two structures look (1a)?

**b.** Does the term *equivalent* adequately describe the four hydrogens of methane (1b)?

**c.** Tilt the model so that only two hydrogens are in contact with the desk and imagine pressing the model flat onto the desktop. Draw the way in which the methane molecule would look in two-dimensional space (1c). This is the usual way that three-dimensional structures are written.

**d.** Using a protractor, measure the angle H—C—H on the model (1d).

2. Replace one of the hydrogens of the methane model with a colored sphere, which represents the halogen chlorine. The new model is chloromethane (methyl chloride), $CH_3Cl$. Position the model so that the three hydrogens are on the desk.

**a.** Grasp the atom representing chlorine and tilt it to the right, keeping two hydrogens on the desk. Write the structure of the projection on the Report Sheet (2a).

**b.** Return the model to its original position and then tilt as before, but this time to the left. Write this projection on the Report Sheet (2b).

**c.** While the projection of the molecule changes, does the structure of chloromethane change (2c)?

3. Now replace a second hydrogen with another chlorine sphere. The new molecule is dichloromethane, $CH_2Cl_2$.

**a.** Examine the model as you twist and turn it in space. Are the projections given below isomers of the molecule $CH_2Cl_2$ or representations of the same structure only seen differently in three dimensions (3a)?

$$
\begin{array}{cccc}
\text{H} & \text{Cl} & \text{H} & \text{Cl} \\
| & | & | & | \\
\text{Cl}-\text{C}-\text{H} & \text{H}-\text{C}-\text{Cl} & \text{Cl}-\text{C}-\text{Cl} & \text{H}-\text{C}-\text{H} \\
| & | & | & | \\
\text{Cl} & \text{H} & \text{H} & \text{Cl}
\end{array}
$$

4. Construct the molecule ethane, $C_2H_6$. Note that you can make ethane from the methane model by removing a hydrogen and replacing the hydrogen with a methyl group, $-CH_3$.

**a.** Write the structural formula for ethane (4a).

**b.** Are all the hydrogens attached to the carbon atoms equivalent (4b)?

**c.** Draw a sawhorse representation of ethane. Draw a staggered and an eclipsed Newman projection of ethane (4c).

**d.** Replace any hydrogen in your model with chlorine. Write the structure of the molecule chloroethane (ethyl chloride), $C_2H_5Cl$ (4d).

**e.** Twist and turn your model. Draw two Newman projections of the chloroethane molecule (4e).

**f.** Do the projections that you drew represent different isomers or conformers of the same compound (4f)?

**5.** Dichloroethane, $C_2H_4Cl_2$

**a.** In your molecule of chloroethane, if you choose to remove another hydrogen note that you now have a choice among the hydrogens. You can either remove a hydrogen from the carbon to which the chlorine is attached, or you can remove a hydrogen from the carbon that has only hydrogens attached. First, remove the hydrogen from the carbon that has the chlorine attached and replace it with a second chlorine. Write its structure on the Report Sheet (5a).

**b.** Compare this structure to the model which would result from removal of a hydrogen from the other carbon and its replacement by chlorine. Write its structure (5b) and compare it to the previous example. One isomer is 1,1-dichloroethane; the other is 1,2-dichloroethane. Label the structures drawn on the Report Sheet with the correct name.

**c.** Are all the hydrogens of chloroethane equivalent? Are some of the hydrogens equivalent? Label those hydrogens which are equivalent to each other (5c).

**6.** Butane

**a.** Butane has the formula $C_4H_{10}$. With help from a partner, construct a model of butane by connecting the four carbons in a series (C—C—C—C) and then adding the needed hydrogens. First, orient the model in such a way that the carbons appear as a straight line. Next, tilt the model so that the carbons appear as a zig-zag line. Then, twist around any of the C—C bonds so that a part of the chain is at an angle to the remainder. Draw each of these structures in the space on the Report Sheet (6a). Note that the structures you draw are for the same molecule but represent a different orientation and projection.

**b.** Sight along the carbon-carbon bond of $\overset{2}{C}$ and $\overset{3}{C}$ on the butane chain: $\overset{1}{C}H_3{-}\overset{2}{C}H_2{-}\overset{3}{C}H_2{-}\overset{4}{C}H_3$. Draw a staggered Newman projection. Rotate the $C_2$ carbon clockwise by 60°; draw the eclipsed Newman projection. Again, rotate the $C_2$ carbon clockwise by 60°; draw the Newman projection. Is the last projection staggered or eclipsed (6b)? Continue rotation of the $C_2$ carbon clockwise by 60° increments and observe the changes that take place.

**c.** Examine the structure of butane for equivalent hydrogens. In the space on the Report Sheet (6c), redraw the structure of butane and label those hydrogens which are equivalent to each other. On the basis of this examination, predict how many monochlorobutane isomers ($C_4H_9Cl$) could be obtained from the structure you drew in 6c (6d). Test your prediction by replacement of hydrogen by chlorine on the models. Draw the structures of these isomers (6e).

**d.** Reconstruct the butane system. First, form a three-carbon chain, then connect the fourth carbon to the center carbon of the three-carbon chain. Add the necessary hydrogens. Draw the structure of 2-methylpropane (isobutane) (6f).

Can any manipulation of the model, by twisting or turning of the model or by rotation of any of the bonds, give you the butane system? If these two, butane and 2-methylpropane (isobutane), are *isomers*, then how may we recognize that any two structures are isomers (6g)?

e. Examine the structure of 2-methylpropane for equivalent hydrogens. In the space on the Report Sheet (6h), redraw the structure of 2-methylpropane and label the equivalent hydrogens. Predict how many monochloroisomers of 2-methylpropane could be formed (6i) and test your prediction by replacement of hydrogen by chlorine on the model. Draw the structures of these isomers (6j).

## 7. $C_2H_6O$

a. There are two isomers with the molecular formula, $C_2H_6O$, ethanol (ethyl alcohol) and dimethyl ether. With your partner, construct both of these isomers. Draw these isomers on the Report Sheet (7a) and name each one.

b. Manipulate each model. Can either be turned into the other by a simple twist or turn (7b)?

c. For each compound, label those hydrogens which are equivalent. How many sets of equivalent hydrogens are there in ethanol (ethyl alcohol) and dimethyl ether (7c)?

## 8. Optional: Butenes

a. If springs are available for the construction of double bonds, construct 2-butene, $CH_3$—$CH$=$CH$—$CH_3$. There are two isomers for compounds of this formulation: the isomer with the two —$CH_3$ groups on the same side of the double bond, *cis*-2-butene; and the isomer with the two —$CH_3$ groups on opposite sides of the double bond, *trans*-2-butene. Draw these two structures on the Report Sheet (8a).

b. Can you twist, turn, or rotate one model into the other? Explain (8b).

c. How many bonds are connected to any single carbon of these structures (8c)?

d. With the protractor, measure the C—C=C angle (8d).

e. Construct methylpropene, $CH_3$— C = $CH_2$.
$$\mid$$
$$CH_3$$

Can you have a *cis*- or a *trans*- isomer in this system (8e)?

## 9. Optional: Butynes

a. If springs are available for the construction of triple bonds, construct 2-butyne, $CH_3$—$C$≡$C$—$CH_3$. Can you have a *cis*- or a *trans*- isomer in this system (9a)?

b. With the protractor, measure the C—C≡C angle (9b).

c. Construct a second butyne with your molecular models and springs. How does this isomer differ from the one in (a) above (9c)?

## Chemicals and Equipment

1. Molecular models (you may substitute other available colors for the spheres):
   - 2 Black spheres
   - 6 Yellow spheres
   - 2 Colored spheres (e.g., green)
   - 1 Blue sphere
   - 8 Sticks
2. Protractor
3. Optional: 3 springs

## Experiment 24

**PRE-LAB QUESTIONS**

1. How many bonds can each of the elements below form with neighboring atoms in a compound?

   C   H   O   N   Br   S   Cl

2. How does a *molecular formula* differ from a *structural formula*?

3. Write structural formulas for all the compounds with the molecular formula of $C_3H_7Br$.

4. Write the molecular formula for each of the compounds shown below.

a.
```
    H  H  H
    |  |  |
H – C– C– C– O – H
    |  |  |
    H  H  H
```

b.
```
    H  H  H
    |  |  |
H – C– C– C – H
    |  |  |
    H  O  H
       |
       H
```

c.
```
    H        H  H
    |        |  |
H – C – O – C– C – H
    |        |  |
    H        H  H
```

How are these compounds related to each other?

5. What information is provided by showing a molecule's stereochemistry?

## Experiment 24

### REPORT SHEET

**1.** Methane

   **a.**

   **b.**

   **c.**

   **d.**

**2.** Chloromethane (methyl chloride)

   **a.**

   **b.**

   **c.**

**3.** Dichloromethane

   **a.**

**4.** Ethane and chloroethane (ethyl chloride)

a.

b.

c.

d.

e.

f.

**5.** Dichloroethane

a.

b.

c.

**6.** Butane

a.

b.

c.

d.

**e.**

**f.**

**g.**

**h.**

**i.**

**j.**

**7.** C$_2$H$_6$O

**a.**

**b.**

**c.** Ethanol (ethyl alcohol) has _____ set(s) of equivalent hydrogens.

Dimethyl ether has _____ set(s) of equivalent hydrogens.

**8.** Butenes

**a.**

**b.**

**c.**

**d.** C—C=C angle

**e.**

**9.** Butynes

    **a.**

    **b.**

    **c.**

## POST-LAB QUESTIONS

**1.** Draw the three (3) isomers with the formula $C_5H_{12}$.

**2.** Draw the structure of propane and identify equivalent hydrogens. Identify equivalent sets by letters, e.g., $H_a$, $H_b$, etc.

**3.** Explain why there is no *cis-* nor *trans-* isomer for the open chain isomer of butane, $CH_3CH_2CH_2CH_3$, but for 2-butene, $CH_3CH=CHCH_3$, there is a *cis-* and a *trans-* isomer.

**4.** Draw a staggered and an eclipsed conformer for propane, $\overset{1}{C}H_3-\overset{2}{C}H_2-\overset{3}{C}H_3$, sighting along the $\overset{1}{C}-\overset{2}{C}$ bond.

# Experiment | 25

## Stereochemistry: use of molecular models. II

In Experiment 24, we looked at some molecular variations that acyclic organic molecules can take:

**1.** *Constitutional isomerism.* Molecules can have the same molecular formula but different arrangements of atoms.

    **a.** *skeletal isomerism*: structural isomers where differences are in the order in which atoms that make up the skeleton are connected; e.g., $C_4H_{10}$

$$CH_3CH_2CH_2CH_3 \qquad CH_3-\overset{\overset{\displaystyle CH_3}{|}}{CH}-CH_3$$

Butane                 2-Methylpropane

    **b.** *positional isomerism*: structural isomers where differences are in the location of a functional group; e.g., $C_3H_7Cl$

$$CH_3CH_2CH_2-Cl \qquad CH_3-\overset{\overset{\displaystyle Cl}{|}}{CH}-CH_3$$

1-Chloropropane            2-Chloropropane

**2.** *Stereoisomerism.* Molecules which have the same order of attachment of atoms but differ in the arrangement of the atoms in three-dimensional space.

    **a.** *cis-/trans- isomerism*: molecules that differ due to the geometry of substitution around a double bond; e.g., $C_4H_8$

$$\underset{\text{H}}{\overset{\text{CH}_3}{\diagdown}}C=C\underset{\text{H}}{\overset{\text{CH}_3}{\diagup}} \qquad \underset{\text{H}}{\overset{\text{CH}_3}{\diagdown}}C=C\underset{\text{CH}_3}{\overset{\text{H}}{\diagup}}$$

*cis*-2-Butene         *trans*-2-Butene

    **b.** *conformational isomerism*: variation in acyclic molecules as a result of a rotation about a single bond; e.g., ethane, $CH_3—CH_3$

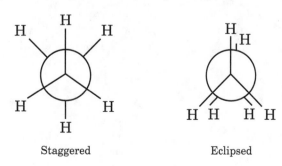

Staggered                    Eclipsed

In this experiment, we will further investigate stereoisomerism by examining a cyclic system, cyclohexane, and several acyclic tetrahedral carbon systems. The latter possess more subtle characteristics as a result of the spatial arrangement of the component atoms. We will do this by building models of representative organic molecules, then studying their properties.

## Objectives

1. To use models to study the conformations of cyclohexane.
2. To use models to distinguish between chiral and achiral systems.
3. To define and illustrate enantiomers, diastereomers, and meso forms.
4. To learn how to represent these systems in two-dimensional space.

## Procedure

You will build models and then you will be asked questions about the models. You will provide answers to these questions in the appropriate places on the Report Sheet. In doing this laboratory, it will be convenient if you tear out the Report Sheet and keep it by the Procedure as you work through the exercises. In this way, you can answer the questions without unnecessarily turning pages back and forth.

### Cyclohexane

Obtain a model set of "atoms" that contain the following:
- 8 Carbon components—model atoms with 4 holes at the tetrahedral angle (e.g., black);
- 2 Substituent components (halogens)—model atoms with 1 hole (e.g., red);
- 18 Hydrogen components—model atoms with 1 hole (optional) (e.g., white);
- 24 Connecting links—bonds.

1. Construct a model of cyclohexane by connecting 6 carbon atoms in a ring; then into each remaining hole insert a connecting link (bond) and, if available, add a hydrogen to each.

   a. Is the ring rigid or flexible, that is, can the ring of atoms move and take various arrangements in space, or is the ring of atoms locked into only one configuration (1a)?

   b. Of the many configurations, which appears best for the ring—a planar or a puckered arrangement (1b)?

   c. Arrange the ring atoms into a *chair* conformation (Fig. 25.1a) and compare it to the picture of the lounge chair (Fig. 25.1b). (Does the term fit the picture?)

**a)** The chair conformation            **b)** A lounge chair

**Figure 25.1** • The chair conformation for a 6-carbon ring.

2. With the model in the chair conformation, rest it on the tabletop.

   **a.** How many hydrogens are in contact with the tabletop (2a)?

   **b.** How many hydrogens point in a direction 180° opposite to these (2b)?

   **c.** Take your pencil and place it into the center of the ring perpendicular to the table. Now, rotate the ring around the pencil; we'll call this an *axis of rotation*. How many hydrogens are on bonds parallel to this axis (2c)? These hydrogens are called the *axial* hydrogens, and the bonds are called the *axial* bonds.

   **d.** If you look at the perimeter of the cyclohexane system, the remaining hydrogens lie roughly in a ring perpendicular to the axis through the center of the molecule. How many hydrogens are on bonds lying in this ring (2d)? These hydrogens are called *equatorial* hydrogens, and the bonds are called the *equatorial* bonds.

   **e.** Compare your model to the diagrams in Fig. 25.2 and be sure you are able to recognize and distinguish between axial and equatorial positions.

**a)** Axial position            **b)** Equatorial position

**Figure 25.2** • Axial and equatorial hydrogens in the chair conformation.

   In the space provided on the Report Sheet (2e), draw the structure of cyclohexane in the chair conformation with all 12 hydrogens attached. Label all the axial hydrogens, $H_a$, and all the equatorial hydrogens, $H_e$. How many hydrogens are labeled $H_a$ (2f)? How many hydrogens are labeled $H_e$ (2g)?

3. Look along any bond connecting any two carbon atoms in the ring. (Rotate the ring and look along a new pair of carbon atoms.) How are the bonds connected to these two carbons arranged? Are they staggered or are they eclipsed (3a)? In the space provided

on the Report Sheet (3b), draw the Newman projection for the view (see Experiment 24 for an explanation of this projection); for the bond connecting a ring carbon, label that group "ring."

4. Pick up the cyclohexane model and view it from the side of the chair. Visualize the "ring" around the perimeter of the system perpendicular to the axis through the center. Of the 12 hydrogens, how many are pointed "up" relative to the plane (4a)? How many are pointed "down" (4b)?

5. Orient your model so that you look at an edge of the ring and it conforms to Fig. 25.3. Are the two axial positions labeled *A cis* or *trans* to each other (5a)? Are the two equatorial positions labeled *B cis* or *trans* to each other (5b)? Are the axial and equatorial positions *A* and *B cis* or *trans* to each other (5c)? Rotate the ring and view new pairs of carbons in the same way. See whether the relationships of positions vary from the above. Position your eye as in Fig. 25.3 and view along the carbon-carbon bond. In the space provided on the Report Sheet (5d), draw the Newman projection. Using this projection, review your answers to 5a, 5b, and 5c.

**Figure 25.3**
Cyclohexane ring
viewed on edge.

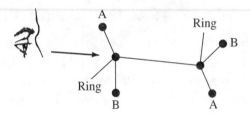

6. Replace one of the axial hydrogens with a colored component atom. Do a "ring flip" by moving one of the carbons *up* and moving the carbon farthest away from it *down* (Fig. 25.4). In what position is the colored component after the ring flip (6a)—axial or equitorial? Do another ring flip. In what position is the colored component now (6b)? Observe all the axial positions and follow them through a ring flip.

**Figure 25.4**
A "ring flip."

7. Refer to Fig. 25.3 and replace both positions labeled *A* by colored component atoms. Are they *cis* or *trans* (7a)? Do a ring flip. Are the two colored components *cis* or *trans* (7b)? Does the geometry change for the two components as the ring undergoes a ring flip (7c)? Repeat the exercise, replacing atoms in positions labeled *A* and *B* and answer the same three questions for this model.

8. Replace one of the colored components with a methyl, —$CH_3$, group. Manipulate the model so that the —$CH_3$ group is in an axial position; examine the model. Do a ring flip placing the —$CH_3$ in an equatorial position; examine the model. Which of the chair conformations, —$CH_3$ axial or —$CH_3$ equatorial, is more crowded (8a)? What would account for one of the conformations being more crowded than the other (8b)? Which would be of higher energy and thus less stable (8c)? In the space provided on the

Report Sheet (8d), draw the two conformations and connect with equilibrium arrows. Given your answers to 8a, 8b, and 8c, toward which conformation will the equilibrium lay (indicate by drawing one arrow bigger and thicker than the other)?

9. *A substituent group in the equatorial position of a chair conformation is more stable than the same substituent group in the axial position.* Do you agree or disagree? Explain your answer (9).

For the exercises in 10–15, although we will *not* be asking you to draw each and every conformation, we encourage you to practice drawing them in order to gain experience and facility in creating drawings on paper. Your instructor may make these exercises optional.

10. Construct *trans*-1,2-dimethylcyclohexane. By means of ring flips, examine the model with the two —CH$_3$ groups axial and the two —CH$_3$ groups equatorial. Which is the more stable conformation? Explain your answer (10).

11. Construct *cis*-1,2-dimethylcyclohexane by placing one —CH$_3$ group axial and the other equatorial. Do ring flips and examine the two chair conformations. Which is the more stable conformation? Explain your answer (11a). Given the two isomers, *trans*-1,2-dimethylcyclohexane and *cis*-1,2-dimethylcyclohexane, which is the more stable isomer? Explain your answer (11b).

12. Construct *cis*-1,3-dimethylcyclohexane by placing both —CH$_3$ groups in the axial positions. Do ring flips and examine the two chair conformations. Which is the more stable conformation? Explain your answer (12).

13. Construct *trans*-1,3-dimethylcyclohexane by placing one —CH$_3$ group axial and the other equatorial. Do ring flips and examine the two chair conformations. Which is the more stable conformation? Explain your answer (13a). Given the two isomers, *trans*-1,3-dimethylcyclohexane and *cis*-1,3-dimethylcyclohexane, which is the more stable isomer? Explain your answer (13b).

14. Construct *trans*-1,4-dimethylcyclohexane by placing both —CH$_3$ groups axial. Do ring flips and examine the two chair conformations. Which is the more stable conformation? Explain your answer (14).

15. Construct *cis*-1,4-dimethylcyclohexane by placing one —CH$_3$ group axial and the other equatorial. Do ring flips and examine the two chair conformations. Which is the more stable conformation? Explain your answer (15a). Given the two isomers, *trans*-1,4-dimethylcyclohexane and *cis*-1,4-dimethylcyclohexane, which is the more stable isomer? Explain your answer (15b).

16. Before we leave the cyclohexane ring system, there are some additional ring conformations we can examine. As we move from one cyclohexane chair conformation to another, the *boat* is a transitional conformation between them (Fig. 25.5). Examine a model of the boat conformation by viewing along a carbon-carbon bond, as shown by Fig. 25.5. In the space provided on the Report Sheet (16a), draw the Newman projection for this view and compare with the Newman projection of 5d. By examining the models and comparing the Newman projections, explain which conformation, the chair or the boat, is more stable (16b). Replace the "flagpole" hydrogens by —CH$_3$ groups. What happens when this is done (16c)? The steric strain can be relieved by twisting the ring and separating the two bulky groups. What results is a *twist boat*.

**Figure 25.5**
The boat conformation.

"Flagpole" positions

**17.** Review the conformations the cyclohexane ring can assume as it moves from one chair conformation to another:

$$\text{chair} \rightleftharpoons \text{twist boat} \rightleftharpoons \text{boat} \rightleftharpoons \text{twist boat} \rightleftharpoons \text{chair}$$

## Chiral Molecules

For this exercise, obtain a small hand mirror and a model set of "atoms" which contain the following:

- 8 Carbon components—model atoms with four holes at the tetrahedral angle (e.g., black);
- 32 Substituent components—model atoms with one hole in four colors (e.g., 8 red; 8 white; 8 blue; 8 green; or any other colors which your set may have);
- 28 Connecting links—bonds.

## Enantiomers

**1.** Construct a model consisting of a tetrahedral carbon center with four different component atoms attached: red, white, blue, green; each color represents a *different* group or atom attached to carbon. Does this model have a *plane of symmetry* (1a)? A plane of symmetry can be described as a cutting plane—a plane that when passed through a model or object *divides it into two equivalent halves*; the elements on one side of the plane are the exact reflection of the elements on the other side. If you are using a pencil to answer these questions, examine the pencil. Does it have a plane of symmetry (1b)?

**2.** Molecules without a plane of symmetry are *chiral*. In the model you constructed in no. 1, the tetrahedral carbon is the stereocenter; the molecule is chiral. A simple test for a stereocenter in a molecule is to look for a stereocenter with four different atoms or groups attached to it; this molecule will have no plane of symmetry. On the Report Sheet (2) are three structures; label the stereocenter in each structure with an asterisk (*).

**3.** Now take the model you constructed in no. 1 and place it in front of a mirror. Construct the model of the image projected in the mirror. You now have two models. If one is the object, what is the other (3a)? Do either have a plane of symmetry (3b)? Are both chiral (3c)? Now try to superimpose one model onto the other, that is, to place one model on top of the other in such a way that all five elements (i.e., the colored atoms) fall exactly one on top of the other. Can you superimpose one model onto the other (3d)? *Enantiomers* are two molecules that are related to each other such that they are *nonsuperimposable mirror images of each other*. Are the two models you have a pair of enantiomers (3e)?

4. Molecules with a plane of symmetry are *achiral*. Replace the blue substituent with a second green one. The model should now have three different substituents attached to the carbon. Does the model now have a plane of symmetry (4a)? Passing the cutting plane through the model, what colored elements does it cut in half (4b)? What is on the left half and right half of the cutting plane (4c)? Place this model in front of the mirror. Construct the model of the image projected in the mirror. You now have two models—an object and its mirror image. Are these two models superimposable on each other (4d)? Are the two models representative of different molecules or identical molecules (4e)?

Each stereoisomer in a pair of enantiomers has the property of being able to rotate monochromatic plane-polarized light. The instrument chemists use to demonstrate this property is called a *polarimeter* (see your text for a further description of the instrument). A pure solution of a single one of the enantiomers (referred to as an *optical isomer*) can rotate the light in either a clockwise (dextrorotatory, +) or a counterclockwise (levorotatory, −) direction. Thus those molecules that are optically active possess a "handedness" or chirality. Achiral molecules are optically inactive and do not rotate the light.

## *Meso* Forms and Diastereomers

5. With your models, construct a pair of enantiomers. From each of the models, remove the same common element (e.g., the white component) and the connecting links (bonds). Reconnect the two central carbons by a bond. What you have constructed is the *meso* form of a molecule, such as *meso*-tartaric acid. How many chiral carbons are there in this compound (5a)?

$$HOOC - C_aH - C_bH - COOH$$
$$\qquad\qquad\ \ | \qquad\ |$$
$$\qquad\qquad OH \quad\ OH$$

Tartaric acid

Is there a plane of symmetry (5b)? Is the molecule chiral or achiral (5c)?

6. In the space provided on the Report Sheet (6), use circles to indicate the four different groups for carbon $C_a$ and squares to indicate the four different groups for carbon $C_b$.

7. Project the model into a mirror and construct a model of the mirror image. Are these two models superimposable or nonsuperimposable (7a)? Are the models identical or different (7b)?

8. Now take one of the models you constructed in no. 7, and on one of the carbon centers exchange any two colored component groups. Does the new model have a plane of symmetry (8a)? Is it chiral or achiral (8b)? How many stereocenters are present (8c)? Take this model and one of the models you constructed in no. 7 and see whether they are superimposable. Are the two models superimposable (8d)? Are the two models identical or different (8e)? Are the two models mirror images of each other (8f)? Here we have a pair of molecular models, each with two stereocenters, that are not mirror images of each other. These two examples represent *diastereomers*, stereoisomers that are not related as mirror images.

9. Take the new model you constructed in no. 8 and project it into a mirror. Construct a model of the image in the mirror. Are the two models superimposable (9a)? What term describes the relationship of the two models (9b)?

Thus if we let these three models represent different isomers of tartaric acid, we find that there are three stereoisomers for tartaric acid—a *meso* form and a pair of enantiomers. A *meso* form with any one of the enantiomers of tartaric acid represents a pair of diastereomers. Although it may not be true for this compound because of the *meso* form, in general, if you have **n** stereocenters, there are $2^n$ stereoisomers possible (see Post-Lab question no. 3).

## Drawing Stereoisomers

This section will deal with conventions for representing these three-dimensional systems in two-dimensional space.

10. Construct models of a pair of enantiomers; use tetrahedral carbon and four differently colored components for the four different groups: red, green, blue, white. Hold one of the models in the following way:

   **a.** Grasp the blue group with your fingers and rotate the model until the green and red groups are pointing toward you (Fig. 25.6a). (Use the model which has the green group on the left and the red group on the right.)

   **b.** Holding the model in this way, the blue and white groups point away from you.

   **c.** If we use a drawing that describes a bond pointing toward you as a wedge and a bond pointing away from you as a dashed line, the model can be drawn as shown in Fig. 25.6b.

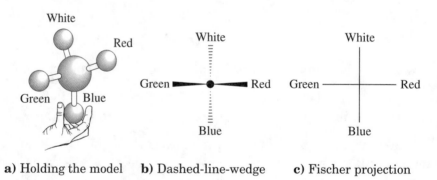

**a)** Holding the model      **b)** Dashed-line-wedge      **c)** Fischer projection

**Figure 25.6** • Projections in two-dimensional space.

If this model were compressed into two-dimensional space, we would get the projection shown in Fig. 25.6c. This is termed a *Fischer projection* and is named after a pioneer in stereochemistry, Emile Fischer. The Fischer projection has the following requirements:

   **(1)** the center of the cross represents the chiral carbon and is in the plane of the paper;

(2) the horizontal line of the cross represents those bonds projecting *out front* from the plane of the paper;

(3) the vertical line of the cross represents bonds projecting *behind* the plane of the paper.

**d.** In the space provided on the Report Sheet (10), use the enantiomer of the model in Fig. 25.6a and draw both the dashed-line-wedge and Fischer projection.

**11.** Take the model shown in Fig. 25.6a and rotate by 180° (turn upside down). Draw the Fischer projection (11a). Does this keep the requirements of the Fischer projection (11b)? Is the projection representative of the same system or of a different system (i.e., the enantiomer) (11c)?

In general, if you have a Fischer projection and rotate it in the plane of the paper by 180°, the resulting projection is of the *same* system. Test this assumption by taking the Fischer projection in Fig. 25.6c, rotating it in the plane of the paper by 180°, and comparing it to the drawing you did for no. 11a.

**12.** Again, take the model shown in Fig. 25.6a. Exchange the red and the green components. Does this exchange give you the enantiomer (12a)? Now exchange the blue and the white components. Does this exchange return you to the original model (12b)?

In general, for a given stereocenter, whether we use the dashed-line wedge or the Fischer projection, an odd-numbered exchange of groups leads to the mirror image of that center; an even-numbered exchange of groups leads back to the original system.

**13.** Test the above by starting with the Fischer projection given below and carrying out the operations directed in a, b, and c; use the space provided on the Report Sheet (13) for the answers.

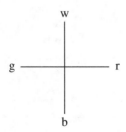

**a.** Exchange *r* and *g*; draw the Fischer projection you obtain; label this new projection as either the same as the starting model or the enantiomer.

**b.** Using the new Fischer projection from above, exchange *b* and *w*; draw the Fischer projection you now have.

**c.** Now rotate the last Fischer projection you obtained by 180°; draw the Fischer projection you now have; label this as either the same as the starting model or the enantiomer.

**14.** Let us examine models with two stereocenters by using tartaric acid as the example, HOOC—CH(OH)—CH(OH)—COOH; use your colored components to represent the various groups. Hold your models so that each stereoisomer is oriented as in Fig. 25.7. In the space provided on the Report Sheet (14), draw each of the corresponding Fischer projections.

**a)** Meso                           **b)** Enantiomers

**Figure 25.7** • The stereoisomers of tartaric acid.

Circle the Fischer projection that shows a plane of symmetry. Underline all the Fischer projections that would be optically active.

**15.** Use the Fischer projection of *meso*-tartaric acid and carry out even and odd exchanges of the groups; follow these exchanges with a model. Does an odd exchange lead to an enantiomer, a diastereomer, or to a system identical to the *meso* form (15a)? Does an even exchange lead to an enantiomer, a diastereomer, or to a system identical to the *meso* form (15b)?

---

**Chemicals and Equipment**

Model kits vary in size and color of components. Use what is available; other colors may be substituted.
1. Cyclohexane model kit: 8 carbons (black, 4 holes); 18 hydrogens (white, 1 hole); 2 substituents (red, 1 hole); 24 bonds.
2. Chiral model kit: 8 carbons (black, 4 holes); 32 substituents (8 red, 1 hole; 8 white, 1 hole; 8 blue, 1 hole; 8 green, 1 hole); 28 bonds.
3. Hand mirror

## Experiment 25

### PRE-LAB QUESTIONS

**1.** What shape or conformation does a cyclohexane ring take? Draw it.

**2.** Look at your hands and your feet. Which term best explains the relationship of the two hands and of the two feet: identical, constitutional, conformational, or enantiomeric?

**3.** What characteristics are present in a molecule that is chiral?

**4.** Label the chiral carbons in the molecules below with an asterisk (*).

**a.**

$$\begin{array}{cc} \text{OH} & \text{Cl} \\ | & | \\ \text{CH}_3-\text{CH}-\text{CH}-\text{CH}_2\text{CH}_3 \end{array}$$

**b.**

**5.** Explain the following statement: (+)-Lactic acid is an optical isomer.

## Experiment 25

### REPORT SHEET

#### *Cyclohexane*

1. a.

   b.


2. a.

   b.

   c.

   d.

   e.


   f.

   g.


3. a.

   b.


4. a.

   b.


5. a.

   b.

c.

d.

6. a.

b.

7.            *Trial 1*                          *Trial 2*

a.

b.

c.

8. a.

b.

c.

d.

9.

10. e,e or a,a

11. a. a,e or e,a

b.

12. a,a or e,e

13. a. a,e or e,a

b.

**14.** a,a or e,e

**15. a.** a,e or e,a

    **b.**

**16. a.**

    **b.**

    **c.**

## Enantiomers

**1. a.**

    **b.**

**2.**

$$CH_3-\overset{\overset{\displaystyle OH}{|}}{CH}-CH_2CH_3 \qquad CH_3-\overset{\overset{\displaystyle OH}{|}}{CH}-COOH \qquad ClCH_2-\overset{\overset{\displaystyle Br}{|}}{CH}-CH_3$$

**3. a.**

    **b.**

    **c.**

    **d.**

    **e.**

**4. a.**

    **b.**

    **c.**

    **d.**

    **e.**

## Meso *forms and diastereomers*

**5. a.**

**b.**

**c.**

6. $HOOC-\underset{\underset{\displaystyle HO}{|}}{\overset{\overset{\displaystyle H}{|}}{C_a}}-\underset{\underset{\displaystyle OH}{|}}{\overset{\overset{\displaystyle H}{|}}{C_b}}-COOH$ 　　　$HOOC-\underset{\underset{\displaystyle HO}{|}}{\overset{\overset{\displaystyle H}{|}}{C_a}}-\underset{\underset{\displaystyle OH}{|}}{\overset{\overset{\displaystyle H}{|}}{C_b}}-COOH$

7. **a.**

　**b.**

8. **a.**

　**b.**

　**c.**

　**d.**

　**e.**

　**f.**

9. **a.**

　**b.**

*Drawing stereoisomers*

10.

11. **a.**

　**b.**

　**c.**

12. **a.**

　**b.**

**13. a.**

**b.**

**c.**

**14.**

**15. a.**

**b.**

## POST-LAB QUESTIONS

**1.** Which position is more stable for the methyl group in methylcyclohexane: an equatorial position or an axial position? Explain your answer.

**2.** Draw the Fischer projections for the pair of enantiomers of lactic acid, $CH_3$—$CH(OH)$—$COOH$.

**3.** For 2,3-dibromobutane:

**a.** How many stereoisomers are possible for this compound?

$$
\begin{array}{cc}
Br & Br \\
| & | \\
\end{array}
$$
$$
CH_3-CH-CH-CH_3
$$

**b.** Draw Fischer projections for each stereoisomer; label enantiomers. Label any *meso* isomers (if there are any).

**4.** Determine the relationship between the following pairs of structures: identical, enantiomers, diastereomers.

**a.**  and

**b.**

$$
\begin{array}{c}
COOH \\
H_2N \blacktriangleright C \blacktriangleleft H \\
CH_3
\end{array}
$$
and
$$
\begin{array}{c}
COOH \\
H \blacktriangleright C \blacktriangleleft NH_2 \\
CH_3
\end{array}
$$

# Identification of hydrocarbons

## Background

The number of known organic compounds totals into the millions. Of these compounds, the simplest types are those which contain only hydrogen and carbon atoms. These are known as *hydrocarbons*. Because of the number and variety of hydrocarbons that can exist, some means of classification is necessary.

One means of classification depends on the way in which carbon atoms are connected. *Chain* aliphatic hydrocarbons are compounds consisting of carbons linked either in a single chain or in a branched chain. *Cyclic* hydrocarbons are aliphatic compounds that have carbon atoms linked in a closed polygon (also referred to as a *ring*). For example, hexane (single) and 2-methylpentane (branched) are chain aliphatic molecules, while cyclohexane is a cyclic aliphatic compound.

$$CH_3CH_2CH_2CH_2CH_2CH_3$$

Hexane

$$CH_3CHCH_2CH_2CH_3 \\ \quad\ | \\ \quad\ CH_3$$

2-Methylpentane

$$\begin{array}{c} H_2 \\ C \\ H_2C \quad CH_2 \\ H_2C \quad CH_2 \\ C \\ H_2 \end{array}$$

Cyclohexane

Another means of classification depends on the type of bonding that exists between carbons. Hydrocarbons which contain only carbon-to-carbon single bonds are called *alkanes*. These are also referred to as *saturated* molecules. Hydrocarbons containing at least one carbon-to-carbon double bond are called *alkenes*, and those compounds with at least one carbon-to-carbon triple bond are called *alkynes*. These are compounds that are referred to as *unsaturated* molecules. Finally, a class of cyclic hydrocarbons that contain a closed loop (sextet) of electrons are called *aromatic* (see Chapter 13 in your text for further details). Table 26.1 distinguishes between the families of hydrocarbons.

With so many compounds possible, identification of the bond type is an important step in establishing the molecular structure. Quick, simple tests on small samples can establish the physical and chemical properties of the compounds by class.

Some of the observed physical properties of hydrocarbons result from the nonpolar character of the compounds. In general, hydrocarbons do not mix with polar solvents such as water or ethyl alcohol. On the other hand, hydrocarbons mix with relatively nonpolar solvents such as ligroin (a mixture of alkanes), carbon tetrachloride ($CCl_4$), or dichloromethane ($CH_2Cl_2$). Since the density of most hydrocarbons is less than that of water, they will float. Crude oil and crude oil products (home heating oil and gasoline) are mixtures of hydrocarbons; these substances, when spilled on water, spread quickly along the surface because they are insoluble in water.

## Table 26.1 Types of Hydrocarbons

| Class | Characteristic Bond Type | | Example | |
|---|---|---|---|---|
| **I. Aliphatic** | | | | |
| 1. Alkane* | $-\overset{\mid}{\underset{\mid}{C}}-\overset{\mid}{\underset{\mid}{C}}-$ | single | $CH_3CH_2CH_2CH_2CH_2CH_3$ | hexane |
| 2. Alkene† | $\overset{}{\underset{}{>}}C=C\overset{}{\underset{}{<}}$ | double | $CH_3CH_2CH_2CH_2CH=CH_2$ | 1-hexene |
| 3. Alkyne† | $-C\equiv C-$ | triple | $CH_3CH_2CH_2CH_2C\equiv CH$ | 1-hexyne |
| **II. Cyclic** | | | | |
| 1. Cycloalkane* | $-\overset{\mid}{\underset{\mid}{C}}-\overset{\mid}{\underset{\mid}{C}}-$ | single | | cyclohexane |
| 2. Cycloalkene† | $\overset{}{\underset{}{>}}C=C\overset{}{\underset{}{<}}$ | double | | cyclohexene |
| 3. Aromatic | | | | benzene |
| | | | | toluene |

*Saturated    †Unsaturated

The chemical reactivity of hydrocarbons is determined by the type of bond in the compound. Although saturated hydrocarbons (alkanes) will burn (undergo *combustion*), they are generally unreactive to most reagents. (Alkanes do undergo a substitution reaction with halogens but require ultraviolet light.) Unsaturated hydrocarbons, alkenes and alkynes, not only burn, but also react by *addition* of reagents to the double or triple bonds. The addition products become saturated, with fragments of the reagent becoming attached to the carbons of the multiple bond. Aromatic compounds, with a higher carbon-to-hydrogen ratio than nonaromatic compounds, burn with a sooty flame as a result of unburned carbon particles being present. These compounds undergo *substitution* in the presence of catalysts rather than an addition reaction.

1. *Combustion*. The major component in "natural gas" is the hydrocarbon methane ($CH_4$). Other hydrocarbons used for heating or cooking purposes are propane ($C_3H_8$) and butane ($C_4H_{10}$). The products from combustion are carbon dioxide and water (heat is evolved, also).

$$CH_4 + 2O_2 \longrightarrow CO_2 + 2H_2O + \text{Heat}$$

$$CH_3CH_2CH_3 + 5O_2 \longrightarrow 3CO_2 + 4H_2O + \text{Heat}$$

**2.** *Reaction with bromine.* Unsaturated hydrocarbons react rapidly with bromine in a solution of carbon tetrachloride or cyclohexane. The reaction is the addition of the elements of bromine to the carbons of the multiple bonds.

$$CH_3CH = CHCH_3 + Br_2 \longrightarrow \underset{\text{Colorless}}{CH_3\underset{\overset{|}{Br}}{CH} - \underset{\overset{|}{Br}}{CH}CH_3}$$

Red

$$CH_3C \equiv CCH_3 + 2Br_2 \longrightarrow CH_3\underset{\underset{Br}{\overset{Br}{|}}}{C} - \underset{\underset{Br}{\overset{Br}{|}}}{C}CH_3$$

Red

Colorless

The bromine solution is red; the product that has the bromine atoms attached to carbon is colorless. Thus a reaction has taken place when there is a loss of color from the bromine solution and a colorless solution remains. Since alkanes have only single C—C bonds present, no reaction with bromine is observed; the red color of the reagent would persist when added. Aromatic compounds resist addition reactions because of their "aromaticity": *the possession of a closed loop (sextet) of electrons.* These compounds react with bromine in the presence of a catalyst such as iron filings or aluminum chloride.

Note that a substitution reaction has taken place and the gas HBr is produced.

**3.** *Reaction with concentrated sulfuric acid.* Alkenes react with cold concentrated sulfuric acid by addition. Alkyl sulfonic acids form as products and are soluble in $H_2SO_4$.

$$CH_3 - CH = CH - CH_3 + \underset{(H_2SO_4)}{HOSO_2OH} \longrightarrow CH_3 - \underset{\overset{|}{H}}{CH} - \underset{\overset{|}{OSO_2OH}}{CH} - CH_3$$

Saturated hydrocarbons are unreactive (additions are not possible); alkynes react slowly and require a catalyst ($HgSO_4$); aromatic compounds also are unreactive since addition reactions are difficult.

**4.** *Reaction with potassium permanganate.* Dilute or alkaline solutions of $KMnO_4$ oxidize unsaturated compounds. Alkanes and aromatic compounds are generally unreactive. Evidence that a reaction has occurred is observed by the loss of the purple color of $KMnO_4$ and the formation of the brown precipitate manganese dioxide, $MnO_2$.

$$3CH_3 - \underset{\text{Purple}}{CH = CH - CH_3} + 2KMnO_4 + 4H_2O \longrightarrow 3CH_3 - \underset{\overset{|}{OH}}{CH} - \underset{\overset{|}{OH}}{CH} - CH_3 + \underset{\text{Brown}}{2MnO_2} + 2KOH$$

Note that the product formed from an alkene is a glycol.

## Objectives

1. To investigate the physical properties, solubility, and density of some hydrocarbons.
2. To compare the chemical reactivity of an alkane, an alkene, and an aromatic compound.
3. To use physical and chemical properties to identify an unknown.

## Procedure

### CAUTION!

Assume the organic compounds are highly flammable. Use only small quantities. Keep away from open flames. Assume the organic compounds are toxic and can be absorbed through the skin. Avoid contact; wash if any chemical spills on your person. Handle concentrated sulfuric acid carefully. Flush with water if any spills on your person. Potassium permanganate and bromine are toxic; bromine solutions are also corrosive. Although the solutions are diluted, they may cause burns to the skin. Wear gloves when working with these chemicals.

### General Instructions

1. The hydrocarbons hexane, cyclohexene, and toluene (alkane, alkene, and aromatic) are available in dropper bottles.
2. The reagents 1% $Br_2$ in cyclohexane, 1% aqueous $KMnO_4$, and concentrated $H_2SO_4$ are available in dropper bottles.
3. Unknowns are in dropper bottles labeled A, B, and C. They may include an alkane, an alkene, or an aromatic compound.
4. Test tubes of $100 \times 13$ mm will be suitable for all the tests. When mixing the components, grip the test tube between thumb and forefinger; it should be held firmly enough to keep from slipping but loosely enough so that when the third and fourth fingers tap it, the contents will be agitated enough to mix.
5. Record all data and observations in the appropriate places on the Report Sheet.
6. Dispose of all organic wastes as directed by the instructor. *Do not pour into the sink!*

### Physical Properties of Hydrocarbons

1. *Water solubility of hydrocarbons.* Label six test tubes with the name of the substance to be tested. Place into each test tube 5 drops of the appropriate hydrocarbon: hexane, cyclohexene, toluene, unknown A, unknown B, unknown C. Add about 5 drops of water dropwise into each test tube. Water is a polar solvent. Is there any separation of components? Which component is on the bottom; which component is on the top? Mix the contents as described above. What happens when the contents are allowed to settle? What do you conclude about the density of the hydrocarbon? Is the hydrocarbon *more*

dense than water or *less* dense than water? Record your observations. Save these solutions for comparison with the next part.

2. *Solubility of hydrocarbons in ligroin.* Label six test tubes with the name of the substance to be tested. Place into each test tube 5 drops of the appropriate hydrocarbon: hexane, cyclohexene, toluene, unknown A, unknown B, unknown C. Add about 5 drops of ligroin dropwise into each test tube. Ligroin is a nonpolar solvent. Is there a separation of components? Is there a bottom layer and top layer? Mix the contents as described above. Is there any change in the appearance of the contents before and after mixing? Compare these test tubes to those from the previous part. Record your observations. Can you make any conclusion about the density of the hydrocarbons from what you actually see?

## Chemical Properties of Hydrocarbons

1. *Combustion.* The instructor will demonstrate this test in the fume hood. Place 5 drops of each hydrocarbon and unknown on separate watch glasses. Carefully ignite each sample with a match. Observe the flame and color of the smoke for each of the samples. Record your observations on the Report Sheet.

2. *Reaction with bromine.* Label six clean, dry test tubes with the name of the substance to be tested. Place into each test tube 5 drops of the appropriate hydrocarbon: hexane, cyclohexene, toluene, unknown A, unknown B, unknown C. Carefully add (dropwise and with shaking) 1% $Br_2$ in cyclohexane. Keep count of the number of drops needed to have the color persist; do not add more than 10 drops. Record your observations. To any sample that gives a negative test after adding 10 drops of bromine solution (i.e., the red color persists), add 5 more drops of 1% $Br_2$ solution and a small quantity of aluminum chloride, the amount on the tip of a micro-spatula; shake the mixture. Hold, with forceps, a piece of moistened blue litmus paper and lower it into the test tube until it is just above the surface of the liquid. Be careful and try not to touch the sides of the test tube or the liquid with the litmus paper. Record any change in the color of the solution and the litmus paper.

---

**CAUTION!**

Use 1% $Br_2$ solution in the hood; wear gloves when using this chemical.
Use the aluminum chloride in the hood; wear gloves when using this chemical. The bottle of aluminum chloride is kept in a desiccator; quickly take your sample and return the bottle to the desiccator. Do not leave the bottle open or outside the desiccator for a long time.

---

3. *Reaction with $KMnO_4$.* Label six clean, dry test tubes with the name of the substance to be tested. Place into each test tube 5 drops of the appropriate hydrocarbon: hexane, cyclohexene, toluene, unknown A, unknown B, unknown C. Carefully add (dropwise) 1% aqueous $KMnO_4$ solution; after each drop, shake to mix the solutions. Keep count of the number of drops needed to have the color of the permanganate solution persist; do not add more than 10 drops. Record your observations.

4. *Reaction with concentrated $H_2SO_4$.* Label six clean, dry test tubes with the name of the substance to be tested. Place into each test tube 5 drops of the appropriate hydrocarbon: hexane, cyclohexene, toluene, unknown A, unknown B, unknown C. Place all of the test tubes in an ice bath. *Wear gloves and carefully* add (with shaking) 3 drops of cold, concentrated sulfuric acid to each test tube. Note whether heat is evolved by feeling the test tube. Note whether the solution has become homogeneous or whether a color is produced. (The evolution of heat or the formation of a homogeneous solution or the appearance of a color is evidence that a reaction has occurred.) Record your observations.

5. *Unknowns.* By comparing the observations you made for your unknowns with that of the known hydrocarbons, you can identify unknowns A, B, and C. Record their identities on your Report Sheet.

---

### Chemicals and Equipment

1. 1% aqueous $KMnO_4$
2. 1% $Br_2$ in cyclohexane
3. Blue litmus paper
4. Concentrated $H_2SO_4$
5. Cyclohexene
6. Hexane
7. Aluminum chloride
8. Test tubes
9. Ligroin
10. Toluene
11. Unknowns A, B, and C
12. Watch glasses
13. Ice

## Experiment 26

### PRE-LAB QUESTIONS

**1.** Give an example of a *saturated* hydrocarbon and an *unsaturated* hydrocarbon.

**2.** Show the structural feature that distinguishes whether a hydrocarbon is an

alkane

alkene

alkyne

aromatic

**3.** Gasoline and water do not mix. Why?

## Experiment 26

## REPORT SHEET

### Physical properties of hydrocarbons

*Solubility:* Does the hydrocarbon mix with the solvent, *soluble,* or not mix with the solvent, *insoluble*?  Use the observations you make for the solubility tests and determine whether the hydrocarbons are polar or non-polar substances.

*Density:* For water, is the density *greater* than water (sinks) or *less* than water (floats)? For ligroin, can you tell anything about the relative densities?

| Hydrocarbon | H₂O Solubility | Density | Ligroin Solubility | Density |
|---|---|---|---|---|
| Hexane | | | | |
| Cyclohexene | | | | |
| Toluene | | | | |
| Unknown A | | | | |
| Unknown B | | | | |
| Unknown C | | | | |

## Chemical properties of hydrocarbons

| Hydrocarbon | Combustion | Bromine Test | KMnO$_4$ Test | H$_2$SO$_4$ Test |
|---|---|---|---|---|
| Hexane | | | | |
| Cyclohexene | | | | |
| Toluene | | | | |
| Unknown A | | | | |
| Unknown B | | | | |
| Unknown C | | | | |

Unknown A is _____.

Unknown B is _____.

Unknown C is _____.

## POST-LAB QUESTIONS

**1.** Write the structure of the major organic product for the following reactions; if no reaction, write NR.

**a.** $CH_3 - CH = CH_2 + Br_2 \longrightarrow$

**b.** $+ \quad KMnO_4 \quad + \quad H_2O \quad \longrightarrow$

**c.** $CH_3 - CH = CH - CH_3 + H_2SO_4 \longrightarrow$

**d.** $+ \quad KMnO_4 \quad \longrightarrow$

**2.** Octane is an unbranched alkane of formula $C_8H_{18}$. Based on your observations in this experiment, predict the following:

    **a.** Solubility in water:

    **b.** Solubility in ligroin:

    **c.** Combustion characteristics:

    **d.** Density versus water:

**3.** 1-Hexene is an alkene. Based on your observations in this experiment, what should you expect to see for this compound in the following tests:

    **a.** Bromine test:

    **b.** $KMnO_4$ test:

    **c.** Combustion:

**4.** An unknown compound, believed to be an hydrocarbon, showed the following behavior: it burned with a yellow sooty flame; it did not react with conc. $H_2SO_4$ or $KMnO_4$ solutions; it reacted with bromine solution only when a catalyst was added. From the compounds below, circle the one which fits the observations.

$CH_3CH_2CH_2CH_2CH_3$      $CH_2=CHCH_2CH_2CH_3$      $H-C\equiv C-CH_2CH_2CH_3$

# Column and paper chromatography: separation of plant pigments

## Background

Chromatography is a widely used experimental technique by which a mixture of compounds can be separated into its individual components. Two kinds of chromatographic experiments will be explored. In column chromatography, a mixture of components dissolved in a solvent is poured over a column of solid adsorbent and is eluted with the same or a different solvent. This is therefore a solid-liquid system; the stationary phase (the adsorbent) is solid and the mobile phase (the eluent) is liquid. In paper chromatography, the paper adsorbs water from the atmosphere of the developing chromatogram. (The water is present in the air as vapor, and it may be supplied as one component in the eluting solution.) The water is the stationary phase. The (other) component of the eluting solvent is the mobile phase and carries with it the components of the mixture. This is a liquid-liquid system.

Column chromatography is used most conveniently for preparative purposes, when one deals with a relatively large amount of the mixture and the components need to be isolated in milligrams or grams quantities. Paper chromatography, on the other hand, is used mostly for analytical purposes. Microgram or even picogram quantities can be separated by this technique, and they can be characterized by their $R_f$ number. This number is an index of how far a certain spot moved on the paper.

$$R_f = \frac{\text{Distance of the center of the sample spot from the origin}}{\text{Distance of the solvent front from the origin}}$$

For example, in Fig. 27.1 the $R_f$ values are as follows:

$R_f$ (substance 1) = 3.1 cm/11.2 cm = 0.28 and

$R_f$ (substance 2) = 8.5 cm/11.2 cm = 0.76

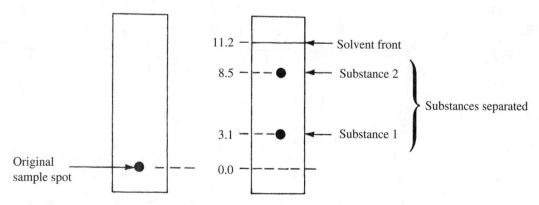

**Figure 27.1** • Illustration of chromatograms before and after elution.

Using the $R_f$ values, one is able to identify the components of the mixture with the individual components. The two main pigment components of tomato paste are β-carotene (yellow-orange) and lycopene (red) pigments. Their structures are given below:

Lycopene

β-Carotene

The colors of these pigments are due to the numerous double bonds in their structure. When bromine is added to double bonds, it saturates them and the color changes accordingly. In the tomato juice "rainbow" experiment, we stir bromine water into the tomato juice. The slow stirring allows the bromine water to penetrate deeper and deeper into the cylinder in which the tomato juice was placed. As the bromine penetrates, more and more double bonds will be saturated. Therefore, you may be able to observe a continuous change, a "rainbow" of colors, starting with the reddish tomato color at the bottom of the cylinder where no reaction occurred (since the bromine did not reach the bottom). Lighter colors will be observed on the top of the cylinder where most of the double bonds have been saturated.

---

### Objectives

1. To compare separation of components of a mixture by two different techniques.
2. To demonstrate the effect of bromination on plant pigments of tomato juice.

## Paper Chromatography

1. Obtain a sheet of Whatman no.1 filter paper, cut to size.

**Figure 27.2** • Preparation of chromatographic paper for spotting.

2. Plan the **spotting** of the samples as illustrated on Fig. 27.2. Five spots will be applied. The first and fifth spots will be β-carotene solutions supplied by your instructor. The second, third, and fourth spots will have your tomato paste extracts in different concentrations. Use a pencil to mark the placement of the spots lightly according to Fig. 27.2.

3. Pigments of tomato paste will be **extracted** in two steps.

   (a) Weigh about 10 g of tomato paste in a 50-mL beaker. Add 15 mL of 95% ethanol. Stir the mixture vigorously with a spatula until the paste will not stick to the stirrer. Place a small amount of glass wool (the size of a pea) in a small funnel, blocking the funnel exit. Place the funnel into a 50-mL Erlenmeyer flask and pour the tomato paste–ethanol mixture into the funnel. When the filtration is completed, squeeze the glass wool lightly with your spatula. In this step, we removed the water from the tomato paste and the aqueous components are in the filtrate, which we discard. The residue in the glass wool will be used to extract the pigments.

   (b) Place the residue from the glass wool in a 50-mL beaker. Add 10 mL petroleum ether and stir the mixture for about 2 min. to extract the pigments. Filter the extract as before through a new funnel with glass wool blocking the exit into a new and clean 50-mL beaker. Place the beaker under the hood on a hot plate (or water bath). No open flame, such as a Bunsen burner, is allowed. Evaporate the solvent to about 1 mL volume. Use low heat and take care not to evaporate all the solvent. After evaporation, cover the beaker with aluminum foil.

**Figure 27.3**
Withdrawing samples
with a capillary tube.

## Spotting

**4.** Place your chromatographic paper on a clean area (another filter paper) in order not to contaminate it. Use separate capillaries for your tomato paste extract and for the β-carotene solution. First, apply your capillary to the extracted pigment by dipping it into the solution as illustrated in Fig. 27.3. Apply the capillary lightly to the chromatographic paper by touching sequentially the spots marked 2, 3, and 4. Make sure you apply only small spots, not larger than 2 mm diameter, by quickly withdrawing the capillary from the paper each time you touch it. (See Fig. 27.4.)

**Figure 27.4**
Spotting.

While allowing the spots to dry, use your second capillary to apply spots of β-carotene in lanes 1 and 5. Return to the first capillary and apply another spot of the extract on top of the spots of lanes 3 and 4. Let them dry (Fig. 27.5). Finally, apply one more spot on top of lane 4. Let the spots dry. The unused extract in your beaker should be covered with aluminum foil. Place it in your drawer in the dark to save it for the second part of your experiment.

**Figure 27.5**
Drying chromatographic spots.

### *Developing the paper chromatogram*

**5.** Curve the paper into a cylinder and staple the edges above the 2-cm line, as is shown in Fig. 27.6.

**Figure 27.6** • Stapling.          **Figure 27.7** • Developing the chromatogram.

**6.** Pour 20 mL of the eluting solvent (petroleum ether : toluene : acetone in 45 : 1 : 4 ratio, supplied by your instructor) into a 600-mL beaker.

**7.** Place the stapled chromatogram into the 600-mL beaker, the spots being at the bottom near the solvent surface but **not covered by it**. Cover the beaker with aluminum foil (Fig. 27.7). Allow the solvent front to migrate up to 0.5–1 cm below the edge of the paper. This may take from 15 min. to 1 hr. Make certain by frequent inspection that the **solvent front does not run over the edge of the paper**. Remove the chromatogram from the beaker when the solvent front reaches 0.5–1 cm from the edge; then **proceed to step 11**.

## Column Chromatography

**8.** While you are waiting for the paper chromatogram to develop (step no. 7), you can perform the column chromatography experiment. Take a 25-mL buret. (You may use a chromatographic column, if available, of 1.6 cm diameter and about 13 cm long; see Fig. 27.8. If you use the column instead of the buret, all subsequent quantities below should be doubled.)

**Figure 27.8**
Chromatographic column.

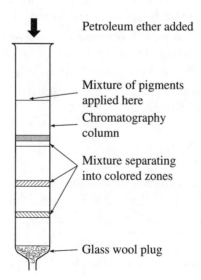

Petroleum ether added

Mixture of pigments applied here

Chromatography column

Mixture separating into colored zones

Glass wool plug

Add a small piece of glass wool and with the aid of a glass rod push it down near the stopcock. Add 15–16 mL of petroleum ether to the buret. Open the stopcock slowly and allow the solvent to fill the tip of the buret. Close the stopcock. You should have 12–13 mL of solvent above the glass wool. Weigh 20 g of aluminum oxide (alumina) in a 100-mL beaker. Place a small funnel on top of your buret. Pour the alumina into the buret. Allow the alumina to settle in order to form a 20-cm column. *Drain* the solvent but *do not allow the column to run dry. Always have at least 0.5 mL of clear solvent above the alumina in the column.* If alumina adheres to the walls of the buret, wash it down with more solvent.

9. Transfer by pipet 0.5–1 mL of the extract you stored in your drawer onto the column. The pipet containing the extract should be placed near the surface of the solvent on top of the column. Touching the walls of the buret with the tip of the pipet, allow the extract to drain slowly on top of the column. Open the stopcock slightly. Allow the sample to enter the column, *but make sure there is a small amount of solvent above the alumina in the column. (The column should never run dry.)* Add 10 or more mL of petroleum ether and wash the sample into the column by opening the stopcock and collecting the eluted solvent in a beaker.

10. As the solvent elutes the sample, you observe the migration of the pigments and their separation into at least two bands. When the fastest-moving pigment band reaches near the bottom of the column, close the stopcock and observe the color of the pigment bands and how far they migrated from the top of the column. Record your observation on the Report Sheet. This concludes the column chromatographic part of the experiment. Discard your solvent in a bottle supplied by your instructor for a later redistillation.

11. Meanwhile your paper chromatogram has developed. You must remove the filter paper from the 600-mL beaker before the solvent front reaches the edges of the paper. *Mark the position of the solvent front with a pencil.* Put the paper standing on its edges under the hood and let it dry.

## Tomato Juice "Rainbow"

12. While waiting for the paper to dry, you can perform the following short experiment. Weigh about 15 g of tomato paste in a beaker. Add about 30 mL of water and stir. Transfer the tomato juice into a 50-mL graduated cylinder and, with the aid of a pipet, add 5 mL of saturated bromine water (dropwise). With a glass rod, stir the solution very gently. Observe the colors and their positions in the cylinder. Record your observations on the Report Sheet.

## Paper Chromatography (continued)

13. Remove the staples from the dried chromatogram. Mark the spots of the pigments by circling with a pencil. Note the colors of the spots. Measure the distance of the center of each spot from its origin. Calculate the $R_f$ values.

14. If the spots on the chromatogram are faded, we can visualize them by exposing the chromatogram to iodine vapor. Place your chromatogram into a wide-mouthed jar containing a few iodine crystals. Cap the jar and warm it slightly on a hot plate to enhance the sublimation of iodine. The iodine vapor will interact with the faded

pigment spots and make them visible. After a few minutes of exposure to iodine vapor, remove the chromatogram and mark the spots **immediately** with a pencil. The spots will fade again with exposure to air. Measure the distance of the center of the spots from the origin and calculate the $R_f$ values.

**15.** Record the results of the paper chromatography on the Report Sheet.

## Chemicals and Equipment

1. Melting point capillaries open at both ends
2. 25-mL buret or chromatographic column
3. Glass wool
4. Whatman no.1 filter paper, 10 × 20 cm, cut to size
5. Heat lamp (optional)
6. Stapler
7. Hot plate (with or without water bath)
8. Tomato paste
9. Aluminum oxide (alumina)
10. Petroleum ether (b.p. 30–60°C)
11. 95% ethanol
12. Toluene
13. Acetone
14. 0.5% β-carotene in petroleum ether
15. Saturated bromine water
16. Iodine crystals
17. Ruler
18. Wide-mouthed jar

## Experiment 27

### PRE-LAB QUESTIONS

1. What marks are placed on your chromatography paper in order to calculate the $R_f$ value of a substance?

2. The structures of the two main pigments, lycopene and $\beta$-carotene, are given in the first part (**Background**):

   (a) What is the basic difference between the structures of these two pigments?

   (b) How many double bonds are in the $\beta$-carotene and in the lycopene structures?

   (c) To what class of hydrocarbons do these pigments belong?

3. Write the structure of $\beta$-carotene after it completely reacts with $Br_2$.

4. In paper chromatography, water is considered to be a stationary phase. Why is water stationary?

## Experiment 27

## REPORT SHEET

### Paper chromatography

| Sample | Distance from origin to solvent front (cm) (a) | Distance from origin to center of spot (cm) (b) | $R_f$ (b)/(a) | Color |
|---|---|---|---|---|
| **β-carotene** lane 1 | | | | |
| lane 5 | | | | |
| **Tomato extract** lane 2 (a) | | | | |
| (b) | | | | |
| (c) | | | | |
| (d) | | | | |
| lane 3 (a) | | | | |
| (b) | | | | |
| (c) | | | | |
| (d) | | | | |
| lane 4 (a) | | | | |
| (b) | | | | |
| (c) | | | | |
| (d) | | | | |

## Column chromatography

| Number of bands | Distance migrated from top of the column (cm) | Color |
|---|---|---|
| 1 | | |
| 2 | | |
| 3 | | |

### "Rainbow"

Describe the colors observed in the tomato juice "rainbow" experiment, starting from the bottom of the cylinder:

**1.** red                 **2.**                 **3.**

**4.**                 **5.**                 **6.**

### POST-LAB QUESTIONS

**1.** Did your tomato paste contain lycopene? What support is there for your answer?

**2.** Did your "rainbow" experiment indicate that the bromine penetrated to the bottom of your cylinder?

**3.** What is the effect of the amount of sample applied to the paper on the separation of the tomato pigments? Compare the results on lanes 2, 3, and 4 of the paper chromatogram.

**4.** Based on the $R_f$ value you obtained for lycopene in lane 3c, calculate how far your pigment would move if the solvent front traveled 25 cm.

**5.** Based on the "rainbow" experiment, which color indicates the presence of the smallest number of double bonds? Explain.

# Identification of alcohols and phenols

## Background

Specific groups of atoms in an organic molecule can determine its physical and chemical properties. These groups are referred to as *functional groups*. Organic compounds which contain the functional group $-OH$, the hydroxyl group, are called *alcohols*.

Alcohols are important commercially and include uses as solvents, drugs, and disinfectants. The most widely used alcohols are methanol or methyl alcohol, $CH_3OH$, ethanol or ethyl alcohol, $CH_3CH_2OH$, and 2-propanol or isopropyl alcohol, $(CH_3)_2CHOH$. Methanol is found in automotive products such as antifreeze and "dry gas." Ethanol is used as a solvent for drugs and chemicals, but is more popularly known for its effects as an alcoholic beverage. 2-Propanol, also known as "rubbing alcohol," is an antiseptic.

Alcohols may be classified as either primary, secondary, or tertiary:

$$R - CH_2 - OH$$
Primary alcohol

$$R - \underset{\underset{OH}{|}}{CH} - R'$$
Secondary alcohol

$$R - \underset{\underset{R''}{|}}{\overset{\overset{R'}{|}}{C}} - OH$$
Tertiary alcohol

Note that the classification depends on the number of carbon-containing groups, R (alkyl or aromatic), attached to the carbon bearing the hydroxyl group. Examples of each type are as follows:

$$CH_3CH_2 - OH$$
Ethanol
(Ethyl alcohol)
a primary alcohol

$$CH_3 - \underset{\underset{CH_3}{|}}{CH} - OH$$
2-Propanol
(Isopropyl alcohol)
a secondary alcohol

$$CH_3 - \underset{\underset{CH_3}{|}}{\overset{\overset{CH_3}{|}}{C}} - OH$$
2-Methyl-2-propanol
(*t*-Butyl alcohol)
a tertiary alcohol

Phenols bear a close resemblance to alcohols structurally since the hydroxyl group is present. However, since the $-OH$ group is bonded directly to a carbon that is part of an aromatic ring, the chemistry is quite different from that of alcohols. Phenols are more acidic than alcohols; concentrated solutions of the compound phenol are quite toxic and can cause severe skin burns. Phenol derivatives are found in medicines; for example, thymol is used to kill fungi and hookworms. (Also see Table 28.1.)

Phenol

Thymol
(2-isopropyl-5-methylphenol)

In this experiment, you will examine physical and chemical properties of representative alcohols and phenols. You will be able to compare the differences in chemical behavior between these compounds and use this information to identify an unknown.

| Table 28.1 | Selected Alcohols and Phenols |
|---|---|
| **Compound** | **Name and Use** |
| $CH_3OH$ | Methanol: solvent for paints, shellacs, and varnishes |
| $CH_3CH_2OH$ | Ethanol: alcoholic beverages; solvent for medicines, perfumes, and varnishes |
| $CH_3-\underset{\underset{OH}{\|}}{CH}-CH_3$ | 2-Propanol (isopropyl alcohol): rubbing alcohol; astringent; solvent for cosmetics, perfumes, and skin creams |
| $\underset{\underset{OH}{\|}}{CH_2}-\underset{\underset{OH}{\|}}{CH_2}$ | Ethylene glycol: antifreeze |
| $\underset{\underset{OH}{\|}}{CH_2}-\underset{\underset{OH}{\|}}{CH}-\underset{\underset{OH}{\|}}{CH_2}$ | Glycerol (glycerin): sweetening agent; solvent for medicines; lubricant; moistening agent |
| | Phenol (carbolic acid): cleans surgical and medical instruments; topical antipruritic (relieves itching) |
| | Vanillin: flavoring agent (vanilla flavor) |
| | Tetrahydrourushiol: irritant in poison ivy |

## Physical Properties

Since the hydroxyl group is present in alcohols and phenols, these compounds are polar. The polarity of the hydroxyl group, coupled with its ability to form hydrogen bonds, enables many alcohols and phenols to mix with water. In general, for alcohols of comparable molecular weight, the greater the number of hydroxyl groups present in the molecule, the greater the number of hydrogen bonds that can form with water.  As a result, the solubility in water increases. Since these compounds also contain nonpolar portions, they show additional solubility in many organic solvents, such as dichloromethane ($CH_2Cl_2$) and diethyl ether ($CH_3CH_2OCH_2CH_3$).

Hydrogen bonding of the hydroxyl group with water.

## Chemical Properties

The chemical behavior of the different classes of alcohols and of phenols can be used as a means of identification. Quick, simple tests that can be carried out in test tubes will be performed.

1. *Lucas test*. This test is used to distinguish between water-soluble primary, secondary, and tertiary alcohols. Lucas reagent is a mixture of zinc chloride, $ZnCl_2$, in concentrated HCl. Upon addition of this reagent, a tertiary alcohol reacts rapidly and immediately gives an insoluble white layer. A secondary alcohol reacts slowly and, after heating slightly, gives the white layer within 10 min. A primary alcohol does not react. Any formation of a heterogeneous phase or appearance of an emulsion is a positive test.

   $CH_3CH_2$—OH  +  HCl  +  $ZnCl_2$  →  no reaction
   primary alcohol

   $(CH_3)_2CH$—OH  +  HCl  +  $ZnCl_2$  →  $(CH_3)_2CH$—Cl + $H_2O$ (10 min. heat)
   secondary alcohol                                          insoluble

   $(CH_3)_3C$—OH  +  HCl  +  $ZnCl_2$  →  $(CH_3)_3C$—Cl  +  $H_2O$ (<5 min.)
   tertiary alcohol                                        insoluble

2. *Chromic acid test*. This test is able to distinguish primary and secondary alcohols from tertiary alcohols. Using acidified dichromate solution, primary alcohols are oxidized to carboxylic acids; secondary alcohols are oxidized to ketones; tertiary alcohols are not oxidized. (Note that in those alcohols which are oxidized, the carbon that has the hydroxyl group *loses a hydrogen*.) In the oxidation, the brown-red color of the chromic acid changes to a blue-green solution. Phenols are oxidized to nondescript brown tarry masses. (Aldehydes are also oxidized under these conditions to carboxylic acids, but ketones remain intact; see Experiment 29 for further discussion.)

$$3CH_3CH_2-OH + 4H_2CrO_4 + 6H_2SO_4 \longrightarrow 3CH_3-\overset{\overset{\displaystyle O}{\|}}{C}-OH + 2Cr_2(SO_4)_3 + 13H_2O$$

primary alcohol          brown-red                              carboxylic acid         blue-green

$$3CH_3-\overset{\overset{\displaystyle OH}{|}}{CH}-CH_3 + 2H_2CrO_4 + 3H_2SO_4 \longrightarrow 3CH_3-\overset{\overset{\displaystyle O}{\|}}{C}-CH_3 + Cr_2(SO_4)_3 + 8H_2O$$

secondary alcohol         brown-red                               ketone         blue-green

$$(CH_3)_3C-OH + H_2CrO_4 + H_2SO_4 \longrightarrow \text{no reaction}$$

tertiary alcohol

3. *Iodoform test.* This test is more specific than the previous two tests. Only ethanol (ethyl alcohol) and alcohols with the part structure $CH_3CH(OH)$ react. These alcohols react with iodine in aqueous sodium hydroxide to give the yellow precipitate iodoform.

$$\overset{\overset{\displaystyle OH}{|}}{RCHCH_3} + 4I_2 + 6NaOH \longrightarrow \overset{\overset{\displaystyle O}{\|}}{RC}-O^- \ Na^+ + 5NaI + 5H_2O + HCI_3 \,(s)$$

                                                      iodoform
                                                      yellow

Phenols also react under these conditions. With phenol, the yellow precipitate triiodophenol forms.

triiodophenol
yellow precipitate

4. *Acidity of phenol.* Phenol is also called carbolic acid. Phenol is an acid and will react with base; thus phenols readily dissolve in base solutions. In contrast, alcohols are not acidic.

5. *Iron(III)chloride test.* Addition of aqueous iron(III) chloride to a phenol gives a colored solution. Depending on the structure of the phenol, the color can vary from green to purple.

light yellow

violet color

1. To learn characteristic chemical reactions of alcohols and phenols.
2. To use these chemical characteristics for identification of an organic compound.

## Procedure

### CAUTION!

Chromic acid is very corrosive. Any spill should be immediately flushed with water. Phenol is toxic. Also, contact with the solid will cause burns to skin; any contact should be thoroughly washed with large quantities of water. Solid phenol should be handled only with a spatula or forceps. Use gloves with these reagents. Dispose of reaction mixtures and excess reagents in proper containers as directed by your instructor.

## Physical Properties of Alcohols and Phenols

1. You will test the alcohols 1-butanol (a primary alcohol), 2-butanol (a secondary alcohol), 2-methyl-2-propanol (a tertiary alcohol), and phenol; you will also have as an unknown one of these compounds (labeled A, B, C, or D). As you run a test on a known, test the unknown at the same time for comparison. Note that the phenol will be provided as an aqueous solution.

2. Into separate test tubes (100 × 13 mm) labeled 1-butanol, 2-butanol, 2-methyl-2-propanol, and unknown, place 10 drops of each sample; dilute by mixing with 3 mL of distilled water. Into a separate test tube, place 2 mL of a prepared water solution of phenol. Are all the solutions homogeneous? Record your observations on the Report Sheet (1).

3. Test the pH of each of the aqueous solutions. Do the test by first dipping a clean glass rod into the solutions and then transferring a drop of liquid to pH paper. Use a broad pH indicator paper (e.g., pH range 1–12) and read the value of the pH by comparing the color to the chart on the dispenser. Record the results on the Report Sheet (2).

## Chemical Properties of Alcohols and Phenols

1. *Iodoform test*. Prepare a water bath of 60°C in a 400-mL beaker. Place into separate clean, dry labeled test tubes (150 x 18 mm), (a) 1-butanol, (b) 2-butanol, (c) 2-methyl-2-propanol, (d) phenol, and (e) unknown, 5 drops of sample to be tested. Add to each test tube 2 mL of water. If the compound is not soluble, add 1,2-dimethoxyethane (dropwise) until the solution is homogeneous. Add to each test tube (dropwise) 2 mL of 10% NaOH; tap the test tube with your finger to mix. Warm the mixtures in the 60°C water bath. Add the prepared solution of $I_2$-KI test reagent, dropwise to each test tube, (with shaking) until each solution becomes brown. Cork each test tube and shake

vigorously; return to the warm water bath. (If the color fades, add more I$_2$-KI test reagent until the dark color persists for 2 min. at 60°C.) Add 10% NaOH (dropwise) until each solution becomes colorless; cork and shake again. Return the test tubes to the warm water bath for 5 min. (*Remember to remove the corks when heating.*) Remove the test tubes from the water and dilute each solution with cold water; leave enough room so you can cork and shake one last time.  Let cool and look for a light yellow precipitate. The formation of the yellow precipitate tends to be slow. Put these test tubes to one side and make your observations when all the other tests are completed. Record your observations on the Report Sheet (3).

2. *Lucas test.* Place 5 drops of each sample into separate clean, dry test tubes (100 × 13 mm), labeled as before. Add 1 mL of Lucas reagent; mix well by stoppering each test tube with a cork, tapping the test tube sharply with your finger for a few seconds to mix; remove the cork after mixing and allow each test tube to stand for 5 min. Look carefully for any cloudiness that may develop during this time period. If there is no cloudiness after 10 min., warm the test tubes that are clear for 15 min. in a 60°C water bath. Record your observations on the Report Sheet (4).

3. *Chromic acid test.* Place into separate clean, dry test tubes (100 × 13 mm), labeled as before, 5 drops of sample to be tested. To each test tube add 10 drops of reagent grade acetone and 2 drops of chromic acid. Place the test tubes in a 60°C water bath for 5 min. Note the color of each solution. (Remember, the loss of the brown-red and the formation of a blue-green color is a positive test.) Record your observations on the Report Sheet (5).

4. *Iron(III) chloride test.* Place into separate clean, dry test tubes (100 × 13 mm), labeled as before, 5 drops of sample to be tested. Add 2 drops of iron(III) chloride solution to each. Note any color changes in each solution. (Remember, a purple color indicates the presence of a phenol.) Record your observations on the Report Sheet (6).

5. From your observations identify your unknown.

---

### Chemicals and Equipment

1. Aqueous phenol
2. Acetone (reagent grade)
3. 1-Butanol
4. 2-Butanol
5. 2-Methyl-2-propanol (*t*-butyl alcohol)
6. Chromic acid solution
7. 1, 2-Dimethoxyethane
8. Iron(III) chloride solution
9. I$_2$-KI solution
10. 10% NaOH, sodium hydroxide
11. Lucas reagent
12. Corks
13. Hot plate
14. pH paper
15. Unknown

## Experiment 28

### PRE-LAB QUESTIONS

1. Below are the structures of four alcohols. Classify each alcohol as either primary, secondary, tertiary, or phenol.


a.   $CH_3CH-OH$
     $\quad\quad\;\;|$
     $\quad\quad CH_3$


b.   $CH_3\,CH_2\,CHCH_2\,CH_2-OH$
     $\quad\quad\quad\quad\;|$
     $\quad\quad\quad\quad CH_3$


c.   $\quad\quad\quad\quad CH_3$
     $\quad\quad\quad\quad\;|$
     $CH_3\,CH_2\,C-OH$
     $\quad\quad\quad\quad\;|$
     $\quad\quad\quad\quad CH_3$


d.   HO
     $\quad\quad CH_3$

**2.** The compound below, thymol (a constituent of garden thyme), is an example of a compound which contains the phenolic functional group. Circle the phenol part of the molecule.

Thymol

**3.** Explain why many alcohols and phenols can mix with water.

## Experiment 28

## REPORT SHEET

| Test | 1-Butanol | 2-Butanol | 2-Methyl-2-propanol | Phenol | Unknown |
|---|---|---|---|---|---|
| **1.** Water | | | | | |
| **2.** pH | | | | | |
| **3.** Iodoform | | | | | |
| **4.** Lucas | | | | | |
| **5.** Chromic acid | | | | | |
| **6.** Iron(III) chloride | | | | | |

Identity of unknown:

Unknown no. _____. The unknown compound is _____.

## POST-LAB QUESTIONS

**1.** Write the structures for the following alcohols. Also, indicate whether the alcohol is primary, secondary, or tertiary.

   **a.** 1-butanol

**b.** 2-butanol

**c.** 2-methyl-2-propanol

2. Write the structure of the major organic product expected from each of the following reactions. If no reaction is expected write "No Reaction."

**a.**

+ NaOH $\longrightarrow$

**b.** $CH_3-CH-CH_2CH_3 + H_2CrO_4 \xrightarrow{H_2SO_4}$
       |
       OH

**c.**

$$CH_3CH_2 - \overset{\overset{\displaystyle CH_3}{|}}{\underset{\underset{\displaystyle CH_3}{|}}{C}} - OH + H_2CrO_4 \xrightarrow{H_2SO_4}$$

**d.**

$$CH_3 - \overset{\overset{\displaystyle CH_3}{|}}{\underset{\underset{\displaystyle CH_3}{|}}{C}} - OH + HCl \xrightarrow{ZnCl_2}$$

3. The compounds below give a purple color with iron(III) chloride solution. Circle the part in each structure responsible for the reaction that gives this test.

Resveratrol
(grapes)

Capsaicin
(hot pepper)

**4.** Ethylene glycol (see Table 28.1) is a liquid at room temperature and is soluble in water in all proportions; it is the liquid that cars use in radiators as an antifreeze. However, butane, $CH_3CH_2CH_2CH_3$, a compound of similar molecular weight and a gas at room temperature, is *insoluble* in water. How do you account for these differences?

**5.** A student had two unknown liquid alcohols. Unknown A was soluble in base and gave a violet color with iron(III) chloride. Unknown B was insoluble in base and gave a blue-green color with chromic acid; it did not form a precipitate with Lucas reagent, even after heating for 30 min. To which classes do alcohols A and B belong?

**6.** Solubility in water *increases* for the following alcohols : slightly soluble 1-pentanol $(CH_3CH_2CH_2CH_2CH_2OH) < 1,5$-pentanediol $(HOCH_2CH_2CH_2CH_2CH_2OH) <$ very soluble 1,3,6-hexanetriol $[HOCH_2CH_2CH(OH)CH_2CH_2CH_2OH]$. Explain this order in solubility, where < means *less than*.

# Identification of aldehydes and ketones

## Background

Aldehydes and ketones are representative of compounds which possess the carbonyl group:

$$
\begin{array}{c}
O \\
\parallel \\
-C- \\
\end{array}
\qquad \text{The carbonyl group}
$$

Aldehydes have at least one hydrogen attached to the carbonyl carbon; in ketones, no hydrogens are directly attached to the carbonyl carbon, only carbon containing R-groups:

$$
\begin{array}{c}
O \\
\parallel \\
R-C-H \\
\text{Aldehyde}
\end{array}
\qquad
\begin{array}{c}
O \\
\parallel \\
R-C-R' \\
\text{Ketone}
\end{array}
\qquad
\begin{array}{l}
\text{(R and R' can be} \\
\text{alkyl or aromatic)}
\end{array}
$$

   Aldehydes and ketones of low molecular weight have commercial importance. Many others occur naturally. Table 29.1 has some representative examples.

| Table **29.1** Representative Aldehydes and Ketones | | |
|---|---|---|
| **Compound** | | **Source and Use** |
| HCH (with O double bonded to C) | Formaldehyde | Oxidation of methanol; plastics; preservative |
| $CH_3CCH_3$ (with O double bonded to C) | Acetone | Oxidation of 2-propanol; solvent |
| Citral structure | Citral | Lemongrass oil; fragrance |
| Jasmone structure | Jasmone | Oil of jasmine; fragrance |

In this experiment you will investigate the chemical properties of representative aldehydes and ketones.

## Classification Tests

1. *Chromic acid test*. Aldehydes are oxidized to carboxylic acids by chromic acid; ketones are not oxidized. A positive test results in the formation of a blue-green solution from the brown-red color of chromic acid.

$$3R-\overset{\overset{\displaystyle O}{\|}}{C}-H \;+\; 2H_2CrO_4 \;+\; 3H_2SO_4 \longrightarrow 3R-\overset{\overset{\displaystyle O}{\|}}{C}-OH \;+\; Cr_4(SO_4)_3 \;+\; 5H_2O$$

aldehyde      brown-red                 blue-green

$$R-\overset{\overset{\displaystyle O}{\|}}{C}-R \;\xrightarrow[\text{H}_2\text{SO}_4]{\text{H}_2\text{CrO}_4}\; \text{no reaction}$$

ketone

2. *Tollens' test*. Most aldehydes reduce Tollens' reagent (ammonia and silver nitrate) to give a precipitate of silver metal. The free silver forms a silver mirror on the sides of the test tube. (This test is sometimes referred to as the "silver mirror" test.) The aldehyde is oxidized to a carboxylic acid.

$$R-\overset{\overset{\displaystyle O}{\|}}{C}-H \;+\; 2Ag(NH_3)_2OH \longrightarrow 2Ag(s) \;+\; R-\overset{\overset{\displaystyle O}{\|}}{C}-O^-NH_4^+ \;+\; H_2O \;+\; 3NH_3$$

aldehyde                           silver<br>mirror

3. *Iodoform test*. Methyl ketones give the yellow precipitate iodoform when reacted with iodine in aqueous sodium hydroxide.

$$R-\overset{\overset{\displaystyle O}{\|}}{C}-CH_3 \;+\; 3I_2 \;+\; 4NaOH \longrightarrow 3NaI \;+\; 3H_2O \;+\; R-\overset{\overset{\displaystyle O}{\|}}{C}-O^-Na^+ \;+\; HCI_3(s)$$

methyl<br>ketone                              iodoform<br>yellow

**4.** *2,4-Dinitrophenylhydrazine test.* All aldehydes and ketones give an immediate precipitate with 2,4-dinitrophenylhydrazine reagent. This reaction is general for both these functional groups. The color of the precipitate varies from yellow to red. (Note that alcohols do not give this test.)

## Identification by forming a derivative

The classification tests (summarized in Table 29.2), when properly done, can distinguish between various types of aldehydes and ketones. However, these tests alone may not allow for the identification of a specific unknown aldehyde or ketone. A way to correctly identify an unknown compound is by using a known chemical reaction to convert it into another compound that is known. The new compound is referred to as a *derivative*. Then, by comparing the physical properties of the unknown and the derivative to the physical properties of known compounds listed in a table, an identification can be made.

| Table **29.2** Summary of Classification Tests | |
|---|---|
| **Compound** | **Reagent for Positive Test** |
| Aldehydes and ketones | 2,4-Dinitrophenylhydrazine |
| Aldehydes | Chromic acid |
| | Tollens' reagent |
| Methyl ketones | Iodoform |

The ideal derivative is a solid. A solid can be easily purified by crystallization and easily characterized by its melting point. Thus two similar aldehydes or two similar ketones usually have derivatives that have *different melting points*. The most frequently formed derivatives for aldehydes and ketones are the 2,4-dinitrophenylhydrazone

(2,4-DNP), oxime, and semicarbazone. Table 29.4 (p. 308) lists some aldehydes and ketones along with melting points of their derivatives. If, for example, we look at the properties of valeraldehyde and crotonaldehyde, though the boiling points are virtually the same, the melting points of the 2,4-DNP, oxime, and semicarbazone are different and provide a basis for identification.

1. *2,4-Dinitrophenylhydrazone*. 2,4-Dinitrophenylhydrazine reacts with aldehydes and ketones to form 2,4-dinitrophenylhydrazones (2,4-DNP).

$$R-\overset{\overset{\displaystyle O}{\|}}{C}-R' \ + \ H_2N-NH-\underset{\text{2,4-dinitrophenylhydrazine}}{\underset{\text{(R'=H or alkyl or aromatic)}}{\boxed{\phantom{NO_2}}}}-NO_2 \ \xrightarrow{-H_2O} \ \underset{R'}{\overset{R}{>}}C=N-NH-\underset{\text{2,4-dinitrophenylhydrazone (2,4-DNP)}}{\boxed{\phantom{NO_2}}}-NO_2$$

The 2,4-DNP product is usually a colored solid (yellow to red) and is easily purified by recrystallization.

2. *Oxime*. Hydroxylamine reacts with aldehydes and ketones to form oximes.

$$R-\overset{\overset{\displaystyle O}{\|}}{C}-R' \ + \ NH_2OH \ \xrightarrow{-H_2O} \ \underset{R'}{\overset{R}{>}}C=NOH$$

hydroxylamine      oxime

(R'=H or alkyl or aromatic)

These are usually derivatives with low melting points.

3. *Semicarbazone*. Semicarbazide, as its hydrochloride salt, reacts with aldehydes and ketones to form semicarbazones.

$$R-\overset{\overset{\displaystyle O}{\|}}{C}-R' \ + \ NH_2NH\overset{\overset{\displaystyle O}{\|}}{C}NH_2 \ \xrightarrow{-H_2O} \ \underset{R'}{\overset{R}{>}}C=NNH\overset{\overset{\displaystyle O}{\|}}{C}NH_2$$

semicarbazide      semicarbazone

(R'=H or alkyl or aromatic)

A pyridine base is used to neutralize the hydrochloride in order to free the semicarbazide so it may react with the carbonyl substrate.

---

### Objectives

1. To learn the chemical characteristics of aldehydes and ketones.
2. To use these chemical characteristics in simple tests to distinguish between examples of aldehydes and ketones.
3. To identify aldehydes and ketones by formation of derivatives.

### Classification Tests

1. Classification tests are to be carried out on four known compounds and one unknown. Any one test should be carried out on all five samples at the same time for comparison. Label test tubes as shown in Table 29.3.

| Table **29.3** | Labelling Test Tubes |
|---|---|
| **Test Tube No.** | **Compound** |
| 1 | Isovaleraldehyde (an aliphatic aldehyde) |
| 2 | Benzaldehyde (an aromatic aldehyde) |
| 3 | Cyclohexanone (a ketone) |
| 4 | Acetone (a methyl ketone) |
| 5 | Unknown |

> **CAUTION!**
>
> Chromic acid is toxic and corrosive. Handle with care and **promptly wash any spill**. Use gloves with this reagent.

2. *Chromic acid test.* Place 5 drops of each substance into separate, labeled test tubes (100 × 13 mm). Dissolve each compound in 20 drops of reagent-grade acetone (to serve as solvent); then add to each test tube 4 drops of chromic acid reagent, one drop at a time; after each drop, mix by sharply tapping the test tube with your finger. Let stand for 10 min. Aliphatic aldehydes should show a change within a minute; aromatic aldehydes take longer. Note the approximate time for any change in color or formation of a precipitate on the Report Sheet.

3. *Tollens' test.*

> **CAUTION!**
>
> This reagent must be freshly prepared before it is to be used and any excess disposed **immediately** after use. Organic residues should be discarded in appropriate waste containers. Unused Tollens' reagent should be collected from every student by the instructor. **Do not store Tollens' reagent since it is explosive when dry.** The instructor should dispose of the excess reagant by adding 1 M $HNO_3$ until acidic, warming on a hot plate. The solution can then be stored in a waste container for heavy metals.

Enough reagent for your use can be prepared in a 25-mL Erlenmeyer flask by mixing 5 mL of Tollens' solution A with 5 mL of Tollens' solution B. To the silver oxide precipitate which forms, add (dropwise, with shaking) 10% ammonia solution until the brown precipitate just dissolves. *Avoid an excess of ammonia.*

Place 5 drops of each substance into separately labeled clean, dry test tubes (100 × 13 mm). Dissolve the compound in *bis*(2-ethoxyethyl)ether by adding this solvent dropwise until a homogeneous solution is obtained. Then, add 2 mL (approx. 40 drops) of the prepared Tollens' reagent and mix by sharply tapping the test tube with your finger. Place the test tube in a 60°C water bath for 5 min. Remove the test tubes from the water and look for a silver mirror. If the tube is clean, a silver mirror will be formed; if not, a black precipitate of finely divided silver will appear. Record your results on the Report Sheet. (Clean your test tubes with 1 M $HNO_3$ and discard the solution in a waste container designated by your instructor.)

4. *Iodoform test.* Prepare a water bath of 60°C in a 400-mL beaker. Place into separately labeled clean, dry test tubes (150 x 18 mm), 5 drops of sample to be tested. Add to each test tube 2 mL of water. If the compound is not soluble, add 1,2-dimethoxyethane (dropwise) until the solution is homogeneous. Add to each test tube (dropwise) 2 mL of 10% NaOH; tap the test tube with your finger to mix. Warm the mixtures in the 60°C water bath. Add the prepared solution of $I_2$-KI test reagent, dropwise to each test tube, (with shaking) until each solution becomes brown. Cork each test tube and shake vigorously; return to the warm water bath. (If the color fades, add more $I_2$-KI test reagent until the dark color persists for 2 min. at 60°C.) Add 10% NaOH (dropwise) until each solution becomes colorless; cork and shake again. Return the test tubes to the warm water bath for 5 min. (*Remember to remove the corks when heating.*) Remove the test tubes from the water and dilute each solution with cold water; leave enough room so you can cork and shake one last time. Let cool and look for a light yellow precipitate. The formation of the yellow precipitate tends to be slow. Put these test tubes to one side and make your observations when all the other tests are completed. Record your observations on the Report Sheet.

5. *2,4-Dinitrophenylhydrazine test.* Place 5 drops of each substance into separately labeled clean, dry test tubes (100 × 13 mm) and add 20 drops of the 2,4-dinitrophenylhydrazine reagent to each. If no precipitate forms immediately, heat for 5 min. in a warm water bath (60°C); cool. Record your observations on the Report Sheet.

## Formation of derivatives

> **CAUTION!**
>
> The chemicals used to prepare derivatives and some of the derivatives are potential carcinogens. Exercise care in using the reagents and in handling the derivatives. Avoid skin contact by wearing gloves.

1. This section is *optional.* Consult your instructor to determine whether this section is to be completed. Your instructor will indicate how many derivatives and which derivatives you should make.

2. *Waste.* Place all the waste solutions from these preparations in designated waste containers for disposal by your instructor.

3. *General procedure for recrystallization.* Heat a small volume (10–20 mL) of solvent to boiling on a steam bath (or carefully on a hot plate). Place crystals into a test tube (100 × 13 mm) and add the hot solvent (dropwise) until the crystals just dissolve

**Figure 29.1** • Vacuum filtration with a Hirsch funnel.

(keep the solution hot, also). Allow the solution to cool to room temperature; then cool further in an ice bath. Collect the crystals on a Hirsch funnel by vacuum filtration. Use a trap between the Hirsch funnel set-up and the aspirator (Fig. 29.1). Wash the crystals with 10 drops of ice cold solvent. Allow the crystals to dry by drawing air through the Hirsch funnel. Take a melting point (see Experiment 15 for a review of the technique).

4. *2,4-Dinitrophenylhydrazone* (2,4-DNP). Place 5 mL of the 2,4-dinitrophenylhydrazine reagent in a test tube (150 × 18 mm). Add 10 drops of the unknown compound; sharply tap the test tube with your finger to mix. If crystals do not form immediately, gently heat in a water bath (60°C) for 5 min. Cool in an ice bath until crystals form. Collect the crystals by vacuum filtration using a Hirsch funnel (Fig. 29.1). Allow the crystals to dry on the Hirsch funnel by drawing air through the crystals. Take a melting point and record on the Report Sheet. (The crystals are usually pure enough to give a good melting point. However, if the melting point range is too large, recrystallize from a minimum volume of ethanol.)

5. *Oxime*. Prepare fresh reagent by dissolving 1.0 g of hydroxylamine hydrochloride and 1.5 g of sodium acetate in 4 mL of distilled water in a test tube (150 × 18 mm). Add 20 drops of unknown and sharply tap the test tube with your finger to mix. Warm in a hot water bath (60°C) for 5 min. Cool in an ice bath until crystals form. (If no crystals form, scratch the inside of the test tube with a glass rod.) Collect the crystals on a Hirsch funnel by vacuum filtration (Fig. 29.1). Allow the crystals to air dry on the Hirsch funnel by drawing air through the crystals. Take a melting point and record on the Report Sheet. (Recrystallize, if necessary, from a minimum volume of ethanol.)

6. *Semicarbazone*. Place 2.0 mL of the semicarbazide reagent in a test tube (150 × 18 mm); add 10 drops of unknown. If the solution is not clear, add methanol (dropwise) until a clear solution results. Add 2.0 mL of pyridine and gently warm in a hot bath (60°C) for 5 min. Crystals should begin to form. (If there are no crystals, place the test tube in an ice bath and scratch the inside of the test tube with a glass rod.) Collect the crystals on a

Table **29.4**   Selection of Aldehydes and Ketones with Derivatives

| Compound | Formula | b.p.°C | 2,4-DNP m.p.°C | Oxime m.p.°C | Semi-carbazone m.p.°C |
|---|---|---|---|---|---|
| **Aldehydes** | | | | | |
| Isovaleraldehyde (3-methylbutanal) | $CH_3-CH-CH_2-C-H$ with $CH_3$ branch and $C=O$ | 93 | 123 | 49 | 107 |
| Valeraldehyde (pentanal) | $CH_3CH_2CH_2CH_2-C-H$ with $C=O$ | 103 | 106 | 52 | — |
| Crotonaldehyde (2-butenal) | $CH_3-CH=CH-C-H$ with $C=O$ | 104 | 190 | 119 | 199 |
| Caprylaldehyde (octanal) | $CH_3CH_2CH_2CH_2CH_2CH_2CH_2-C-H$ with $C=O$ | 171 | 106 | 60 | 101 |
| Benzaldehyde | $C_6H_5-C-H$ with $C=O$ | 178 | 237 | 35 | 222 |
| **Ketones** | | | | | |
| Acetone (2-propanone) | $CH_3-C-CH_3$ with $C=O$ | 56 | 126 | 59 | 187 |
| 2-Pentanone | $CH_3-C-CH_2CH_2CH_3$ with $C=O$ | 102 | 144 | 58 (b.p. 167°C) | 112 |
| 3-Pentanone | $CH_3CH_2-C-CH_2CH_3$ with $C=O$ | 102 | 156 | 69 (b.p. 165°C) | 139 |
| Cyclopentanone | cyclopentanone ring with $C=O$ | 131 | 146 | 56 | 210 |
| Cyclohexanone | cyclohexanone ring with $C=O$ | 156 | 162 | 90 | 166 |
| Acetophenone | $C_6H_5-C-CH_3$ with $C=O$ | 202 | 238 | 60 | 198 |

*Source*: Compiled by Zvi Rappoport, *CRC Handbook of Tables for Organic Compound Identification, 3rd ed.*, The Chemical Rubber Co.: Cleveland (1967).

Hirsch funnel by vacuum filtration (Fig. 29.1). Allow the crystals to air dry on the Hirsch funnel by drawing air through the crystals. Take a melting point and record on the Report Sheet. (Recrystallize, if necessary, from a minimum volume of ethanol.)

7. Based on the observations you recorded on the Report Sheet, and by comparing the melting points of the derivatives for your unknown to the knowns listed in Table 29.4, identify your unknown.

---

## Chemicals and Equipment

1. Acetone (reagent grade)
2. 10% ammonia solution
3. Benzaldehyde
4. *Bis*(2-ethoxyethyl) ether
5. Chromic acid reagent
6. Cyclohexane
7. 2,4-Dinitrophenylhydrazine reagent
8. 1, 2-Dimethoxyethane
9. Ethanol
10. Hydroxylamine hydrochloride
11. $I_2$-KI test solution
12. Isovaleraldehyde
13. Methanol
14. 10% NaOH, sodium hydroxide
15. Pyridine
16. Semicarbazide reagent
17. Sodium acetate
18. Tollens' reagent (solution A and solution B)
19. Hirsch funnel
20. Hot plate
21. Neoprene adapter (no. 2)
22. Rubber stopper (no. 6, one-hole), with glass tubing
23. 50-mL side-arm filter flask
24. 250-mL side-arm filter flask
25. Vacuum tubing (heavy walled)

# Experiment 29

## PRE-LAB QUESTIONS

**1.** Write a general structure for the following functional groups:

    **a.** an aldehyde

    **b.** an aromatic aldehyde

    **c.** a ketone

    **d.** a methyl ketone

**2.** How does chromic acid distinguish between an aldehyde and a ketone?

**3.** Which functional group gives a reaction with Tollens' reagent? What do you observe when there is a positive Tollens' test?

**4.** How can you identify a specific unknown aldehyde or ketone?

## Experiment 29

## REPORT SHEET

| Test | Isovaler-aldehyde | Benz-aldehyde | Cyclo-hexanone | Acetone | Unknown |
|---|---|---|---|---|---|
| Chromic acid | | | | | |
| Tollens' | | | | | |
| Iodoform | | | | | |
| 2,4-Dinitro-phenyl-hydrazine | | | | | |

### Optional

| Derivative | Observed m.p. | Literature m.p. |
|---|---|---|
| 2,4-DNP | | |
| Oxime | | |
| Semicarbazone | | |

Unknown no. _____. The unknown compound is _____.

## POST-LAB QUESTIONS

1. What test tube reaction will distinguish between 2-pentanone and 3-pentanone?

2. What kind of results do you see when the following compounds are mixed together with the given test solution?

   a. with 2,4-dinitrophenylhydrazine

   b. $CH_3CH_2-\overset{\overset{\displaystyle O}{\|}}{C}-H$   with Tollens' reagent

   c. $-\overset{\overset{\displaystyle O}{\|}}{C}-CH_3$   with $I_2$-KI reagent

   d. $-\overset{\overset{\displaystyle O}{\|}}{C}-H$   with chromic acid

3. The boiling points of valeraldehyde and crotonaldehyde are too close to make a clear identification (refer to Table 29.4). How can the two aldehydes be identified?

4. Using the laboratory tests of this experiment, show how you could distinguish between the following compounds (Hint: use tests that would give a clear, positive test result for a listed compound that is unique to it and thus different from the others):

| Test | Cyclohexane | Cyclohexanone | Hexanal |
|---|---|---|---|
|  |  |  |  |
|  |  |  |  |
|  |  |  |  |

# Properties of carboxylic acids and esters

## Background

Carboxylic acids are structurally like aldehydes and ketones in that they contain the carbonyl group. However, an important difference is that carboxylic acids contain an hydroxyl group attached to the carbonyl carbon.

$$\overset{\displaystyle \overset{O}{\|}}{-C-OH} \quad \text{The carboxylic acid group}$$

This combination gives the group its most important characteristic; it behaves as an acid.

As a family, carboxylic acids are weak acids that ionize only slightly in water. As aqueous solutions, typical carboxylic acids ionize to the extent of only one percent or less.

$$\overset{\displaystyle \overset{O}{\|}}{R-C-OH} + H_2O \rightleftharpoons \overset{\displaystyle \overset{O}{\|}}{R-C-O^-} + H_3O^+$$

At equilibrium, most of the acid is present as un-ionized molecules. Dissociation constants, $K_a$, of carboxylic acids, where R is an alkyl group, are $10^{-5}$ or less. Water solubility depends to a large extent on the size of the R-group. Only a few low-molecular-weight acids (up to four carbons) are very soluble in water.

Although carboxylic acids are weak, they are capable of reacting with bases stronger than water. Thus while benzoic acid shows limited water solubility, it reacts with sodium hydroxide to form the soluble salt sodium benzoate. (Sodium benzoate is a preservative in some soft drinks.)

| Benzoic acid | Sodium benzoate |
|---|---|
| Insoluble | Soluble |

Sodium carbonate, $Na_2CO_3$, and sodium bicarbonate, $NaHCO_3$, solutions can neutralize carboxylic acids also.

The combination of a carboxylic acid and an alcohol gives an ester; water is eliminated. Ester formation is an equilibrium process, catalyzed by an acid catalyst.

$$CH_3CH_2CH_2\overset{\overset{\displaystyle O}{\|}}{C}-OH \ + \ CH_3CH_2OH \ \underset{}{\overset{H^+}{\rightleftharpoons}} \ H_2O \ + \ CH_3CH_2CH_2\overset{\overset{\displaystyle O}{\|}}{C}-OCH_2CH_3$$

| Butyric acid | Ethyl alcohol | Ethyl butyrate (Ester) |

$$\text{Esterification} \longrightarrow$$
$$\longleftarrow \text{Hydrolysis}$$

The reaction typically gives 60% to 70% of the maximum yield. The reaction is a reversible process. An ester reacting with water, giving the carboxylic acid and alcohol, is called *hydrolysis*; it is acid catalyzed. The base-promoted decomposition of esters yields an alcohol and a salt of the carboxylic acid; this process is called *saponification*. Saponification means "soap making," and the sodium salt of a fatty acid (e.g., sodium stearate) is a soap.

$$CH_3CH_2CH_2\overset{\overset{\displaystyle O}{\|}}{C}-OCH_2CH_3 \ + \ NaOH \ \longrightarrow \ CH_3CH_2CH_2\overset{\overset{\displaystyle O}{\|}}{C}-O^-Na^+ \ + \ CH_3CH_2OH$$

Saponification

A distinctive difference between carboxylic acids and esters is in their characteristic odors. Carboxylic acids are noted for their sour, disagreeable odors. On the other hand, esters have sweet and pleasant odors often associated with fruits, and fruits smell the way they do because they contain esters. These compounds are used in the food industry as fragrances and flavoring agents. For example, the putrid odor of rancid butter is due to the presence of butyric acid, while the odor of pineapple is due to the presence of the ester, ethyl butyrate. Only those carboxylic acids of low molecular weight have odor at room temperature. Higher-molecular-weight carboxylic acids form strong hydrogen bonds, are solid, and have a low vapor pressure. Thus few molecules reach our noses. Esters, however, do not form hydrogen bonds among themselves; they are liquid at room temperature, even when the molecular weight is high. Thus they have high vapor pressure and many molecules can reach our noses, providing odor.

---

### Objectives

1. To study the physical and chemical properties of carboxylic acids: solubility, acidity, aroma.
2. To prepare a variety of esters and note their odors.
3. To demonstrate saponification.

---

### Carboxylic Acids and Their Salts

*Characteristics of acetic acid*

1. Place into a clean, dry test tube (100 × 13 mm) 2 mL of water and 10 drops of glacial acetic acid. Note its odor by wafting (moving your hand quickly over the open end of the test tube) the vapors toward your nose. Of what does it remind you?

2. Take a glass rod and dip it into the solution. Using wide-range indicator paper (pH 1–12), test the pH of the solution by touching the paper with the wet glass rod. Determine the value of the pH by comparing the color of the paper with the chart on the dispenser.

3. Now, add 2 mL of 2 M NaOH to the solution. Cork the test tube and sharply tap it with your finger. Remove the cork and determine the pH of the solution as before; if not basic, continue to add more base (dropwise) until the solution is basic. Note the odor and compare to the odor of the solution before the addition of base.

4. By dropwise addition of 3 M HCl, carefully reacidify the solution from step no. 3 (above); test the solution as before with pH paper until the solution tests acid. Does the original odor return?

*Characteristics of benzoic acid*

1. Your instructor will weigh out 0.1 g of benzoic acid for sample size comparison. With your microspatula, take some sample equivalent to the preweighed sample (an exact quantity is not important here). Add the solid to a test tube (100 × 13 mm) along with 2 mL of water. Is there any odor? Mix the solution by sharply tapping the test tube with your finger. How soluble is the benzoic acid?

2. Now add 1 mL of 2 M NaOH to the solution from step no. 1 (above), cork, and mix by sharply tapping the test tube with your finger. What happens to the solid benzoic acid? Is there any odor?

3. By dropwise addition of 3 M HCl, carefully reacidify the solution from step no. 2 (above); test as before with pH paper until acidic. As the solution becomes acidic, what do you observe?

### Esterification

1. Into five clean, dry test tubes (100 × 13 mm), add 10 drops of liquid carboxylic acid or 0.1 g of solid carboxylic acid and 10 drops of alcohol according to the scheme in Table 30.1. Note the odor of each reactant.

| Table 30.1 | Acids and Alcohols | |
| --- | --- | --- |
| Test Tube No. | Carboxylic Acid | Alcohol |
| 1 | Formic | Isobutyl |
| 2 | Acetic | Benzyl |
| 3 | Acetic | Isopentyl |
| 4 | Acetic | Ethyl |
| 5 | Salicylic | Methyl |

2. Add 5 drops of concentrated sulfuric acid to each test tube and mix the contents thoroughly by sharply tapping the test tube with your finger.

---

## CAUTION!

Sulfuric acid causes severe burns. Flush any spill with lots of water. Use gloves with this reagent.

---

3. Place the test tubes in a warm water bath at 60°C for 15 min. Remove the test tubes from the water bath, cool, and add 2 mL of water to each. Note that there is a layer on top of the water in each test tube. With a Pasteur pipet, take a few drops from this top layer and place on a watch glass. Note the odor. Match the ester from each test tube with one of the following odors: banana, peach, raspberry, nail polish remover, wintergreen.

### Saponification

This part of the experiment can be done while the esterification reactions are being heated.

1. Place into a test tube (150 × 18 mm) 10 drops of methyl salicylate and 5 mL of 6 M NaOH. Heat the contents in a boiling water bath for 30 min. Record on the Report Sheet what has happened to the ester layer (1).

2. Cool the test tube to room temperature by placing it in an ice water bath. Determine the odor of the solution and record your observation on the Report Sheet (2).

3. Carefully add 6 M HCl to the solution, 1 mL at a time, until the solution is acidic. After each addition, mix the contents and test the solution with litmus. When the solution is acidic, what do you observe? What is the name of the compound formed? Answer these questions on the Report Sheet (3).

## Chemicals and Equipment

1. Glacial acetic acid
2. Benzoic acid
3. Formic acid
4. Salicylic acid
5. Benzyl alcohol
6. Ethanol (ethyl alcohol)
7. 2-Methyl-1-propanol (isobutyl alcohol)
8. 3-Methyl-1-butanol (isopentyl alcohol)
9. Methanol (methyl alcohol)
10. Methyl salicylate
11. 3 M HCl
12. 6 M HCl
13. 2 M NaOH
14. 6 M NaOH
15. Concentrated $H_2SO_4$
16. pH paper (broad range pH 1–12)
17. Litmus paper
18. Pasteur pipet
19. Hot plate

## Experiment 30

### PRE-LAB QUESTIONS

1. Write the structures of the following carboxylic acids:

   a. acetic acid

   b. formic acid

   c. salicylic acid

2. Write the products from the reaction of acetic acid and sodium hydroxide.

3. Methyl butyrate has the flavor of apples. Draw the structure and give the name of the alcohol and the carboxylic acid needed to synthesize this ester.

4. What is a saponification?

5. Give an example of a soap.

## Experiment 30

**REPORT SHEET**

*Carboxylic acids and their salts*

| Characteristics of Acetic Acid | | | |
|---|---|---|---|
| **Property** | **Water Solution** | **NaOH Solution** | **HCl Solution** |
| Odor | | | |
| Solubility | | | |
| pH | | | |

| Characteristics of Benzoic Acid | | | |
|---|---|---|---|
| **Property** | **Water Solution** | **NaOH Solution** | **HCl Solution** |
| Odor | | | |
| Solubility | | | |
| pH | | | |

## Esterification

| Test Tube | Acid | Odor | Alcohol | Odor | Ester | Odor |
|-----------|----------|------|----------|------|-------|------|
| 1 | Formic | | Isobutyl | | | |
| 2 | Acetic | | Benzyl | | | |
| 3 | Acetic | | Isopentyl | | | |
| 4 | Acetic | | Ethyl | | | |
| 5 | Salicylic | | Methyl | | | |

## Saponification

1. What has happened to the ester layer?

2. What has happened to the odor of the ester?

3. What forms on reacidification of the solution? Name the compound.

4. Write the chemical equation for the saponification of methyl salicylate.

## POST-LAB QUESTIONS

1. (a.) Acetic acid is completely soluble in water. Draw the structural formula of the molecule as it is in the water. (b) Sodium hydroxide, NaOH, is added and the pH raised to 12. What happens to the acid in this solution? Draw the structural formula now. (c) Hydrochloric acid, HCl, is now added and the pH lowered to 2. Draw the structural formula of the acetic acid in this solution.

2. Write equations for each of the five esterification reactions you performed in this experiment.

   a.

   b.

   c.

   d.

   e.

3. Benzoic acid and diphenyl ketone are both insoluble in water. Suggest a method for separating a mixture of these compounds.

   Benzoic acid          Diphenyl ketone

# Experiment 31

## Properties of amines and amides

### Background

Amines and amides are two classes of organic compounds which contain nitrogen. Amines behave as organic bases and may be considered as derivatives of ammonia. Amides are compounds which have a carbonyl group connected to a nitrogen atom and are neutral. In this experiment, you will learn about the physical and chemical properties of some members of the amine and amide families.

If the hydrogens of ammonia are replaced by alkyl or aryl groups, amines result. Depending on the number of carbon atoms bonded directly to nitrogen, amines are classified as either primary (one carbon atom), secondary (two carbon atoms), or tertiary (three carbon atoms) (Table 31.1).

**Table 31.1  Types of Amines**

| | **Primary Amines** | **Secondary Amines** | **Tertiary Amines** |
|---|---|---|---|
| $NH_3$ <br> Ammonia | $CH_3NH_2$ <br> Methylamine | $(CH_3)_2NH$ <br> Dimethylamine | $(CH_3)_3N$ <br> Trimethylamine |
| | Aniline | N-Methylaniline | N,N-Dimethylaniline |

There are a number of similarities between ammonia and amines that carry beyond the structure. Consider odor. The smell of amines resembles that of ammonia but is not as sharp. However, amines can be quite pungent. Anyone handling or working with raw fish knows how strong the amine odor can be, since raw fish contains low-molecular-weight amines such as dimethylamine and trimethylamine. Other amines associated with decaying flesh have names suggestive of their odors: putrescine and cadaverine.

$$NH_2CH_2CH_2CH_2CH_2NH_2 \qquad NH_2CH_2CH_2CH_2CH_2CH_2NH_2$$

Putrescine                                        Cadaverine
(1,4-Diaminobutane)                        (1,5-Diaminopentane)

The solubility of low-molecular-weight amines in water is high. In general, if the total number of carbons attached to nitrogen is four or less, the amine is water soluble; amines with a carbon content greater than four are water insoluble. However, all amines are soluble in organic solvents such as diethyl ether or dichloromethane.

Since amines are organic bases, water solutions show weakly basic properties. If the basicity of aliphatic amines and aromatic amines are compared to ammonia, aliphatic amines are stronger than ammonia, while aromatic amines are weaker. Amines characteristically react with acids to form ammonium salts; the nonbonded electron pair on nitrogen bonds the hydrogen ion.

$$R\overset{\cdot\cdot}{N}H_2 + HCl \longrightarrow RNH_3^+Cl^-$$

Amine            Ammonium Salt

If an amine is insoluble, reaction with an acid produces a water-soluble salt. Since ammonium salts are water soluble, many drugs containing amines are prepared as ammonium salts. After working with fish in the kitchen, a convenient way to rid one's hands of fish odor is to rub a freshly cut lemon over the hands. The citric acid found in the lemon reacts with the amines found on the fish; a salt forms which can be easily rinsed away with water.

Amides are carboxylic acid derivatives. The amide group is recognized by the nitrogen connected to the carbonyl group. Amides are neutral compounds; the electrons are delocalized into the carbonyl (resonance) and thus, are not available to bond to a hydrogen ion.

Amide group (resonance)       Acetamide       Benzamide

Under suitable conditions, amide formation can take place between an amine and a carboxylic acid, an acyl halide, or an acid anhydride. Along with ammonia, primary and secondary amines yield amides with carboxylic acids or derivatives. Table 31.2 relates the nitrogen base with the amide class (based on the number of alkyl or aryl groups on the nitrogen of the amide).

| Table 31.2 | Classes of Amides | |
|---|---|---|
| **Nitrogen Base** | **(reacts to form)** | **Amide** ($-\overset{O}{\overset{\|}{C}}-\overset{\|}{N}-$) |
| Ammonia | | Primary amide (no R groups) |
| Primary amine | | Secondary amide (one R group) |
| Secondary amine | | Tertiary amide (two R groups) |

Hydrolysis of amides can take place in either acid or base. Primary amides hydrolyze in acid to ammonium salts and carboxylic acids. Neutralization of the acid and ammonium salts releases ammonia which can be detected by odor or by litmus.

$$\underset{\substack{O\\ \|}}{R-C-NH_2} + HCl + H_2O \longrightarrow \underset{\substack{O\\ \|}}{R-C-OH} + NH_4Cl$$

$$NH_4Cl + NaOH \longrightarrow NH_3 + NaCl + H_2O$$

Secondary and tertiary amides would release the corresponding alkyl ammonium salts which, when neutralized, would yield the amine.

In base, primary amides hydrolyze to carboxylic acid salts and ammonia. The presence of ammonia (or amine from corresponding amides) can be detected similarly by odor or litmus. The carboxylic acid would be generated by neutralization with acid.

$$\underset{\substack{O\\ \|}}{R-C-NH_2} + NaOH \longrightarrow \underset{\substack{O\\ \|}}{R-C-O^-Na^+} + NH_3$$

$$\underset{\substack{O\\ \|}}{R-C-O^-Na^+} + HCl \longrightarrow \underset{\substack{O\\ \|}}{R-C-OH} + NaCl$$

## Objectives

1. To show some physical and chemical properties of amines and amides.
2. To demonstrate the hydrolysis of amides.

## Procedure

### CAUTION!

Amines are toxic chemicals. Avoid excessive inhaling of the vapors and use gloves to avoid direct skin contact. Anilines are more toxic than aliphatic amines and are readily absorbed through the skin. Wash any amine or aniline spill with large quantities of water. Diethyl ether (ether) is extremely flammable. Be certain there are **NO** open flames in the immediate area. Discard all solutions in properly labeled organic waste containers.

### Properties of Amines

1. Place 5 drops of liquid or 0.1 g of solid from the compounds listed in the following table into labeled clean, dry test tubes (100 × 13 mm).

| Test Tube No. | Nitrogen Compound |
|---|---|
| 1 | 6 M $NH_3$ |
| 2 | Triethylamine |
| 3 | Aniline |
| 4 | N,N-Dimethylaniline |
| 5 | Acetamide |

2. Carefully note the odors of each compound. **Do not inhale deeply. Merely wave your hand across the mouth of the test tube toward your nose in order to note the odor.** Record your observations on the Report Sheet.

3. Add 2 mL of distilled water to each of the labeled test tubes. Mix thoroughly by sharply tapping the test tube with your finger. Note on the Report Sheet whether the amines are soluble or insoluble.

4. Take a glass rod, and test each solution for its pH. Carefully dip one end of the glass rod into a solution and touch a piece of pH paper. Between each test, be sure to clean and dry the glass rod. Record the pH by comparing the color of the paper with the chart on the dispenser.

5. Carefully add 2 mL of 6 M HCl to each test tube. Mix thoroughly by sharply tapping the test tube with your finger. Compare the odor and solubility of this solution to previous observations.

6. Place 5 drops of liquid or 0.1 g of solid from the compounds listed in the table into labeled clean, dry test tubes (100 × 13 mm). Add 2 mL of diethyl ether (ether) to each test tube. Stopper with a cork and mix thoroughly by shaking. Record the observed solubilities.

7. Carefully place on a watch glass, side-by-side, without touching, a drop of triethylamine and a drop of concentrated HCl. Record your observations. (Do this in the hood.)

## Hydrolysis of Acetamide

1. Dissolve 0.5 g of acetamide in 5 mL of 6 M $H_2SO_4$ in a large test tube (150 × 18 mm). Heat the solution in a boiling water bath for 5 min.

2. Hold a small strip of moist pH paper over the mouth of the test tube; note any changes in color; record the pH reading. Remove the test tube from the water bath, holding it in a test tube holder. Carefully note any odor.

3. Place the test tube in an ice water bath until cool to the touch. Now *carefully add, dropwise with shaking*, 6 M NaOH to the cool solution until basic. (You will need more than 7 mL of base.) Hold a piece of moist pH paper over the mouth. Record the pH reading. Carefully note any odor.

1. Acetamide
2. 6 M $NH_3$, ammonia water
3. Aniline
4. N,N-Dimethylaniline
5. Triethylamine
6. Diethyl ether (ether)
7. 6 M NaOH
8. Concentrated HCl
9. 6 M HCl
10. 6 M $H_2SO_4$
11. pH papers
12. Hot plate

## Experiment 31

### PRE-LAB QUESTIONS

**1.** How do primary, secondary, and tertiary amines differ?

**2.** Ethylamine, $CH_3CH_2NH_2$, is soluble in water but pentylamine, $CH_3CH_2CH_2CH_2CH_2NH_2$, is not. Explain this observation.

**3.** N,N-Dimethylaniline is insoluble in water (see Table 31.1 for structure). What would you do to get it to dissolve in water?

**4.** For the compounds below, tell whether it is an amine or an amide:

$$CH_3CH_2-NH_2 \qquad \underset{\underset{\displaystyle CH_2CH_3}{|}}{CH_3-\overset{\overset{\displaystyle CH_3}{|}}{N}} \qquad \text{(phenyl)}-NH-\overset{\overset{\displaystyle O}{\|}}{C}-CH_3$$

## Experiment 31

### REPORT SHEET

*Properties of amines*

| | Odor | | Solubility | | | pH |
| --- | --- | --- | --- | --- | --- | --- |
| | Original Sol. | with HCl | $H_2O$ | Ether | HCl | $H_2O$ |
| 6 M $NH_3$ | | | | | | |
| Triethylamine | | | | | | |
| Aniline | | | | | | |
| N,N-Dimethylaniline | | | | | | |
| Acetamide | | | | | | |

Triethylamine and concentrated hydrochloric acid observation:

### Hydrolysis of acetamide

1. Acid solution

   **a.** pH reading:

   **b.** Odor noted:

2. Base solution

   **a.** pH reading:

   **b.** Odor noted:

## POST-LAB QUESTIONS

1. The prescribed drug, Benzedrine, is the sulfate salt of amphetamine. Why is this stimulant administered as the salt and not as the amine?

Amphetamine

2. What accounts for the odor in raw fish and shellfish? How can you remove this smell from your hands after handling seafood? Why does this work?

3. Write the chemical equation for the reaction of triethylamine with concentrated hydrochloric acid.

4. Write the equations that account for what happens in the hydrolysis of the acetamide solution in (a) acid and in (b) base.

   a.

   b.

# Polymerization reactions

## Background

Polymers are giant molecules made of many (poly-) small units. The starting material, which is a single unit, is called the monomer. Many of the most important biological compounds are polymers. Cellulose and starch are polymers of glucose units, proteins are made of amino acids, and nucleic acids are polymers of nucleotides. Since the 1930s, a large number of synthetic polymers have been manufactured. They contribute to our comfort and gave rise to the previous slogan of DuPont Co.: "Better living through chemistry." Synthetic fibers such as nylon and polyesters, plastics such as the packaging materials made of polyethylene and polypropylene films, polystyrene, and polyvinyl chloride, just to name a few, all became household words. Synthetic polymers are parts of buildings, automobiles, machinery, toys, appliances, etc.; we encounter them daily in our life.

We focus our attention in this experiment on synthetic polymers and the basic mechanism by which some of them are formed. The two most important types of reactions that are employed in polymer manufacturing are the addition and condensation polymerization reactions. The first is represented by the polymerization of styrene and the second by the formation of nylon.

Styrene is a simple organic monomer which, by its virtue of containing a double bond, can undergo addition polymerization.

$$H_2C=CH \ + \ H_2C=CH \longrightarrow H_3C-CH-CH=CH$$

The reaction is called an *addition reaction* because two monomers are added to each other with the elimination of a double bond. This is also called a chain growth polymerization reaction. However, the reaction as such does not go without the help of an unstable molecule, called an *initiator*, that starts the reaction. Benzoyl peroxide or *t*-butyl benzoyl peroxide are such initiators. Benzoyl peroxide splits into two halves under the influence of heat or ultraviolet light and thus produces two free radicals. A *free radical* is a molecular fragment that has one unpaired electron. Thus, when the central bond was broken in the benzoyl peroxide, each of the shared pair of electrons went with one half of the molecule, each containing an unpaired electron.

benzoyl peroxide

Similarly, *t*-butyl benzoyl peroxide also gives two free radicals:

*t*-butyl benzoyl peroxide

The dot indicates the unpaired electron. The free radical reacts with styrene and initiates the reaction:

styrene

After this, the styrene monomers are added to the growing chain one by one until giant molecules containing hundreds and thousands of styrene-repeating units are formed. Please note the distinction between the monomer and the repeating unit. The monomer is the starting material, and the repeating unit is part of the polymer chain. Chemically they are not identical. In the case of styrene, the monomer contains a double bond, while the repeating unit (in the brackets in the following structure) does not.

$$CH_3-CH-\left[CH_2-CH-\right]_n CH_2-CH_2$$

polystyrene

Since the initiators are unstable compounds, care should be taken not to keep them near flames or heat them directly. If a bottle containing a peroxide initiator is dropped, a minor explosion can even occur.

The second type of reaction is called a *condensation reaction* because we condense two monomers into a longer unit, and at the same time we eliminate—expel—a small molecule. This is also called a step growth polymerization reaction. Nylon 6–6 is made of adipoyl chloride and hexamethylene diamine:

$$n \; Cl-\overset{\overset{\displaystyle O}{\|}}{C}-(CH_2)_4-\overset{\overset{\displaystyle O}{\|}}{C}-Cl \; + \; n \; H_2N-(CH_2)_6-NH_2 \longrightarrow$$

adipoyl chloride             hexamethylene diamine

$$Cl-\overset{\overset{\displaystyle O}{\|}}{C}-(CH_2)_4-\overset{\overset{\displaystyle O}{\|}}{C}-\left[NH-(CH_2)_6-NH-\overset{\overset{\displaystyle O}{\|}}{C}-(CH_2)_4-\overset{\overset{\displaystyle O}{\|}}{C}-\right]_n NH-(CH_2)_6-NH_2 \; + \; n \; HCl$$

repeating unit

We form an amide linkage between the adipoyl chloride and the amine with the elimination of HCl. The polymer is called nylon 6–6 because there are six carbon atoms in the acyl chloride and six carbon atoms in the diamine. Other nylons, such as nylon 6–10, are made of sebacoyl chloride (a 10-carbon atom containing acyl chloride) and hexamethylene diamine (a six carbon atom containing diamine). We use an acyl chloride rather than a carboxylic acid to form the amide bond because the former is more reactive. NaOH is added to the polymerization reaction in order to neutralize the HCl that is released every time an amide bond is formed.

The length of the polymer chain formed in both reactions depends on environmental conditions. Usually the chains formed can be made longer by heating the products longer. This process is called *curing*.

## Objectives

1. To acquaint students with the conceptual and physical distinction between monomer and polymer.
2. To perform addition and condensation polymerization and solvent casting of films.

## Procedure

### Preparation of Polystyrene

1. Set up your hot plate in the hood. Place 50 g of sea sand in a 150-mL beaker. Position a thermometer (0–200°C) in the sand bath so that it does not touch the bottom of the beaker. Heat the sand bath to 140°C.

2. Place approximately 2.5 mL styrene in a 16/18 × 150-mm Pyrex test tube. Add 3 drops of *t*-butyl benzoyl peroxide (*t*-butyl peroxide benzoate) initiator. Mix the solution.

3. Place the test tube in a test tube holder. Immerse the test tube in the sand bath. **Make sure the test tube points away from your face. (Caution: Do not touch either the test tube or the beaker with your hand.)** Heat the mixture in the test tube to about 140°C. The mixture will turn yellow.

4. When bubbles appear, remove the test tube from the sand bath. The polymerization reaction is exothermic and thus it generates its own heat. Overheating would create sudden boiling. When the bubbles disappear, put the test tube back in the sand bath. But every time the mixture starts to boil you must remove the test tube.

5. Continue the heating until the mixture in the test tube has a syrupy consistency.

6. Immerse a glass rod in the hot mixture. Swirl it around a few times. Remove the glass rod immediately. A chunk of polystyrene will be attached to the glass rod which will solidify upon cooling. The remaining polystyrene will solidify on the walls of the test tube when you remove it from the sand bath. Turn off the hot plate and let the sand bath cool to room temperature. While it is cooling, add a few drops of xylene to the test tube and dissolve some of the polystyrene by warming it in the sand bath.

7. Pour a few drops of the warm xylene solution on a microscopic slide and let the solvent evaporate. A thin film of polystyrene will be obtained. This is one of the techniques—the so-called solvent-casting technique—used to make films from bulk polymers.

8. Discard the remaining xylene solution into a special jar labeled "Waste." Discard the test tube with the polystyrene in a special box labeled "Glass."

9. Investigate the consistency of the solidified polystyrene on your glass rod, removing the solid mass by prying it off with a spatula.

## Preparation of Nylon

1. Set up a 50-mL reaction beaker and clamp above it a cylindrical paper roll (from toilet paper) or a stick.

2. Add 2.0 mL of 20% NaOH solution and 10 mL of a 5% aqueous solution of hexamethylene diamine.

3. Take 10 mL of 5% adipoyl chloride solution in cyclohexane with a pipet or syringe. Layer the cyclohexane solution slowly on top of the aqueous solution in the beaker. Two layers will form and nylon will be produced at the interface (Fig. 32.1).

4. With a bent wire first scrape off the nylon formed on the walls of the beaker.

**Figure 32.1**
Preparation of nylon.

copper hook

collapsed film

diacid chloride in organic solvent

polyamide film forming at interface

diamine in water

5. Slowly lift and pull the film from the center. If you pull it too fast, the nylon rope will break.

6. Wind it around the paper roll or stick two to three times. Do not touch it with your hands.

7. Slowly rotate the roll or the stick and wind at least a 1-m nylon rope.

8. Cut the rope and transfer the wound rope into a beaker filled with water (or 50% ethanol). Watch as the thickness of the rope collapses. Dry the rope between two filter papers.

9. There are still monomers left in the beaker. Mix the contents vigorously with a glass rod. Observe the beads of nylon that have formed.

10. Pour the mixture into a cold water bath and wash it. Dry the nylon between two filter papers. Note the consistency of your products.

11. Dissolve a small amount of nylon in 80% formic acid. Place a few drops of the solution onto a microscope slide and evaporate the solvent under the hood.

12. Compare the appearance of the solvent-cast nylon film with that of the polystyrene.

## Chemicals and Equipment

1. Styrene
2. Hexamethylene diamine solution
3. Adipoyl chloride solution
4. Sodium hydroxide solution
5. Xylene
6. Formic acid solution
7. *t*-Butyl peroxide benzoate initiator
8. Sea sand
9. Hot plate
10. Test tube (16/18 × 150-mm) Pyrex No. 9820
11. Test tube holder
12. Paper roll or stick
13. Bent wires
14. 10-mL pipets
15. Spectroline pipet filler
16. Beaker tongs

## Experiment 32

**PRE-LAB QUESTIONS**

**1.** Why should you *not* expose *t*-butyl peroxide to direct heat?

**2.** What would be the product of the reaction between benzoyl peroxide radical and vinylchloride, $CH_2=CHCl$?

**3.** Write the reaction for the polymerization of vinyl chloride (chloroethene). Show the repeating unit of the resulting polymer (Polyvinyl chloride, PVC).

**4.** Write the structure of the monomers and that of the repeating unit in nylon 10-6. (In numbering nylons, the first number indicates the number of carbon atoms in the diamine and the second number refers to the number of carbon atoms in the acyl chloride.)

**5.** Think of the LeChatelier's principle. Why is it advantageous to add NaOH to the aqueous phase?

## Experiment 32

**REPORT SHEET**

1. Describe the appearance of polystyrene and nylon.

2. Describe the difference in physical characteristics between polystyrene and nylon.

3. Is there any difference in the appearance of the solvent cast films of nylon and polystyrene?

## POST-LAB QUESTIONS

**1.** A polyester is made of sebacoyl chloride and ethylene glycol,

$$\text{Cl-}\overset{\displaystyle O}{\overset{\displaystyle \|}{\text{C}}}\text{-(CH}_2)_8\text{-}\overset{\displaystyle O}{\overset{\displaystyle \|}{\text{C}}}\text{-Cl} \quad \text{and} \quad \overset{\displaystyle \text{CH}_2\text{OH}}{\underset{\displaystyle \text{CH}_2\text{OH}}{|}}$$

**a.** Draw the structure of the polyester formed.

**b.** What molecules have been eliminated in this condensation reaction?

**2.** Two compounds, $\text{Cl-}\overset{\displaystyle O}{\overset{\displaystyle \|}{\text{C}}}\text{-(CH}_2)_3\text{-}\overset{\displaystyle O}{\overset{\displaystyle \|}{\text{C}}}\text{-Cl}$ in cyclohexane and $\text{H}_2\text{N-(CH}_2)_4\text{-NH}_2$ in water, are reacted. Write the structure of the polyamide rope formed.

**3.** Did you have any evidence that the polymerization of styrene is exothermic?

**4.** Since the polymerization of styrene is an exothermic reaction, why do you need to heat the mixture to 140°C?

**5.** Distinguish between the polarities of the solvents that solubilize polystyrene and nylon, respectively.

# Experiment 33

## Preparation of acetylsalicylic acid (aspirin)

### Background

One of the most widely used nonprescription drugs is aspirin. In the United States, more than 15,000 pounds are sold each year. It is no wonder there is such wide use when one considers the medicinal applications for aspirin. It is an effective analgesic (pain killer) that can reduce the mild pain of headaches, toothache, neuralgia (nerve pain), muscle pain, and joint pain (from arthritis and rheumatism). Aspirin behaves as an antipyretic drug (it reduces fever) and an antiinflammatory agent capable of reducing the swelling and redness associated with inflammation. It is an effective agent in preventing strokes and heart attacks due to its ability to act as an anticoagulant by preventing platelet aggregation.

Early studies showed the active agent that gave these properties to be salicylic acid. However, salicylic acid contains the phenolic and the carboxylic acid groups. As a result, the compound was too harsh to the linings of the mouth, esophagus, and stomach. Contact with the stomach lining caused some hemorrhaging. The Bayer Company in Germany patented the ester acetylsalicylic acid and marketed the product as "aspirin" in 1899. Their studies showed that this material was less of an irritant; the acetylsalicylic acid was hydrolyzed in the small intestine to salicylic acid, which then was absorbed into the bloodstream. The relationship between salicylic acid and aspirin is shown in the following formulas:

Salicylic acid

Acetylsalicylic acid (Aspirin)

Aspirin still has side effects. Hemorrhaging of the stomach walls can occur even with normal dosages. These side effects can be reduced through the addition of coatings or through the use of buffering agents. Magnesium hydroxide, magnesium carbonate, and aluminum glycinate, when mixed into the formulation of the aspirin (e.g., Bufferin), reduce the irritation.

This experiment will acquaint you with a simple synthetic problem in the preparation of aspirin. The preparative method uses acetic anhydride and an acid catalyst, like sulfuric or phosphoric acid, to speed up the reaction with salicylic acid.

$$\text{Salicylic acid} + \text{Acetic anhydride} \xrightarrow[\Delta]{H^+} \text{Aspirin} + CH_3COOH$$

| Salicylic acid | Acetic anhydride | Aspirin | Acetic acid |

If any salicylic acid remains unreacted, its presence can be detected with a 1% iron(III) chloride solution. Salicylic acid has a phenol group in the molecule. The iron(III) chloride gives a violet color with any molecule possessing a phenol group (see Experiment 28). Notice the aspirin no longer has the phenol group. Thus a pure sample of aspirin will not give a purple color with 1% iron(III) chloride solution.

## Objectives

1. To illustrate the synthesis of the drug aspirin.
2. To use a chemical test to determine the purity of the preparation.

## Procedure

### Preparation of Aspirin

1. Prepare a bath using a 400-mL beaker filled about half way with water. Heat to boiling.

2. Take 2.0 g of salicylic acid and place it in a 125-mL Erlenmeyer flask. Use this quantity of salicylic acid to calculate the theoretical or expected yield of aspirin (1). Carefully add 3 mL of acetic anhydride to the flask and, while swirling, add 3 drops of concentrated phosphoric acid.

> **CAUTION!**
>
> Acetic anhydride will irritate your eyes. Phosphoric acid will cause burns to the skin. Use gloves with these reagents. Handle both chemicals with care. Dispense in the hood.

3. Mix the reagents and then place the flask in the boiling water bath; heat for 15 min. (Fig. 33.1). The solid will completely dissolve. Swirl the solution occasionally.

**Figure 33.1**
Assembly for the
synthesis of aspirin.

4. Remove the Erlenmeyer flask from the bath and let it cool to approximately room temperature. Then, slowly pour the solution into a 150-mL beaker containing 20 mL of ice water, mix thoroughly, and place the beaker in an ice bath. Use a glass rod to mix the solution while in the ice bath, vigorously rubbing the glass rod (called scratching) along the bottom of the beaker (be careful not to poke a hole through the beaker). The water destroys any unreacted acetic anhydride and will cause the insoluble aspirin to precipitate from the solution. Scratching the beaker helps to precipitate the aspirin, also.

5. Collect the crystals by filtering under suction with a Büchner funnel. The assembly is shown in Fig. 33.2. (Also see Fig. 29.1, p. 307.)

**Figure 33.2**
Filtering using the
Büchner funnel.

6. Obtain a 250-mL filter flask and connect the side arm of the filter flask to a water aspirator with heavy wall vacuum rubber tubing. (The thick walls of the tubing will not collapse when the water is turned on and the pressure is reduced.)

7. The Büchner funnel is inserted into the filter flask through either a filtervac, a neoprene adapter, or a one-hole rubber stopper, whichever is available. Filter paper is then placed into the Büchner funnel. Be sure that the paper lies flat and covers all the holes. Wet the filter paper with water.

8. Turn on the water aspirator to maximum water flow. Pour the solution into the Büchner funnel.

9. Wash the crystals with two 5-mL portions of cold water, followed by one 10-mL portion of cold ethyl acetate.

10. Continue suction through the crystals for several minutes to help dry them. Disconnect the rubber tubing from the filter flask before turning off the water aspirator.

11. Using a spatula, place the crystals between several sheets of paper toweling or filter paper and press dry the solid.

12. Weigh a 50-mL beaker (2). Add the crystals and reweigh (3). Calculate the weight of crude aspirin (4). Determine the percent yield (5).

## Determine the Purity of the Aspirin

1. The aspirin you prepared is not pure enough for use as a drug and is *not* suitable for ingestion. The purity of the sample will be tested with 1% iron(III) chloride solution and compared with a commercial aspirin and salicylic acid.

2. Label three test tubes (100 × 13 mm) 1, 2, and 3; place a few crystals of salicylic acid into test tube no. 1, a small sample of your aspirin into test tube no. 2, and a small sample of a crushed commercial aspirin into test tube no. 3. Add 5 mL of distilled water to each test tube and swirl to dissolve the crystals.

3. Add 10 drops of 1% aqueous iron(III) chloride to each test tube.

4. Compare and record your observations. The formation of a purple color indicates the presence of salicylic acid. The intensity of the color qualitatively tells how much salicylic acid is present.

## Chemicals and Equipment

1. Acetic anhydride
2. Concentrated phosphoric acid, $H_3PO_4$
3. Commercial aspirin tablets
4. Ethyl acetate
5. 1% Iron(III) chloride
6. Salicylic acid
7. Boiling chips
8. Büchner funnel, small
9. 250-mL filter flask
10. Filter paper
11. Filtervac or neoprene adapter
12. Hot plate

## Experiment 33

### PRE-LAB QUESTIONS

**1.** Are the medicinal properties of aspirin due to acetylsalicylic acid or salicylic acid?

**2.** Draw the structure of aspirin. Should this compound test positive with 1% iron(III) chloride solution? Explain your answer.

**3.** Salicylic acid can irritate the stomach. Draw the structure of the molecule. Circle and name the functional group (or groups) responsible for this effect.

**4.** Magnesium hydroxide, $Mg(OH)_2$, and magnesium carbonate, $MgCO_3$, are often used as buffering agents with aspirin. What is the purpose of the buffering agents? How do these chemicals act as buffering agents?

## Experiment 33

### REPORT SHEET

1. Theoretical yield:

_____ g salicylic acid $\times \dfrac{180 \text{ g aspirin}}{1 \text{ mole}} \times \dfrac{1 \text{ mole}}{138 \text{ g salicylic acid}}$

$= $ _____ g aspirin

2. Weight of 50-mL beaker _____ g

3. Weight of your aspirin and beaker _____ g

4. Weight of your aspirin: (3) − (2) _____ g

5. Percent yield: [(4)/(1)] × 100 = % _____ %

6. Iron(III) chloride test

| No. | Sample | Color | Intensity |
|-----|--------|-------|-----------|
| 1 | Salicylic acid | | |
| 2 | Your aspirin | | |
| 3 | Commercial aspirin | | |

### POST-LAB QUESTIONS

1. The experiment required 3 drops of concentrated phosphoric acid. Why was this acid added? What would happen if this reagent were omitted?

2. What would happen to your percent yield if in step no. 11 of the procedure you failed to dry completely your aspirin preparation by omitting the drying between filter paper?

3. A student expected 3.05 g of acetylsalicyclic acid, but obtained only 2.53 g. What is the percentage yield?

4. Some nonprescription pain relievers available as substitutes for aspirin are acetaminophen, ibuprofen, and naproxen. Which drug would give a positive test for phenol?

Acetaminophen        Ibuprofen                              Naproxen

5. Bottles containing old aspirin tablets often smell of vinegar. The presence of what chemical causes this odor? Where does this chemical come from?

6. A tablet of aspirin usually contains 5 grains. How many mg is this? (1 grain = 0.0648 g)

# Experiment 34

## Measurement of the active ingredient in aspirin pills

### Background

Medication delivered in the form of a pill contains an active ingredient or ingredients. Beside the drug itself, the pill also contains fillers. The task of the filler is many fold. Sometimes it is there to mask the bitter or otherwise unpleasant taste of the drug. Other times the filler is necessary because the prescribed dose of the drug is so small in mass that it would be difficult to handle. Drugs that have the same generic name contain the same active ingredient. The dosage of the active ingredient must be listed as specified by law. On the other hand, neither the quantity of the filler nor its chemical nature appears on the label. That does not mean that the fillers are completely inactive. They usually affect the rate of drug delivery. In order to deliver the active ingredient, the pill must fall apart in the stomach. For this reason, many fillers are polysaccharides, for example, starch, that either are partially soluble in stomach acid or swell, allowing the drug to be delivered in the stomach or in the intestines.

In the present experiment, we measure the amount of the active ingredient, acetylsalicylic acid (see also Experiment 33), in common aspirin pills. Companies use different fillers and in different amounts, but the active ingredient, acetylsalicylic acid, must be the same in every aspirin tablet. We separate the acetylsalicylic acid from the filler based on their different solubilities. Acetylsalicylic acid is very soluble in ethanol, while neither starch, nor other polysaccharides, or even mono- and disaccharides used as fillers, are soluble in ethanol. Some companies may use inorganic salts as fillers but these too are not soluble in ethanol. On the other hand, some specially formulated aspirin tablets may contain small amounts of ethanol-soluble substances such as stearic acid or vegetable oil. Thus the ethanol extracts of aspirin tablets may contain small amounts of substances other than acetylsalicylic acid.

### Objectives

1. To appreciate the ratio of filler to active ingredients in common aspirin tablets.
2. To learn techniques of quantitative separations.

### Procedure

1. Weigh approximately 10 g of aspirin tablets. Record the actual weight on your Report Sheet (1). Count the number of tablets and record it on your Report Sheet (2).

2. Place the weighed aspirin tablets in a mortar of approximately 100 mL capacity. Before starting to grind, place the mortar on a white sheet of paper and loosely cover it with a filter paper. The purpose of this procedure is to catch small fragments of the tablets

that may fly out of the mortar during the grinding process. Break up the aspirin tablets by gently hammering them with the pestle. Recover and place back in the mortar any fragments that flew out during the hammering. With a twisting motion of your wrist, grind the aspirin pieces into a fine powder with the aid of the pestle.

3. Add 10 mL of 95% ethanol to the mortar and continue to grind for 2 min. Set up a vacuum filtration apparatus (see Fig. 29.1, p. 307) using a Büchner funnel and a 250-mL filter flask. Turn on the water. With the aid of a rubber policeman, transfer the supernatant liquid and solid from the mortar to the Büchner funnel. When about 1 mL of clear filtrate has been collected in the filter flask, turn off the water, lift the Büchner funnel, and allow a drop of filtrate to fall on a clean microscope slide. Return the funnel to the filter flask, turn on the water, and continue the filtration. The drop on the microscope slide will rapidly evaporate, leaving behind crystals of acetylsalicylic acid. This is a qualitative test, showing that the extraction of the active ingredient is successful. Report what you see on the microscope slide on your Report Sheet (3).

4. Add another 10 mL of 95% ethanol to the Büchner funnel and turn on the water. Then collect a drop of filtrate as described above in no.3.

5. Repeat procedure no. 4 two more times; you will use a total of 40 mL of ethanol in the four extractions. Report after each extraction if the extract carries acetylsalicylic acid. Enter these observations on your Report Sheet (4, 5, and 6).

6. When the filtration is completed and only the white, moist solid is left in the filter, transfer the filter paper with its contents into a 100-mL beaker and place the beaker into a drying oven set at 110°C. Dry for 10 min.

7. Carefully remove the beaker from the oven. (**CAUTION!** The beaker is hot.) Allow it to come to room temperature. Weigh a clean and dry 25-mL beaker on your balance. Report the weight on your Report Sheet (7). With the aid of a spatula, **carefully** transfer the dried filler from the filter paper into the 25-mL beaker. Make sure that you do not spill any of the powder. Some of the dried filler may stick to the paper a bit, and you may have to scrape the paper with the spatula. Weigh the 25-mL beaker with its contents on your balance. Report the weight on your Report Sheet (8).

8. Test the dried filler with a drop of Hanus iodine solution. A blue coloration will indicate that it contains starch. Report your findings on your Report Sheet (14).

---

### Chemicals and Equipment

1. Aspirin tablets
2. Mortar and pestle (100-mL capacity)
3. 95% ethanol
4. Filter paper (Whatman no. 2, 7 cm)
5. Balance
6. Drying oven at 110°C
7. Hanus iodine solution
8. Microscope slides
9. 100-mL beaker
10. 25-ml beaker
11. Rubber policeman
12. Vacuum filtration setup (Büchner funnel; 250-mL filter flask)

---

## EXPERIMENT 34

### PRE-LAB QUESTIONS

**1.** What method is used to separate acetylsalicylic acid from starch?

**2.** An adult aspirin pill lists 5.0 grains aspirin content. Another pill recommended to take to prevent strokes and heart attacks contains only 1.25 grains. If a grain is 64.8 mg, how many mg of active ingredient are contained in each each pill?

**3.** An aspirin tablet weighs 135 mg and contains 81 mg active ingredient. Calculate the percentage of fillers in the tablet.

## Experiment 34

### REPORT SHEET

1. Weight of aspirin tablets        _____ g

2. Number of aspirin tablets in your sample        _____

3. Does your first extract contain acetylsalicylic acid?        _____

4. Does your second extract contain acetylsalicylic acid?        _____

5. Does your third extract contain acetylsalicylic acid?        _____

6. Does your fourth extract contain acetylsalicylic acid?        _____

7. Weight of the empty 25-mL beaker        _____ g

8. Weight of the 25-mL beaker and filler        _____ g

9. Weight of the filler: (8) − (7)        _____ g

10. Percent of filler in tablets: $[(9)/(1)] \times 100 = \%$        _____ %

11. Weight of one tablet: (1)/(2)        _____ g

12. Weight of filler per tablet: $(11) \times [(10)/100]$        _____ g

13. Weight of acetylsalicylic acid per tablet: (11) − (12)        _____ g

14. Does your filler contain starch?        _____

### POST-LAB QUESTIONS

1. According to your calculations, does your aspirin tablet contain more, the same, or less active ingredients than the average adult dosage (5.0 grains)?

2. If your ethanol extract contained a filler in addition to the active ingredient, acetylsalicylic acid, how would that affect your calculations of the dosage of the aspirin tablet?

3. Aspirin is completely soluble in 95% ethanol, but only partially soluble in 50% ethanol.

   a. If you have only 50% ethanol available for extraction, how would it affect the yield?

   b. Knowing the fact given above, what would you change in the extraction procedure to get correct results?

   c. What control is present in the extraction protocol that let's you know if the extraction is complete or not?

4. On the basis of the Hanus iodine test performed, what can you say about the nature of the filler in your aspirin tablet?

5. If you did not properly grind your aspirin tablet to a fine powder, would you need more, less, or an equal amount of ethanol extraction to remove most of the aspirin?

6. You obtained a painkiller pill containing acetaminophen. After extracting the active ingredient from a 400-mg pill, you ended up with the following data: beaker: 5.38 g; beaker plus filler: 5.66 g. What was the percent of filler in the painkiller tablet?

# Isolation of caffeine from tea leaves

## Background

Many organic compounds are obtained from natural sources through extraction. This method takes advantage of the solubility characteristics of a particular organic substance with a given solvent. In the experiment here, caffeine is readily soluble in hot water and is thus separated from the tea leaves. Caffeine is one of the main substances that make up the water solution called tea. Besides being found in tea leaves, caffeine is present in coffee, kola nuts, and cocoa beans. As much as 5% by weight of the leaf material in tea plants consists of caffeine.

The caffeine structure is shown below. It is classed as an alkaloid, meaning that with the nitrogen present, the molecule has base characteristics (alkali-like). In addition, the molecule has the purine ring system, a framework which plays an important role in living systems.

Purine            Caffeine

Caffeine is the most widely used of all the stimulants. Small doses of this chemical (50 to 200 mg) can increase alertness and reduce drowsiness and fatigue. The "No-Doz" tablet lists caffeine as the main ingredient. In addition, it affects blood circulation since the heart is stimulated and blood vessels are relaxed (vasodilation). It also acts as a diuretic. There are side effects. Large doses of over 200 mg can result in insomnia, restlessness, headaches, and muscle tremors ("coffee nerves"). Continued, heavy use may bring on physical dependence. (How many of you know somebody who cannot function in the morning until they have that first cup of coffee?)

Tea leaves consist primarily of cellulose; this is the principle structural material of all plant cells. Fortunately, the cellulose is insoluble in water, so that by using a hot water extraction, more soluble caffeine can be separated. Also dissolved in water are complex substances called tannins. These are colored phenolic compounds of high molecular weight (500 to 3000) that have acidic behavior. If a basic salt such as $Na_2CO_3$ is added to the water solution, the tannins can react to form a salt. These salts are insoluble in organic solvents, such as chloroform or dichloromethane, but are soluble in water.

Although caffeine is soluble in water (2 g/100 g of cold water), it is more soluble in the organic solvent dichloromethane (14 g/100 g). Thus caffeine can be extracted from the basic tea solution with dichloromethane, but the sodium salts of the tannins remain behind in the aqueous solution. Evaporation of the dichloromethane yields crude caffeine; the crude material can be purified by sublimation (see Experiment 15).

## Objectives

1. To demonstrate the isolation of a natural product.
2. To learn the techniques of extraction.
3. To use sublimation as a purification technique.

## Procedure

The isolation of caffeine from tea leaves follows the scheme below:

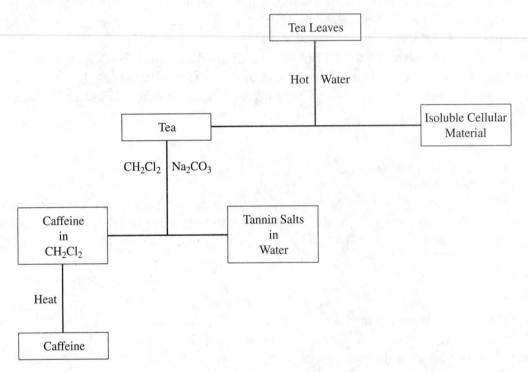

1. Carefully open two commercial tea bags (try not to tear the paper) and weigh the contents to the nearest 0.001 g. Record this weight (1). Place the tea leaves back into the bags, close, and secure the bags with staples.

2. Into a 150-mL beaker, place the tea bags so that they lie flat on the bottom. Add 30 mL of distilled water and 2.0 g of anhydrous $Na_2CO_3$; heat the contents with a hot plate, keeping a *gentle* boil, for 20 min. While the mixture is boiling, keep a watch glass on the beaker. Hold the tea bags under water by occasionally pushing them down with a glass rod.

3. Decant the hot liquid into a 50-mL Erlenmeyer flask. Wash the tea bags with 10 mL of hot water, carefully pressing the tea bag with a glass rod; add this wash water to the tea

extract. (If any solids are present in the tea extract, filter them by gravity to remove.) Cool the combined tea extract to room temperature. The tea bags may be discarded.

4. Transfer the cool tea extract to a 125-mL separatory funnel that is supported on a ring stand with a ring clamp.

5. Carefully add 5.0 mL of dichloromethane to the separatory funnel. Stopper the funnel and lift it from the ring clamp; hold the funnel with two hands as shown in Fig. 35.1. By holding the stopper in place with one hand, invert the funnel. *Make certain the stopper is held tightly and no liquid is spilled; make sure the liquid is not in contact with the stop-cock; open the stop-cock, being sure to point the opening away from you and your neighbors.* Built-up pressure caused by gases accumulating inside will be released. Now, close the stop-cock and gently mix the contents by inverting the funnel two or three times. Again, release any pressure by opening the stop-cock as before.

**Figure 35.1**
Using the separatory funnel.

6. Return the separatory funnel to the ring clamp, remove the stopper, and allow the aqueous layer to separate from the dichloromethane layer (Fig. 35.2). You should see two distinct layers form after a few minutes, with the dichloromethane layer at the bottom. Sometimes an emulsion may form at the juncture of the two layers. The emulsion often can be broken by gently swirling the contents or by gently stirring the emulsion with a glass rod.

**Figure 35.2**
Separation of the aqueous layer and the dichloromethane layer in the separatory funnel.

Water

Caffeine in $CH_2Cl_2$

7. Carefully drain the lower layer into a 25-mL Erlenmeyer flask. Try not to include any water with the dichloromethane layer; careful manipulation of the stop-cock will prevent this.

8. Repeat the extraction with an additional 5.0 mL of dichloromethane. Combine the separated bottom layer with the dichloromethane layer obtained from step no. 7.

9. Add 0.5 g of anhydrous $Na_2SO_4$ to the combined dichloromethane extracts. Swirl the flask. The anhydrous salt is a drying agent and will remove any water that may still be present.

10. Weigh a 25-mL side-arm filter flask containing one or two boiling stones. Record this weight (2). By means of a gravity filtration, filter the dichloromethane–salt mixture into the pre-weighed flask. Rinse the salt on the filter paper with an additional 2.0 mL of dichloromethane.

11. Remove the dichloromethane by evaporation *in the hood*. Be careful not to overheat the solvent, since it may foam over. The solid residue which remains after the solvent is gone is the crude caffeine. Reweigh the cooled flask (3). Calculate the weight of the crude caffeine by subtraction (4) and determine the percent yield (5).

12. Take a melting point of your solid. First, scrape the caffeine from the bottom and sides of the flask with a microspatula and collect a sample of the solid in a capillary tube (review Experiment 15 for the technique). Pure caffeine melts at 238°C. Compare your melting point (6) to the literature value.

## Optional

13. At the option of your instructor, the caffeine may be purified further. The caffeine may be sublimed directly from the flask with a cold finger condenser (Fig. 35.3). Carefully insert the cold finger condenser into a no. 2 neoprene adapter (use a drop of glycerine as a lubricant). Adjust the tip of the cold finger to 1 cm from the bottom of the flask. Clean any glycerine remaining on the cold finger with a Kimwipe and acetone; the cold finger surface must be clean and dry. Connect the cold finger to a faucet by latex tubing (water *in* the upper tube; water *out* the lower tube). Connect the side-arm filter flask to a water aspirator with vacuum tubing, installing a trap between the aspirator and the sublimation set-up (Fig. 35.3). When you turn the water on, press the cold finger into the filter flask until a good seal is made. Gently heat the bottom of the filter flask which holds the caffeine with a microburner (hold the *base* of the microburner); move the flame back and forth and along the sides of the flask. **Do not allow the sample to melt.** If the sample melts, *stop* heating and allow to cool before continuing. When the sublimation is complete, disconnect the heat and allow the system to cool; leave the aspirator connected and the water running.

**Figure 35.3** • Sublimation apparatus connected to an aspirator.

**14.** When the system has reached room temperature, carefully disconnect the aspirator from the side-arm filter flask by removing the vacuum tubing from the side-arm. Turn off the water to the cold finger. *Carefully* remove the cold finger from the flask along with the neoprene adapter without dislodging any crystals. Scrape the sublimed caffeine onto a pre-weighed piece of weighing paper (7). Reweigh (8); determine the weight of caffeine (9). Calculate the percent recovery (10). Determine the melting point (11).

**15.** Collect the caffeine in a sample vial, and submit it to your instructor.

## Chemicals and Equipment

1. Boiling chips
2. Cold finger condenser
3. Filter paper (Whatman no. 7.0), fast flow
4. Hot plate
5. 125-mL separatory funnel with stopper
6. Melting point capillaries
7. No. 2 neoprene adapter
8. 25-mL side-arm filter flask
9. Small sample vials
10. Tea bags
11. Tubing: latex, 2 ft.; vacuum, 2 ft.
12. 250-mL trap: 250-mL side-arm filter flask fitted with a no. 6 one-hole rubber stopper containing a piece of glass tubing (10 cm long $\times$ 37 mm OD)
13. Anhydrous sodium sulfate, $Na_2SO_4$
14. Anhydrous sodium carbonate, $Na_2CO_3$
15. Dichloromethane, $CH_2Cl_2$
16. Stapler

## Experiment 35

### PRE-LAB QUESTIONS

**1.** What are some of the substances found in the water solution of tea?

**2.** What gives caffeine its base characteristics?

**3.** Draw the caffeine molecule and circle the purine nucleus only.

**4.** What are some of the physiological effects of caffeine?

**5.** How does extraction work to remove caffeine from tea leaves?

## Experiment 35

## REPORT SHEET

1. Weight of tea in 2 tea bags      _____ g

2. Weight of 25-mL side-arm filter flask and boiling stones      _____ g

3. Weight of flask, boiling stones, and crude caffeine      _____ g

4. Weight of caffeine: (3) − (2)      _____ g

5. Percent yield: [(4)/(1)] × 100 = %      _____ %

6. Melting point of your crude caffeine      _____ °C

7. Weight of weighing paper      _____ g

8. Weight of sublimed caffeine and paper      _____ g

9. Weight of caffeine: (8) − (7)      _____ g

10. Percent recovery: [(9)/(4)] × 100 = %      _____ %

11. Melting point of sublimed caffeine      _____ °C

## POST-LAB QUESTIONS

1. A student used 6.23 g of tea leaves in the experiment. How much caffeine is expected, assuming all the caffeine is extracted? (*Hint:* see **Background.**)

2. Recovered samples of caffeine from this experiment usually melt lower than the literature value. What would account for the difference?

3. How are the tannins removed from the water extract?

4. If the step using the anhydrous sodium sulfate were omitted, how would the yield and the melting point of the caffeine be affected?

# Carbohydrates

Carbohydrates are polyhydroxy aldehydes, ketones, or compounds that yield polyhydroxy aldehydes or ketones upon hydrolysis. Rice, potatoes, bread, corn, candy, and fruits are rich in carbohydrates. A carbohydrate can be classified as a monosaccharide (glucose or fructose); a disaccharide (sucrose or lactose), which consists of two joined monosaccharides; or a polysaccharide (starch or cellulose), which consists of thousands of monosaccharide units linked together. Monosaccharides exist mostly as cyclic structures containing hemiacetal (or hemiketal) groups. These structures in solutions are in equilibrium with the corresponding open chain structures bearing aldehyde or ketone groups. Glucose, blood sugar, is an example of a polyhydroxy aldehyde (Fig. 36.1).

**Figure 36.1** • The structures of D-glucose.

Disaccharides and polysaccharides exist as cyclic structures containing functional groups such as hydroxyl groups, acetal (or ketal), and hemiacetal (or hemiketal). Most of the di-, oligo-, or polysaccharides have two distinct ends. The one end which has a hemiacetal (or hemiketal) on its terminal is called the reducing end, and the one which does not contain a hemiacetal (or hemiketal) terminal is the nonreducing end. The name "reducing" is given because hemiacetals (and to a lesser extent hemiketals) can reduce an oxidizing agent such as Benedict's reagent.

Fig. 36.2 is an example:

**Figure 36.2**
The structure of maltose, a disaccharide.

Not all disaccharides or polysaccharides contain a reducing end. An example is sucrose, which does not have a hemiacetal (or hemiketal) group on either of its ends (Fig. 36.3).

**Figure 36.3**
The structure of sucrose.

Polysaccharides, such as amylose or amylopectin, do have a hemiacetal group on one of their terminal ends, but practically they are nonreducing substances because there is only one reducing group for every 2,000–10,000 monosaccharidic units. In such a low concentration, the reducing group does not give a positive test with Benedict's or Fehling's reagent.

On the other hand, when a nonreducing disaccharide (sucrose) or a polysaccharide such as amylose is hydrolyzed the glycosidic linkages (acetal) are broken and reducing ends are created. Hydrolyzed sucrose (a mixture of D-glucose and D-fructose) will give a positive test with Benedict's or Fehling's reagent as well as hydrolyzed amylose (a mixture of glucose and glucose containing oligosaccharides). The hydrolysis of sucrose or amylose can be achieved by using a strong acid such as HCl or with the aid of biological catalysts (enzymes).

Starch can form an intense, brilliant, dark blue-, or violet-colored complex with iodine. The straight chain component of starch, the amylose, gives a blue color while the branched component, the amylopectin, yields a purple color. In the presence of iodine, the amylose forms helixes inside of which the iodine molecules assemble as long polyiodide chains. The helix-forming branches of amylopectin are much shorter than those of amylose. Therefore, the polyiodide chains are also much shorter in the amylopectin-iodine complex than in the amylose-iodine complex. The result is a different color (purple). When starch is hydrolyzed and broken down to small carbohydrate units, the iodine will not give a dark blue (or purple) color. The iodine test is used in this experiment to indicate the completion of the hydrolysis.

In this experiment, you will investigate some chemical properties of carbohydrates in terms of their functional groups.

1. *Reducing and nonreducing properties of carbohydrates*

   a. **Aldoses (polyhydroxy aldehydes).** All aldoses are reducing sugars because they contain free aldehyde functional groups. The aldehydes are oxidized by mild oxidizing agents (e.g., Benedict's or Fehling's reagent) to the corresponding carboxylates. For example,

$$R\!-\!CHO + 2Cu^{2+} \xrightarrow{\text{NaOH}} R\!-\!COO^-Na^+ + Cu_2O \downarrow$$
$$\text{(from Fehling's reagent)} \qquad\qquad \text{Red precipitate}$$

**b. Ketoses (polyhydroxy ketones).** All ketoses are reducing sugars because they have a ketone functional group next to an alcohol functional group. The reactivity of this specific ketone (also called $\alpha$-hydroxyketone) is attributed to its ability to form an $\alpha$-hydroxyaldehyde in basic media according to the following equilibrium equations:

$$
\begin{array}{ccccc}
\text{CH}_2\text{OH} & & \text{CHOH} & & \text{CHO} \\
| & \xrightleftharpoons{\text{Base}} & || & \xrightleftharpoons{\text{Base}} & | \\
\text{C}=\text{O} & & \text{C}-\text{OH} & & \text{H}-\text{C}-\text{OH} \\
| & & | & & | \\
\text{H}-\text{C}-\text{OH} & & \text{H}-\text{C}-\text{OH} & & \text{H}-\text{C}-\text{OH} \\
| & & | & & | \\
\text{Ketose} & & \text{Enediol} & & \text{Aldose}
\end{array}
$$

**c. Hemiacetal functional group (potential aldehydes).** Carbohydrates with hemiacetal functional groups can reduce mild oxidizing agents such as Benedict's reagent because hemiacetals can easily form aldehydes through the following equilibrium equation:

$$
\begin{array}{c}
\text{H} \quad \text{OR}' \\
\diagdown \diagup \\
\text{C} \\
\diagup \diagdown \\
\text{R} \quad \text{OH}
\end{array}
\quad \rightleftharpoons \quad
\text{R}-\overset{\overset{\textstyle O}{||}}{\text{C}}-\text{H} \ + \ \text{R}'\text{OH}
$$

Sucrose is, on the other hand, a nonreducing sugar because it does not contain a hemiacetal functional group. Although starch has a hemiacetal functional group at one end of its molecule, it is, however, considered as a nonreducing sugar because the effect of the hemiacetal group in a very large starch molecule becomes insignificant to give a positive Benedict's test.

**2.** *Hydrolysis of acetal groups.* Disaccharides and polysaccharides can be converted into monosaccharides by hydrolysis. The following is an example:

$$
\underset{\substack{\text{Lactose} \\ \text{(milk sugar)}}}{\text{C}_{12}\text{H}_{22}\text{O}_{11}} + \text{H}_2\text{O} \xrightarrow{\text{catalyst}} \underset{\text{Glucose}}{\text{C}_6\text{H}_{12}\text{O}_6} + \underset{\text{Galactose}}{\text{C}_6\text{H}_{12}\text{O}_6}
$$

## Objectives

**1.** To become familiar with the reducing or nonreducing nature of carbohydrates.
**2.** To experience the enzyme-catalyzed and acid-catalyzed hydrolysis of acetal groups.

## Reducing or Nonreducing Carbohydrates

Place approximately 2 mL (approximately 40 drops) of Fehling's solution (20 drops each of solution part A and solution part B) into each of five labeled tubes. Add 10 drops of each of the following carbohydrates to the corresponding test tubes as shown in the following table.

| Test tube no. | Name of carbohydrate |
|:---:|:---:|
| 1 | Glucose |
| 2 | Fructose |
| 3 | Sucrose |
| 4 | Lactose |
| 5 | Starch |

Place the test tubes in a boiling water bath for 5 min. A 600-mL beaker containing about 200 mL of tap water with a few boiling chips is used as the bath. Record your results on your Report Sheet. Which of those carbohydrates are reducing carbohydrates?

## Hydrolysis of Carbohydrates

### Hydrolysis of sucrose (acid versus base catalysis)

Place 3 mL of 2% sucrose solution in each of two labeled test tubes. To the first test tube (no. 1), add 3 mL of water and 3 drops of dilute sulfuric acid solution (3 M $H_2SO_4$). To the second test tube (no. 2), add 3 mL of water and 3 drops of dilute sodium hydroxide solution (3 M NaOH). Heat the test tubes in a boiling water bath for about 5 min. Cool both solutions to room temperature. To the contents of test tube no. 1, add dilute sodium hydroxide solution (3 M NaOH) (about 10 drops) until red litmus paper turns blue. Test a few drops of each of the two solutions (test tube nos. 1 and 2) with Fehling's reagent as described before. Record your results on your Report Sheet.

### Hydrolysis of starch (enzyme versus acid catalysis)

Place 2 mL of 2% starch solution in each of two labeled test tubes. To the first test tube (no. 1), add 2 mL of your own saliva. (Use a 10-mL graduated cylinder to collect your saliva.) To the second test tube (no. 2), add 2 mL of dilute sulfuric acid (3 M $H_2SO_4$). Place both test tubes in a water bath that has been previously heated to 45°C. Allow the test tubes with their contents to stand in the warm water bath for 30 min. Transfer a few drops of each solution into separate depressions of a spot plate or two separately labeled microtest tubes. (Use two clean, separate medicine droppers for transferring.) To each sample (in microtest tubes or on a spot plate), add 2 drops of iodine solution. Record the color of the solutions on your Report Sheet.

## Acid catalyzed hydrolysis of starch

Place 5.0 mL of starch solution in a 15 × 150 mm test tube and add 1.0 mL of dilute sulfuric acid (3 M $H_2SO_4$). Mix it by gently shaking the test tube. Heat the solution in a boiling water bath for about 5 min. Using a clean medicine dropper, transfer about 3 drops of the starch solution into a spot plate or a microtest tube and then add 2 drops of iodine solution. Observe the color of the solution. If the solution gives a positive test with iodine solution (the solution should turn blue), continue heating. Transfer about 3 drops of the boiling solution at 5-min. intervals for an iodine test. (*Note: Rinse the medicine dropper very thoroughly before each test.*) When the solution no longer gives a blue color with iodine solution, stop heating and record the time needed for the completion of hydrolysis.

| Chemicals and Equipment |
| --- |

1. Bunsen burner
2. Medicine droppers
3. Microtest tubes or a white spot plate
4. Boiling chips
5. Fehling's reagent
6. 3 M NaOH
7. 2% starch solution
8. 2% sucrose
9. 2% fructose
10. 2% glucose
11. 2% lactose
12. 3 M $H_2SO_4$
13. 0.01 M iodine in KI

## Experiment 36

### PRE-LAB QUESTIONS

1. Circle and label the hemiacetal functional group and the acetal functional group in the following carbohydrates:

**a.** sucrose          **b.** lactose

2. Which carbohydrate in question 1 is a reducing sugar?

3. Write an equation to indicate the formation of the products resulting from the dilute acid hydrolysis of each of the following carbohydrates.
   (a) maltose

   (b) sucrose

## Experiment 36

### REPORT SHEET

*Reducing or nonreducing carbohydrates*

| Test tube no. | Substance | Observation | Reducing or nonreducing carbohydrates |
|:---:|:---:|:---:|:---:|
| 1 | Glucose | | |
| 2 | Fructose | | |
| 3 | Sucrose | | |
| 4 | Lactose | | |
| 5 | Starch | | |

*Hydrolysis of carbohydrates*

| Hydrolysis of sucrose (acid versus base catalysis) | | | |
|:---:|:---:|:---:|:---:|
| Sample | Condition of hydrolysis | Observation | Fehling's reagent (positive or negative) |
| 1 | Acidic ($H_2SO_4$) | | |
| 2 | Basic (NaOH) | | |

| Hydrolysis of starch (enzyme versus acid catalysis) | | | |
|:---:|:---:|:---:|:---:|
| Sample | Condition of hydrolysis | Observation | Iodine test (positive or negative) |
| 1 | Enzymatic (saliva) | | |
| 2 | Acidic ($H_2SO_4$) | | |

| Acid catalyzed hydrolysis of starch | | |
| --- | --- | --- |
| Sample no. | Heating time (min.) | Iodine test (positive or negative) |
| 1 | 5 | |
| 2 | 10 | |
| 3 | 15 | |
| 4 | 20 | |
| 5 | 25 | |
| 6 | 30 | |

## POST-LAB QUESTIONS

1. Fructose is a ketose. Ketones usually do not give a positive reaction with Fehling's reagent. Why is fructose oxidized by Fehling's reagent? (Consult your textbook.)

2. What chemical test is performed to indicate the completion of the hydrolysis of starch by acid?

3. Which hydrolysis of starch is faster: the acid or the enzyme catalyzed reaction?

4. Amylose gives a positive iodine test and a negative test with Fehling's reagent. On the basis of the structure of amylose, explain the results of the two tests.

# Fermentation of a carbohydrate;
# ethanol from sucrose

Ethanol is the least toxic of the alcohols, and as such, has been used as a beverage. It is an alcohol that has been known since ancient times. How ancient people first discovered the process of making ethanol (ethyl alcohol or grain alcohol) by fermentation is lost in time. However, it is a process known for more than 5000 years; evidence exists from Egyptian tombs that beer preparation can be dated to the third millennium B.C. No doubt the basic chemistry of alcohol fermentation came about completely by accident.

An impure form of ethanol results from the fermented juice of fruits, such as grapes and apples, and grains, such as hops and barley. These beverages are known as wine, cider, beer, and ale. Alchemists further learned that from these juices relatively concentrated ethanol solutions could be had by using the technique of distillation.

In fermentation, any carbohydrate, either a complex polysaccharide or simple sugar, can be used as a starting material. In the present experiment, sucrose, a disaccharide, is converted to ethanol by the action of yeast. Sucrose has the formula $C_{12}H_{22}O_{11}$ and is composed of a glucose molecule and a fructose molecule. Enzymes in yeast convert these sugars to ethanol and carbon dioxide. First, *invertase* catalyzes the hydrolysis of sucrose to glucose and fructose; second, *zymase* converts the hydrolyzed sugars to ethanol and $CO_2$. Addition of phosphate salts to the nutrient medium increases the rate of fermentation.

The fermentation process produces solutions of only 10–15% ethanol. This is because the end-product ethanol inhibits the action of the enzymes in the yeast. In order to get more concentrated solutions, distillation is necessary. It is a method that is based on differences in boiling points of the components of a mixture. If one component is nonvolatile, or is a liquid which boils much higher or lower than the component desired, a simple distillation is used. (See Experiment 5 for an application of this technique.) Further purification can be done by fractional distillation. Here the liquids to be distilled are heated in a boiling flask and the vapors are passed through a fractioning column. This column allows for many condensations and vaporizations to take place before pure vapors finally enter a cooling condenser. The purified material is collected in a clean container.

The ethanol prepared in this way is a constant boiling mixture of 95% ethanol and 5% water. This constant boiling mixture is called an *azeotrope*. Distillation is unable to concentrate and purify further. Pure, or 100%, ethanol is called *absolute* ethanol and is obtained by adding benzene to the water-ethanol mixture. Benzene-water-ethanol forms a low-boiling azeotrope that distills before pure ethanol. Thus, after this three-component azeotrope distills, a 100% solution of ethanol is left. Treatment with calcium oxide removes the last traces of water, and a final distillation gives pure ethanol.

*Proof* is the usual designation of the alcohol content of wines and liquors. Proof is twice the percentage of alcohol. Thus, absolute ethanol is 200 proof, and eighty proof vodka has a 40% alcohol content.

## Objectives

1. To demonstrate a fermentation process.
2. To isolate the ethanol produced.
3. To determine the percent composition of the ethanol solution recovered.

## Procedure

This is a two-part procedure that is to be carried out in two successive weeks. Part A, **Fermentation,** is to be done the first week. Part B, **Distillation,** is to be done the second week.

### Part A. Fermentation

1. Weigh out 20.0 g of sucrose and place it into a 250-mL Erlenmeyer flask.

2. Add 100 mL of water and gently shake until all the sucrose has dissolved.

3. To this solution add 0.50 g disodium hydrogen phosphate ($Na_2HPO_4$), 1.0 g potassium phosphate ($K_3PO_4$), and 2.0 g *dried* baker's yeast. Vigorously shake the contents to mix them thoroughly.

4. Figure 37.1 shows the setup for the fermentation experiment. The Erlenmeyer flask is fitted with a one-hole rubber stopper containing a short piece of glass tubing. Latex tubing (approx. 12 in.) is attached to the glass tubing; at the other end is a short length of glass tubing or a Pasteur pipet. The opened end is placed into a 50-mL Erlenmeyer flask [or a test-tube (15 x 150 mm)] containing 30 mL of calcium hydroxide [$Ca(OH)_2$] solution; be sure the glass opening is at least 1 cm below the surface of the liquid. Carefully cover the top of the calcium hydroxide solution with 1 cm of mineral (paraffin) oil. Hold the glass tubing vertically in place with a clamp attached to a ring stand; plug

**Figure 37.1** • Fermentation setup.

the opening of the Erlenmeyer loosely with a small piece of glass wool or cotton. This setup excludes air (and oxygen) from the system (which allows anaerobic oxidation) and prevents further oxidation (by aerobic oxidation) of the ethanol (to acetic acid).

5. Label the fermentation setup with your name. Your laboratory instructor will instruct you where to place your experiment for the week. The fermentation process requires from 5–7 days at 30–35°C to be complete.

## Part B.  Isolation by Fractional Distillation

1. Carefully remove the rubber stopper from the 250-mL Erlenmeyer flask. Do not shake the flask in order to avoid disturbing the sediment on the bottom.

2. Prepare a vacuum filtration assembly using two 250-mL side-arm filter flasks, a 5.5-cm Büchner funnel (with a Filtervac or neoprene adapter), and two lengths (each 12 in.) of vacuum tubing as assembled in Fig. 37.2. Place a piece of filter paper into the Büchner funnel so that it *covers all the holes and lies flat*. Into a 250-mL beaker, place 100 mL of water and one tablespoon of Celite (Filter Aid, which is diatomaceous earth). Stir vigorously and pour the mixture into the Büchner funnel while the water is running and a vacuum is applied. A thin layer of the Celite Filter Aid will form on the filter paper. Discard the water collected in the filter flask.

3. Carefully decant the liquid in the fermentation flask above the sediment through the Celite Filter Aid, using suction. This technique traps the small yeast particles in the Celite Filter Aid but lets through water, ethanol, and any other liquid impurities. This liquid filtrate will be distilled.

**Figure 37.2** • Vacuum filtration with a Büchner funnel.

4. Obtain a distillation setup and assemble the glassware as shown in Fig. 37.3. Use a small dab of silicone grease on all the standard-taper joints as you connect them.

   **a.** Use a round-bottom distilling flask that will be filled approximately one-half to two-thirds full; a 250-mL round bottom flask should do. Add 2–3 boiling stones to the flask.

   **b.** Use a heating mantle for the heat source and a Variac to control the heat.

   **c.** Pack the fractionating column *loosely,* but uniformly, with soap-free stainless steel cleaning pad material.

5. Gradually turn up the heat until the liquid in the distillation flask begins to boil. As the vapors rise in the column, you will see liquid condensing; this ring of condensate will rise in the column. Control the setting on the Variac so that the condensate rises slowly through the column and at an even rate. (If the rate is too fast, the column will flood.) The temperature readings at the distillation head will rise; when the temperature reaches 78°C, begin to collect the liquid that distills. Discard any liquid distilling before this temperature is reached.

6. Collect liquid distilling between 78 and 90°C. Collect 10 to 15 mL of distillate.

7. Turn off the heat source and remove the heating mantle from the distillation flask.

8. Weigh a 50-mL beaker to the nearest 0.001 g (3). With a 10-mL volumetric pipet, transfer 10 mL of distillate to the beaker ($V_{distillate}$). Reweigh the beaker and liquid (5), and by difference, determine the weight of the distillate (6). Determine the density (7), and by referring to Fig. 37.4, determine the percent composition of the ethanol.

**Figure 37.3** • Fractional distillation apparatus.

**Figure 37.4**
Percent composition of aqueous ethanol as a function of density.

9. Clean up: solid waste can be disposed in solid waste containers; liquid waste can be flushed down the sink with plenty of water.

## Chemicals and Equipment

1. Calcium hydroxide solution, $Ca(OH)_2$
2. Disodium hydrogen phosphate, $Na_2HPO_4$
3. Dried baker's yeast
4. Mineral oil
5. Potassium phosphate, $K_2PO_4$
6. Sucrose
7. Boiling stones
8. Celite (Filter Aid)
9. Clamps
10. Heating mantle (or Thermowell)
11. Kit for fractional distillation
12. Ring stands
13. Silicone grease
14. Steel wool, coarse
15. Variac

## Experiment 37

**PRE-LAB QUESTIONS**

**1.** Define the term *azeotrope*.

**2.** Wine usually contains about 12% ethanol. What is its *proof*? How is *proof* defined?

**3.** Give the complete, balanced equation for the conversion of glucose ($C_6H_{12}O_6$) to ethanol.

**4.** The fermentation process requires the presence of catalysts. What are the names of these fermentation catalysts?

## Experiment 37

### REPORT SHEET

1. Distilling temperature range, °C

   _____

2. Volume of distillate collected

   _____

3. Weight of 50-mL beaker: $W_1$

   _____

4. Volume of distillate: $V_{distillate}$

   10.00 mL

5. Weight of beaker and distillate: $W_2$

   _____

6. Weight of distillate: (5) − (3)

   _____

7. Density of distillate
   $d = (W_2 − W_1)/V_{distillate} = [(5) − (3)]/(4)$

   _____

8. Percentage ethanol

   _____

## POST-LAB QUESTIONS

1. If the air trap was not used in the fermentation setup, would ethanol be the end product? If not, then what would be the end product?

2. During fermentation, bubbles can be seen emerging from the glass tubing. Identify the gas causing the bubbles.

3. A solid forms in the calcium hydroxide trap during the fermentation. What is the solid? Write a balanced equation for the formation of this solid.

4. From the percentage ethanol in your experiment, determine the *proof*.

5. Write the balanced equation for the conversion of sucrose to ethanol. Using the quantity of sucrose from this experiment, calculate the number of grams of ethanol and carbon dioxide you expect. How many liters of carbon dioxide would this be at 20°C and a pressure of 1 atmosphere?

# Experiment 38

## Preparation and properties of a soap

### Background

A soap is the sodium or potassium salt of a long-chain fatty acid. The fatty acid usually contains 12 to 18 carbon atoms. Solid soaps usually consist of sodium salts of fatty acids, whereas liquid soaps consist of the potassium salts of fatty acids.

A soap such as sodium stearate consists of a nonpolar end (the hydrocarbon chain of the fatty acid) and a polar end (the ionic carboxylate).

$$CH_3CH_2CH_2CH_2CH_2CH_2CH_2CH_2CH_2CH_2CH_2CH_2CH_2CH_2CH_2CH_2CH_2 - \overset{\overset{\displaystyle O}{\|}}{C} - O^-Na^+$$

Nonpolar              Polar<br>
(Dissolves in oils)       (Dissolves in water)

Because "like dissolves like," the nonpolar end (hydrophobic or water-hating part) of the soap molecule can dissolve the greasy dirt, and the polar or ionic end (hydrophilic or water-loving part) of the molecule is attracted to water molecules. Therefore, the dirt from the surface being cleaned will be pulled away and suspended in water. Thus soap acts as an *emulsifying agent*, a substance used to disperse one liquid (oil molecules) in the form of finely suspended particles or droplets in another liquid (water molecules).

Treatment of fats or oils with strong bases such as lye (NaOH) or potash (KOH) causes them to undergo hydrolysis (saponification) to form glycerol and the salt of a long-chain fatty acid (soap).

$$
\begin{array}{l}
CH_2 - O - \overset{\overset{\displaystyle O}{\|}}{C} - C_{17}H_{35} \\[1em]
\quad | \qquad\quad \overset{\displaystyle O}{\|} \\
CH - O - \overset{\|}{C} - C_{17}H_{35} \quad + \quad 3NaOH \\[1em]
\quad | \qquad\quad \overset{\displaystyle O}{\|} \\
CH_2 - O - \overset{\|}{C} - C_{17}H_{35}
\end{array}
\quad \xrightarrow{\Delta} \quad
\begin{array}{l}
CH_2OH \\
\quad | \\
CHOH \quad + \quad 3C_{17}H_{35}\overset{\overset{\displaystyle O}{\|}}{C} - O^-Na^+ \\
\quad | \\
CH_2OH
\end{array}
$$

Tristearin             Glycerol     Sodium stearate<br>
(a soap)

Because soaps are salts of strong bases and weak acids, they should be weakly alkaline in aqueous solution. However, a soap with free alkali can cause damage to skin, silk, or wool. Therefore, a test for basicity of the soap is quite important.

Soap has been largely replaced by synthetic detergents during the last two decades, because soap has two serious drawbacks. One is that soap becomes ineffective in hard water. Hard water contains appreciable amounts of $Ca^{2+}$ or $Mg^{2+}$ salts.

$$2C_{17}H_{35}COO^-Na^+ + M^{2+} \longrightarrow [C_{17}H_{35}COO^-]_2\, M^{2+} \downarrow + 2Na^+$$

Soap                                            Scum

$$M = (Ca^{2+} \text{ or } Mg^{2+})$$

The other is that, in an acidic solution, soap is converted to free fatty acid and therefore loses its cleansing action.

$$C_{17}H_{35}COO^-Na^+ + H^+ \longrightarrow C_{17}H_{35}COOH \downarrow + Na^+$$

Soap                                  Fatty acid

## Objectives

1. To prepare a simple soap.
2. To investigate some properties of a soap.

## Procedure

### Preparation of a Soap

Measure 23 mL of a vegetable oil into a 250-mL Erlenmeyer flask. Add 20 mL of ethyl alcohol (to act as a solvent) and 20 mL of 25% sodium hydroxide solution (25% NaOH). While stirring the mixture constantly with a glass rod, the flask with its contents is heated gently in a boiling water bath. A 600-mL beaker containing about 200 mL of tap water and a few boiling chips can serve as a water bath (Fig. 38.1).

**Figure 38.1**
Experimental setup for
soap preparation.

After being heated for about 20 min., the odor of alcohol will disappear, indicating the completion of the reaction. A pasty mass containing a mixture of the soap, glycerol, and excess sodium hydroxide is obtained. Use an ice water bath to cool the flask with its contents. To precipitate or "salt out" the soap, add 150 mL of a saturated sodium chloride solution to the soap mixture while stirring vigorously. This process increases the density of the aqueous solution; therefore, soap will float out from the aqueous solution. Filter the precipitated soap with the aid of suction and wash it with 10 mL of ice cold water. Observe the appearance of your soap and record your observation on the Report Sheet.

## Properties of a Soap

1. *Emulsifying properties*. Shake 5 drops of mineral oil in a test tube containing 5 mL of water. A temporary emulsion of tiny oil droplets in water will be formed. Repeat the same test, but this time add a small piece of the soap you have prepared before shaking. Allow both solutions to stand for a short time. Compare the appearance and the relative stabilities of the two emulsions. Record your observations on the Report Sheet.

2. *Hard water reactions*. Place about one-third spatula full of the soap you have prepared in a 50-mL beaker containing 25 mL of water. Warm the beaker with its contents to dissolve the soap. Pour 5 mL of the soap solution into each of five test tubes (nos. 1, 2, 3, 4, and 5). Test no. 1 with 2 drops of a 5% solution of calcium chloride (5% $CaCl_2$), no. 2 with 2 drops of a 5% solution of magnesium chloride (5% $MgCl_2$), no. 3 with 2 drops of a 5% solution of iron(III) chloride (5% $FeCl_3$), and no. 4 with tap water. Tube no. 5 will be used for a basicity test, which will be performed later. Record your observations on the Report Sheet.

3. *Alkalinity (basicity)*. Test soap solution no. 5 with a wide-range pH paper. What is the approximate pH of your soap solution? Record your answer on the Report Sheet.

## Chemicals and Equipment

1. Hot plate
2. Ice cubes
3. Büchner funnel in no. 7 one-hole rubber stopper
4. 500-mL filter flask
5. Filter paper, 7 cm diameter
6. pHydrion paper
7. Boiling chips
8. 95% ethanol
9. Saturated sodium chloride solution
10. 25% NaOH
11. Vegetable oil
12. 5% $FeCl_3$
13. 5% $CaCl_2$
14. Mineral oil
15. 5% $MgCl_2$

## Experiment 38

### PRE-LAB QUESTIONS

**1.** What chemical process is called saponification (soap making)? Why?

**2.** Consult Table 20.2 of your textbook. If coconut oil is used to make soap, it would be the same as if corn oil was used. List the chemical formula of the most abundant soap you formed in each case.

**3.** If you attempted to make soap from butanoic acid by forming sodium butanoate, would your preparation have the properties to act as an emulsifying agent (soap)?

**4.** Stearic acid is insoluble in water, and sodium stearate (a soap) is soluble. What causes the difference in solubility? Explain.

███████████

## Preparation

Appearance of your soap _____

### Properties

### Emulsifying Properties

Which mixture, oil–water or oil–water–soap, forms a more stable emulsion?




### Hard Water Reaction

No. 1 + $CaCl_2$ _____




No. 2 + $MgCl_2$ _____




No. 3 + $FeCl_3$ _____




No. 4 + tap water _____




### Alkalinity

pH of your soap solution (no. 5) _____

### POST-LAB QUESTIONS

1. When you made soap, first you dissolved vegetable oil in ethanol. What happened to the ethanol during the reaction?

2. Write a chemical equation for the reaction in which you added a few drops of $MgCl_2$ solution to a soap solution.

3. Soaps that have a pH above 8.0 tend to irritate some sensitive skins. Was your soap good enough to compete with commercial preparations?

4. What would happen to the pH of your soap if you
   a. washed the precipitated soap in twice as much volume of cold water as directed on p. 395?

   b. used 300 mL of saturated salt solution rather than the recommended 150 mL?

   c. mixed some vegetable oil with the final product?

# Preparation of a hand cream

## Background

Hand creams are formulated to carry out a variety of cosmetic functions. Among these are softening the skin and preventing dryness; elimination of natural waste products (oils) by emulsification; cooling the skin by radiation thus helping to maintain body temperature. In addition, hand creams must have certain ingredients that aid spreadibility and provide body. In many cases added fragrance improves the odor, and in some special cases medications combat assorted ills.

The basic hand cream formulations all contain water to provide moisture and lanolin which helps its absorption by the skin. The latter is a yellowish wax. Chemically, wax is made of esters of long chain fatty acids and long chain alcohols. Lanolin is usually obtained from sheep wool; it has the ability to absorb 25–30% of its own weight of water and to form a fine emulsion. Mineral oil, which consists of high-molecular-weight hydrocarbons, provides spreadibility. In order to allow nonpolar substances, such as lanolin and mineral oil, to be uniformly dispersed in a polar medium, water, one needs strong emulsifying agents. An emulsifying agent must have nonpolar, hydrophobic portions to interact with the oil and also polar, hydrophilic portions to interact with water. A mixture of stearic acid and triethanolamine, through acid–base reaction, yields the salt that has the requirements to act as an emulsifying agent.

Besides the above five basic ingredients, some hand creams also contain alcohols such as propylene glycol (1,2-propanediol), and esters such as methyl stearate, to provide the desired texture of the hand cream.

In this experiment you will prepare four hand creams using the combination of ingredients as shown in Table 39.1.

### Objectives

1. To learn the method of preparing a hand cream.
2. To appraise the function of the ingredients in the hand cream.

## Procedure

### Preparation of the Hand Creams

For each sample in Table 39.1, assemble the ingredients in two beakers. Beaker 1 contains the polar ingredients, and beaker 2 contains the nonpolar contents.

| Table 39.1 | Recipes to Prepare Hand Creams | | | | |
|---|---|---|---|---|---|
| **Ingredients** | **Sample 1** | **Sample 2** | **Sample 3** | **Sample 4** | |
| Water | 25 mL | 25 mL | 25 mL | 25 mL | |
| Triethanolamine | 1 mL | 1 mL | 1 mL | — | Beaker 1 |
| Propylene glycol | 0.5 mL | 0.5 mL | — | 0.5 mL | |
| | | | | | |
| Stearic acid | 5 g | 5 g | 5 g | 5 g | |
| Methyl stearate | 0.5 g | 0.5 g | — | 0.5 g | |
| Lanolin | 4 g | 4 g | 4 g | 4 g | Beaker 2 |
| Mineral oil | 5 mL | — | 5 mL | 5 mL | |

1. To prepare sample 1, put the nonpolar ingredients in a 50-mL beaker (beaker 2) and heat it in a water bath. The water bath can be a 400-mL beaker half-filled with tap water and heated with a Bunsen burner (Fig. 39.1). Carefully hold the beaker with crucible tongs in the boiling water until all ingredients melt. As an alternative, use hot plates instead of a Bunsen burner.

**Figure 39.1**
Heating ingredients.

2. In the same water bath, heat the 100-mL beaker (beaker 1) containing the polar ingredients for about 5 min. Remove the beaker and set it on the bench top.

3. Into the 100-mL beaker containing polar ingredients, pour slowly the contents of the 50-mL beaker that holds the molten nonpolar ingredients (Fig. 39.2). Stir the mixture for 5 min. until you have a smooth uniform paste.

4. Repeat the same procedure in preparing the other three samples.

**Figure 39.2**
Mixing hand cream
ingredients.

## Characterization of the Hand Cream Preparations

1. Test the pH of the hand creams prepared using a wide-range pH paper.

2. Rubbing a small amount of the hand cream between your fingers, test for smoothness and homogeneity. Also note the appearance. Record your observations on the Report Sheet.

3. Dispose of your hand cream preparations in the waste containers provided. **DO NOT** place in sink.

| Chemicals and Equipment |
| --- |

1. Bunsen burner
2. Lanolin
3. Stearic acid
4. Methyl stearate
5. Mineral oil
6. Triethanolamine
7. Propylene glycol
8. pHydrion paper

## Experiment 39

### PRE-LAB QUESTIONS

1. Write the chemical formula of a wax made of an 18-carbon saturated fatty acid and a 10-carbon saturated straight chain alcohol.

2. The emulsifying agent was prepared from stearic acid and triethanolamine. Give the name of this salt. Write its formula.

3. What functional groups of the emulsifying agent provide the hydrophilic character?

4. Methyl stearate is used in some recipes of hand cream. Write the structure of methyl stearate. Can this act as an emulsifying agent?

## Experiment 39

## REPORT SHEET

### Characterization of the hand cream samples

| Properties | Sample 1 | Sample 2 | Sample 3 | Sample 4 |
|---|---|---|---|---|
| pH | | | | |
| Smoothness | | | | |
| Homogeneity | | | | |
| Appearance | | | | |

## POST-LAB QUESTIONS

1. In comparing the properties of the hand creams you produced, ascertain the function of each of the missing ingredients in the hand cream:

   a. Mineral oil

   b. Triethanolamine

   c. Methyl stearate and propylene glycol

2. A hand cream appears smooth and uniform after you prepared it, but in a week of storage most of the water settles on the bottom and most of the oil separates on the top. What do you think may have gone wrong with the hand cream preparation?

3. Was the pH of all your hand cream preparation the same? If not, explain the differences.

4. In one of your hand cream formulations there was no mineral oil. What characteristic feature was observed in the absence of mineral oil? Explain.

5. If you have omitted water from your preparation, could you still call it a hand cream?

## Extraction and identification of fatty acids from corn oil

**Background**

Fats are esters of glycerol and fatty acids. Liquid fats are often called oils. Whether a fat is solid or liquid depends on the nature of the fatty acids. Solid animal fats contain mostly saturated fatty acids, while vegetable oils contain high amounts of unsaturated fatty acids. To avoid arteriosclerosis, hardening of the arteries, diets which are low in saturated fatty acids as well as in cholesterol are recommended.

Note that even solid fats contain some unsaturated fatty acids, and oils contain saturated fatty acids as well. Besides the degree of unsaturation, the length of the fatty acid chain also influences whether a fat is solid or liquid. Short chain fatty acids, such as found in coconut oil, convey liquid consistency in spite of the low unsaturated fatty acid content. Two of the unsaturated fatty acids, linoleic and linolenic acids, are essential fatty acids because the body cannot synthesize them from precursors; they must be included in the diet.

The four unsaturated fatty acids most frequently found in vegetable oils are:

Oleic acid: $CH_3(CH_2)_7CH = CH(CH_2)_7COOH$

Linoleic acid: $CH_3(CH_2)_4CH = CHCH_2CH = CH(CH_2)_7COOH$

Linolenic acid: $CH_3CH_2CH = CHCH_2CH = CHCH_2CH = CH(CH_2)_7COOH$

Arachidonic acid:
$CH_3(CH_2)_4CH = CHCH_2CH = CHCH_2CH = CHCH_2CH = CH(CH_2)_3COOH$

All the C = C double bonds in the unsaturated fatty acids are *cis* double bonds, which interrupt the regular packing of the aliphatic chains, and thereby convey a liquid consistency at room temperature. This physical property of the unsaturated fatty acid is carried over to the physical properties of triglycerides (oils).

In order to extract and isolate fatty acids from corn oil, first, the ester linkages must be broken. This is achieved in the saponification reaction in which a triglyceride is converted to glycerol and the potassium salt of its fatty acids:

$$CH_2-O-\overset{\overset{O}{\|}}{C}-C_{17}H_{35}$$
$$CH-O-\overset{\overset{O}{\|}}{C}-C_{17}H_{35} + 3KOH \longrightarrow CHOH + 3C_{17}H_{35}\overset{\overset{O}{\|}}{C}-O^-K^+$$
$$CH_2-O-\overset{\overset{O}{\|}}{C}-C_{17}H_{35}$$

with $CH_2OH$, $CHOH$, $CH_2OH$ forming glycerol.

In order to separate the potassium salts of fatty acids from glycerol, the products of the saponification mixture must be acidified. Subsequently, the fatty acids can be extracted by petroleum ether. To identify the fatty acids that were isolated, they must be converted to their respective methyl ester by a perchloric acid catalyzed reaction:

$$C_{17}H_{35}COOH + CH_3OH \xrightarrow{HClO_4} C_{17}H_{35}\overset{\displaystyle O}{\overset{\|}{C}}{-}O{-}CH_3 + H_2O$$

The methyl esters of fatty acids can be separated by thin-layer chromatography (TLC). They can be identified by comparison of their rate of migration ($R_f$ values) to the $R_f$ values of authentic samples of methyl esters of different fatty acids (Fig. 40.1).

**Figure 40.1**
TLC chromatogram.

$$9.0 \text{ cm} \qquad R_f = \frac{4.5}{9.0} = 0.50$$

$R_f$ = distance travelled by fatty acid/distance travelled by the solvent front.

---

**Objectives**

1. To extract fatty acids from neutral fats.
2. To convert them to their methyl esters.
3. To identify them by thin-layer chromatography.

---

**Procedure**

### Part A. Extraction of Fatty Acids

1. Weigh a 50-mL Erlenmeyer flask and record the weight on your Report Sheet (1).

2. Add 2 mL of corn oil and weigh it again. Record the weight on your Report Sheet (2).

3. Add 5 mL of 0.5 M KOH in ethanol to the Erlenmeyer flask. Stopper it. Place the flask in a water bath at 55°C for 20 min.

**CAUTION!**

Strong acid; use gloves with concentrated HCl.

4. When the saponification is completed, add 2.5 mL of the concentrated HCl. Mix it by swirling the Erlenmeyer flask. Transfer the contents into a 50-mL separatory funnel. Add 5 mL of petroleum ether. Mix it thoroughly (see Fig. 35.1). Drain the lower aqueous layer into a flask and the upper petroleum ether layer into a glass-stoppered test tube. Repeat the process by adding back the aqueous layer into the separatory funnel and extracting it with another portion of 5 mL of petroleum ether. Combine the extracts.

## Part B. Preparation of Methyl Esters

1. Place a plug of glass wool (the size of a pea) into the upper stem of a funnel, fitting it loosely. Add 10 g of anhydrous $Na_2SO_4$. Rinse the salt on to the glass wool with 5 mL of petroleum ether; discard the wash. Pour the combined petroleum ether extracts into the funnel and collect the filtrate in an evaporating dish. Add another portion (2 mL) of petroleum ether to the funnel and collect this wash, also in the evaporating dish.

2. Evaporate the petroleum ether under the hood by placing the evaporating dish on a water bath at 60°C. (Alternatively, if dry $N_2$ gas is available, the evaporation could be achieved by bubbling nitrogen through the extract. This also must be done under the hood.)

3. When dry, add 10 mL of the $CH_3OH:HClO_4$ mixture (95:5). Place the evaporating dish on the water bath at 55°C for 10 min.

## Part C. Identification of Fatty Acids

1. Transfer the methyl esters prepared above into a separatory funnel. Extract twice with 5 mL of petroleum ether. Combine the extracts.

2. Prepare another funnel with anhydrous $Na_2SO_4$ on top of the glass wool. Filter the combined petroleum ether extracts through the salt into a dry, clean evaporating dish. Evaporate the petroleum ether on the water bath at 60°C, as before. When dry, add 0.2 mL of petroleum ether and transfer the solution to a clean and dry test tube.

3. Take a 15 × 6.5 cm TLC plate. Make sure you do not touch the TLC plate with your fingers. Preferably use plastic gloves, or handle the plate by holding it only at the edges. This precaution must be observed throughout the whole operation because your fingers may contaminate the sample. With a pencil, lightly draw a line parallel to the 6.5 edge about 1 cm from the edge. Mark the positions of the five spots, equally spaced, where you will spot your samples (Fig. 40.2).

**Figure 40.2**
Spotting.

4. For spots no. 1 and no. 5, use your isolated methyl esters obtained from corn oil. For spot no. 2, use methyl oleate; for spot no. 3, methyl linoleate; and for spot no. 4, methyl palmitate. For each sample use a separate capillary tube. In spotting, apply each sample in the capillary to the plate until it spreads to a spot of 1 mm diameter. Dry the spots with a heat lamp. Pour about 15 mL of solvent (hexane:diethyl ether; 4:1) into a 500-mL beaker. Place the spotted TLC plate diagonally for ascending chromatography. Make certain that the spots applied are **above** the surface of the eluting solvent. Cover the beaker lightly with aluminum foil to avoid excessive solvent evaporation.

5. When the solvent front has risen to about 1–2 cm from the top edge, remove the plate from the beaker. Mark the advance of the solvent front with a pencil. Dry the plate with a heat lamp under the hood. Place the dried plate in a beaker containing a few iodine crystals. Cover the beaker tightly with aluminum foil. Place the beaker in a 110°C oven for 3–4 min. Remove the beaker and let it cool to room temperature. **This part is essential to avoid inhaling iodine vapors.** Remove the TLC plate from the beaker and mark the spots with a pencil.

6. Record the distance the solvent front advanced on your Report Sheet (4). Record on your Report Sheet (5–9) the distance of each iodine-stained spot from its origin. Calculate the $R_f$ values of your samples (10–14).

---

### Chemicals and Equipment

1. Corn oil
2. Methyl palmitate
3. Methyl oleate
4. Methyl linoleate
5. Petroleum ether (b. p. 30–60°C)
6. 0.5 M KOH in ethanol
7. Concentrated HCl
8. Anhydrous $Na_2SO_4$
9. Methanol:perchloric acid mixture (95:5)
10. Hexane:diethyl ether mixture (4:1)
11. Iodine crystals, $I_2$
12. Aluminum foil
13. Polyethylene gloves
14. 15 × 6.5 cm silica gel TLC plate
16. Capillary tubes open on both ends
17. Heat lamp
18. Water bath
19. Ruler
20. Drying oven, 110°C

## Experiment 40

### PRE-LAB QUESTIONS

1. Fatty acids can be extracted by petroleum ether. Salts of fatty acids cannot; they are water soluble. Explain why.

2. Why do you have to wear gloves when handling TLC plates?

3. How can one convert the potassium salt of a fatty acid (i.e., potassium oleate) into a fatty acid (oleic acid)?

4. Why do you have to cool the iodine chamber (the beaker containing the chromatogram and iodine vapor) from 110°C to room temperature?

5. Esterification is an equilibrium reaction. Considering the LeChatelier's principle, propose two processes that will improve the yield of methyl esters of fatty acids.

## Experiment 40

### REPORT SHEET

1. Weight of beaker        _____ g
2. Weight of beaker and oil        _____ g
3. Weight of oil        _____ g

*Distances on the chromatogram in cm*

4. The solvent front        _____
5. Spot no. 1 a, b, c, d, e        a_____ b_____ c_____ d_____ e_____
6. Spot no. 2        _____
7. Spot no. 3        _____
8. Spot no. 4        _____
9. Spot no. 5 a, b, c, d, e        a_____ b_____ c_____ d_____ e_____

*Calculated $R_f$ values*

10. For spot no.1 [(5)/(4)] a, b, c, d, e        a_____ b_____ c_____ d_____ e_____
11. For spot no. 2 [(6)/(4)]        _____
12. For spot no. 3 [(7)/(4)]        _____
13. For spot no. 4 [(8)/(4)]        _____
14. For spot no. 5 [(9)/(4)] a, b, c, d, e        a_____ b_____ c_____ d_____ e_____
15. How many fatty acids were present in your corn oil?

16. How many fatty acids could you identify? Name the identifiable fatty acids in the corn oil.

## POST-LAB QUESTIONS

1. Which of the identifiable fatty acids of your corn oil was a saturated fatty acid?

2. Judging from the iodine spots of samples 2, 3, and 4, which fatty acid reacts most strongly with iodine? Why?

3. What was the role of the **anhydrous** $Na_2SO_4$ in the preparation of the methyl esters of fatty acids?

4. Given two saturated fatty acids, one a short chain of 10 carbons and the other a long chain of 20 carbons, which would move faster on the TLC plate? Explain.

5. Considering the $R_f$ values you obtained for the three methyl esters of the fatty acids in your experiment, how could you achieve a better separation of the spots?

6. Which of your pure methyl ester samples moved the fastest on TLC? Considering only the number of carbons in each sample, what makes the sample move fast?

# Analysis of lipids

## Background

Lipids are chemically heterogeneous mixtures. The only common property they have is their insolubility in water. We can test for the presence of various lipids by analyzing their chemical constituents. Foods contain a variety of lipids, most important among them are fats, complex lipids, and steroids. Fats are triglycerides, esters of fatty acids and glycerol. Complex lipids also contain fatty acids, but their alcohol may be either glycerol or sphingosine. They also contain other constituents such as phosphate, choline, or ethanolamine or mono- to oligo-saccharides. An important representative of this group is lecithin, a glycerophospholipid, containing fatty acids, glycerol, phosphate, and choline. The most important steroid in foods is cholesterol. Different foods contain different proportions of these three groups of lipids.

Structurally, cholesterol contains the steroid nucleus that is the common core of all steroids.

Steroid nucleus                              Cholesterol

There is a special colorimetric test, the Lieberman-Burchard reaction, which uses acetic anhydride and sulfuric acid as reagents, that gives a characteristic green color in the presence of cholesterol. This color is due to the —OH group of cholesterol and the unsaturation found in the adjacent fused ring. The color change is gradual: first it appears as a pink coloration, changing later to lilac, and finally to deep green.

When lecithin is hydrolyzed in acidic medium, both the fatty acid ester bonds and the phosphate ester bonds are broken and free fatty acids and inorganic phosphate are released. Using a molybdate test, we can detect the presence of phosphate in the hydrolysate by the appearance of a purple color. Although this test is not specific for lecithin (other phosphate containing lipids will give a positive molybdate test), it differentiates clearly between fat and cholesterol on the one hand (negative test), and phospholipid on the other (positive test).

A second test that differentiates between cholesterol and lecithin is the acrolein reaction. When lipids containing glycerol are heated in the presence of potassium hydrogen sulfate, the glycerol is dehydrated, forming acrolein, which has an unpleasant odor. Further heating results in polymerization of acrolein, which is indicated by the slight blackening of the reaction mixture. Both the pungent smell and the black color indicate the presence of glycerol, and thereby fat and/or lecithin. Cholesterol gives a negative acrolein test.

$$
\begin{array}{c}
\text{CH}_2\text{OH} \\
| \\
\text{CHOH} \\
| \\
\text{CH}_2\text{OH}
\end{array}
\xrightarrow{\Delta}
\begin{array}{c}
\text{O} \\
\| \\
\text{C}-\text{H} \\
| \\
\text{CH} \\
\| \\
\text{CH}_2
\end{array}
+ 2\text{H}_2\text{O}
$$

## Objectives

To investigate the lipid composition of common foods such as corn oil, butter, and egg yolk.

## Procedure

Use six samples for each test: (1) pure cholesterol, (2) pure glycerol, (3) lecithin preparation, (4) corn oil, (5) butter, (6) egg yolk.

### Phosphate Test

### CAUTION!

6 M nitric acid is a strong acid. Handle it with care. Use gloves.

1. Take six clean and dry test tubes. Label them. Add about 0.2 g of sample to each test tube. Hydrolyze the compounds by adding 3 mL of 6 M nitric acid to each test tube.

2. Prepare a water bath by boiling about 100 mL of tap water in a 250-mL beaker on a hot plate. Place the test tubes in the boiling water bath for 5 min. Do not inhale the vapors. Cool the test tubes. Neutralize the acid by adding 3 mL of 6 M NaOH. Mix. During the hydrolysis, a precipitate may form, especially in the egg yolk sample. The samples in

which a precipitate appeared must be filtered. Place a piece of cheese cloth on top of a 25-mL Erlenmeyer flask. Pour the turbid hydrolysate in the test tube on the cheese cloth and filter it.

3. Transfer 2 mL of each neutralized (and filtered) sample into clean and labeled test tubes. Add 3 mL of a molybdate solution to each test tube and mix the contents. **(Be careful. The molybdate solution contains sulfuric acid.)** Heat the test tubes in a boiling water bath for 5 min. Cool them to room temperature.

4. Add 0.5 mL of an ascorbic acid solution and mix the contents thoroughly. Wait 20 min. for the development of the purple color. Record your observations on the Report Sheet. While you wait, you can perform the rest of the colorimetric tests.

## The Acrolein Test for Glycerol

1. Place 1 g of potassium hydrogen sulfate, $KHSO_4$, in each of seven clean and dry test tubes. Label them. Add a few grains of your pure preparations, lecithin and cholesterol, to two of the test tubes. Add a drop, about 0.1 g, from each, glycerol, corn oil, butter, and egg yolk to the other four test tubes. To the seventh test tube add a few crystals of sucrose.

2. Set up your Bunsen burner in the hood. **It is important that this test be performed under the hood because of the pungent odor of the acrolein.**

3. Gently heat each test tube, one at a time, over the Bunsen burner flame, shaking it continuously from side to side. When the mixture melts it slightly blackens, and you will notice the evolution of fumes. Stop the heating. **Smell the test tubes by moving them sideways under your nose or waft the vapors. Do not inhale the fumes directly.** A pungent odor, resembling burnt hamburgers, is the positive test for glycerol. Sucrose in the seventh test tube also will be dehydrated and will give a black color. However, its smell is different, and thus is not a positive test for acrolein. Do not overheat the test tubes, for the residue will become hard, making it difficult to clean the test tubes. Record your observations on the Report Sheet.

## Lieberman-Burchard Test for Cholesterol

1. Place a few grains of your cholesterol and lecithin preparations in labeled clean and dry test tubes. Similarly, add about 0.1-g samples of glycerol, corn oil, butter, and egg yolk to the other four labeled clean and dry test tubes. **(The next step should be done in the hood.)**

2. Transfer 3 mL of chloroform and 1 mL of acetic anhydride to each test tube. Finally, add 1 drop of concentrated sulfuric acid to each mixture. Mix the contents and record the color changes, if any. Wait 5 min. Record again the color of your solutions. Record your observations on the Report Sheet.

## Chemicals and Equipment

1. 6 M NaOH
2. 6 M $HNO_3$
3. Molybdate reagent
4. Ascorbic acid solution
5. $KHSO_4$
6. Chloroform
7. Acetic anhydride
8. Sulfuric acid, $H_2SO_4$
9. Cholesterol
10. Lecithin
11. Glycerol
12. Corn oil
13. Butter
14. Egg yolk
15. Hot plate
16. Cheese cloth

## Experiment 41

### PRE-LAB QUESTIONS

1. Cephalins are glycerophospholipids present in foods. They differ from lecithins by having ethanolamine or serine instead of choline in their structure. Could you differentiate between lecithins and cephalins on the basis of the three tests to be performed in this experiment?

2. Cholesterol has an alcohol group. One could also dehydrate cholesterol (removing one water molecule by heating). Show the structure you would expect from the dehydration of cholesterol.

3. Would the compound with the structure in question 2 give a positive Lieberman-Burchard test?

4. Choleterol in tissues is sometimes esterified by fatty acids. (a) Draw the structure of cholesteryl oleate. (b) Would this ester give a positive Lieberman-Burchard test?

5. Which of the three tests would you expect to give a positive test with phosphatidyl inositols?

## Experiment 41

### REPORT SHEET

| Tests | Cholesterol | Lecithin | Corn Glycerol | Oil | Butter | Egg yolk | Sucrose |
|---|---|---|---|---|---|---|---|
| **1.** Phosphate<br>  **a.** Color | | | | | | | |
|   **b.** Conclusions | | | | | | | |
| **2.** Acrolein<br>  **a.** Odor | | | | | | | |
|   **b.** Color | | | | | | | |
|   **c.** Conclusions | | | | | | | |
| **3.** Lieberman-Burchard<br>  **a.** Initial color | | | | | | | |
|   **b.** Color after 5 min. | | | | | | | |
|   **c.** Conclusions | | | | | | | |

## POST-LAB QUESTIONS

1. What is your overall conclusion regarding the composition of your corn oil? Was it pure triglyceride?

2. Based on the intensity of color developed in your test for cholesterol, which food contained the most and which contained the least cholesterol?

3. Besides the lecithin and other glycerophospholipids, two more classes of complex lipds are given in your textbook: (a) sphingolipids and (b) glycolipids. (Look up their structures in your textbook.) Would any of these compounds give you a positive test with molybdate solution?

4. Assume that the lecithin still contained all its ester linkages. Would you get a positive acrolein test? Explain.

5. When sucrose is dehydrated by heating it with $KHSO_4$, you can observe the black residue (carbon) and water. This is the origin of the name *carbohydrate*. Can you detect the presence of acrolein by its smell in the dehydration of sucrose?

6. Why was it necessary to hydrolyze the samples with nitric acid before performing the molybdate test?

# Separation of amino acids by paper chromatography

## Background

Amino acids are the building blocks of peptides and proteins. They possess two functional groups—the carboxylic acid group gives the acidic character, and the amino group provides the basic character. The common structure of all amino acids is

$$\begin{array}{c} H \\ | \\ R-C-COOH \\ | \\ NH_2 \end{array}$$

The $R$ represents the side chain that is different for each of the amino acids that are commonly found in proteins. However, all 20 amino acids have a free carboxylic acid group and a free amino (primary amine) group, except proline which has a cyclic side chain and a secondary amino group.

Proline

We use the properties provided by these groups to characterize the amino acids. The common carboxylic acid and amino groups provide the acid–base nature of the amino acids. The different side chains, and the solubilities provided by these side chains, can be utilized to identify the different amino acids by their rate of migration in paper chromatography.

In this experiment, we use paper chromatography to identify aspartame, an artificial sweetener, and its hydrolysis products from certain foods.

Aspartame

Aspartame is the methyl ester of the dipeptide aspartylphenylalanine. Upon hydrolysis with HCl it yields aspartic acid, phenylalanine, and methyl alcohol. When this artificial sweetener was approved by the Food and Drug Administration, opponents of aspartame claimed that it is a health hazard, because aspartame would be hydrolyzed and would yield poisonous methyl alcohol in soft drinks that are stored over long periods of time. The Food and Drug Administration ruled, however, that aspartame is sufficiently stable and fit for human consumption. Only a warning must be put on the labels of foods containing aspartame. This warning is for patients suffering from phenylketonurea who cannot tolerate phenylalanine.

To run a paper chromatography, we use Whatman no.1 chromatography paper. We apply the sample (aspartame or amino acids) as a spot to a strip of chromatography paper. The paper is dipped into a mixture of solvents. The solvent moves up the paper by capillary action and carries the sample with it. Each amino acid may have a different migration rate depending on the solubility of the side chain in the solvent. Amino acids with similar side chains are expected to move with similar, though not identical, rates; those that have quite different side chains are expected to migrate with different velocities. Depending on the solvent system used, almost all amino acids and dipeptides can be separated from each other by paper chromatography.

We actually do not measure the rate of migration of an amino acid or a dipeptide, but rather, how far a particular amino acid travels on the paper relative to the migration of the solvent. This ratio is called the $R_f$ value. In order to calculate the $R_f$ values, one must be able to visualize the position of the amino acid or dipeptide. This is done by spraying the chromatography paper with a ninhydrin solution that reacts with the amino group of the amino acid. A purple color is produced when the paper is heated. (The proline not having a primary amine gives a yellow color with ninhydrin.) For example, if the purple spot of an amino acid appears on the paper 4.5 cm away from the origin and the solvent front migrates 9.0 cm (Fig. 42.1), the $R_f$ value for the amino acid is calculated

$$R_f = \frac{\text{distance traveled by the amino acid}}{\text{distance traveled by the solvent front}} = \frac{4.5 \text{ cm}}{9.0 \text{ cm}} = 0.50$$

In the present experiment you will determine the $R_f$ values of three amino acids: phenylalanine, aspartic acid, and leucine. You will also measure the $R_f$ value of aspartame.

The aspartame you will analyze is actually a commercial sweetener, Equal by the

**Figure 42.1**
Paper chromatogram.

NutraSweet Co., that contains silicon dioxide, glucose, cellulose, and calcium phosphate in addition to the aspartame. None of these other ingredients of Equal will give a purple or any other colored spot with ninhydrin. Other generic aspartame sweeteners may contain other nonsweetening ingredients. Occasionally, some sweeteners may contain a small amount of leucine which can be detected by the ninhydrin test. You will also hydrolyze aspartame using HCl as a catalyst to see if the hydrolysis products will prove that the sweetener is truly aspartame. Finally, you will analyze some commercial soft drinks supplied by your instructor. The analysis of the soft drink can tell you if the aspartame was hydrolyzed at all during the processing and storing of the soft drink.

---

## Objectives

1. To separate amino acids and a dipeptide by paper chromatograhy.
2. To identify hydrolysis products of aspartame.
3. To analyze the state of aspartame in soft drinks.

---

## Procedure

1. Dissolve about 10 mg of the sweetener Equal in 1 mL of 3 M HCl in a test tube. Heat it with a Bunsen burner to a boil for 30 sec., but make sure that the liquid does not completely evaporate. Cool the test tube and label it "Hydrolyzed Aspartame."

2. Label five small test tubes, respectively, for aspartic acid, phenylalanine, leucine, aspartame, and Diet Coke. Place about 0.5-mL samples in each test tube.

3. Use plastic gloves throughout in order not to contaminate the paper chromatogram. Take a strip of Whatman no. 1 chromatographic paper, 15 x 8 cm and 0.016 thickness. With a pencil, lightly draw a line parallel to the 8 cm edge about 1 cm from the edge. Mark the positions of 6 spots, placed equally, where you will spot your samples (Fig. 42.2)

**Figure 42.2**
Spotting.

4. Spotting. For each sample, use a separate capillary tube. Apply a drop of sample to the paper until it spreads to a spot of 1 mm diameter. Dry the spots. (If a heat lamp is available, use it for drying.) Do the spotting in the following order:

| | |
|---|---|
| (1) phenylalanine | one drop |
| (2) aspartic acid | one drop |
| (3) leucine | one drop |
| (4) aspartame (in Equal) | one drop |
| (5) hydrolyzed aspartame | six drops |
| (6) Diet Coca Cola | ten drops |

When you do multiple drops, (5) and (6), allow the paper to dry after each drop before applying the next drop. Do not allow the spots to spread to larger than 2 mm in diameter.

5. Pour about 15 mL of solvent mixture (butanol : acetic acid : water) into a large (1 L) beaker. Place a glass rod over the beaker. Using clear tape, affix the chromatographic paper to the rod so that when you lower the rod over the beaker, the paper will dip into the solvent **but the pencil marks and the spots will be above the solvent surface.** Cover the rod and the beaker with aluminum foil. Place the beaker on a hot plate. Turn the setting to no. 3 (out of 10) and heat the beaker to about 35°C. Allow the solvent front to advance at least 6 cm (about 40–50 min.), but do not allow it to get closer than 1 cm from the edge.

6. When the solvent front has advanced at least 6 cm, remove the rod and the paper from the beaker. You must not allow the solvent front to advance up to or beyond the edge of the paper. Mark immediately *with a pencil* the position of the solvent front. Under a hood, dry the paper with the aid of a heat lamp or hair dryer. Using polyethylene gloves, spray the dry paper with ninhydrin solution. *Be careful not to spray ninhydrin on your hand and not to touch the sprayed areas with bare hands. If the ninhydrin spray touches your skin (which contains amino acids) your fingers will be discolored for a few days.* Place the sprayed paper into a drying oven at 105–110°C for 2–3 min.

7. Remove the paper from the oven. Mark the center of the spots and calculate the $R_f$ values of each spot. Record your observations on the Report Sheet.

8. If the spots on the chromatogram are faded, we can visualize them by exposing the chromatogram to iodine vapor. Place your chromatogram into a wide-mouthed jar containing a few iodine crystals. Cap the jar and warm it slightly on a hot plate to enhance the sublimation of iodine. The iodine vapor will interact with the faded pigment spots and make them visible. After a few minutes' exposure to iodine vapor, remove the chromatogram and mark the spots **immediately** with a pencil. The spots will fade again with exposure to air. Measure the distance of the center of the spots from the origin and calculate the $R_f$ values.

## Chemicals and Equipment

1. 0.1% solutions of aspartic acid, phenylalanine, and leucine
2. 0.5% solution of aspartame (Equal)
3. Diet Coke
4. 3 M HCl
5. 0.2% ninhydrin spray
6. Butanol:acetic acid:water–solvent mixture
7. Equal sweetener
8. Aluminum foil
9. Whatman no. 1 chromatographic paper, 15 x 8 cm
10. Ruler
11. Polyethylene gloves
12. Capillary tubes open on both ends
13. Heat lamp or hair dryer
14. Drying oven, 110°C
15. Wide-mouthed jar
16. Iodine crystals

## Experiment 42

### PRE-LAB QUESTIONS

1. If an amino acid moved 1.5 cm on the chromatogram and the solvent front reached 6.5 cm, what is the $R_f$ value of the amino acid?

2. All amino acids give a purple color when stained with ninhydrin. Only proline gives a yellowish color. Can you give a reason why this amino acid stains differently?

3. What happens if you don't use gloves and your finger comes in contact with the ninhydrin spray?

4. Write the structure of the mono- and dimethyl ester of aspartic acid.

5. Why must you use a pencil and not ink to mark the origin of your spot on the paper?

## Experiment 42

**REPORT SHEET**

**1.**

| Sample | Distance traveled (mm) | Solvent front (mm) | $R_f$ |
|---|---|---|---|
| Phenylalanine | | | |
| Aspartic acid | | | |
| Leucine | | | |
| Aspartame | | | |
| Hydrolyzed aspartame | | | |
| Diet Coke | | | |

**2.** Identification

    **a.** Name the amino acids you found in the hydrolysate of the sweetener Equal.

    **b.** How many spots were stained with ninhydrin (1) in Equal and (2) in Diet Coke samples?

## POST-LAB QUESTIONS

1. Could you identify all the spots in the hydrolyzed aspartame sample with those corresponding to known samples? If your answer is no, how would you guess the identity of your unknown spot?

2. In testing the hydrolysate of aspartame, you forgot to mark the position of the solvent front on your paper. Could you

   a. determine how many amino acids were in the aspartame;

   b. identify those amino acids?

3. Do you have any evidence that the aspartame was hydrolyzed during the processing and storage of the Diet Coke sample? Explain.

4. The difference between aspartic acid and phenylalanine is twofold. Aspartic acid has a polar, acidic side chain, while phenylalanine has a nonpolar side chain. The molecular weight of aspartic acid is smaller than the molecular weight of phenylalanine. Based on the $R_f$ values you obtained for these two amino acids in the solvent employed, which property influenced the rate of migration?

5. The $R_f$ value of leucine is somewhat smaller than that of phenylalanine. Both are nonpolar amino acids. Leucine has a smaller molecular mass than alanine so you would expect it to move faster. Yet it is moving slower. How could you explain your results?

# Experiment 43

## Acid–base properties of amino acids

### Background

In the body, amino acids exist as zwitterions.

$$\begin{array}{c} H \\ | \\ R-C-COO^- \\ | \\ NH_3^+ \end{array}$$

This is an amphoteric compound because it behaves as both an acid and a base in the Brønsted definition. As an acid, it can donate an $H^+$ and becomes the conjugate base:

$$\begin{array}{c} H \\ | \\ R-C-COO^- \\ | \\ NH_3^+ \end{array} + OH^- \rightleftharpoons \begin{array}{c} H \\ | \\ R-C-COO^- \\ | \\ NH_2 \end{array} + H_2O$$

Acid       Base       Conj. base    Conj. acid

As a base, it can accept an $H^+$ ion and becomes the conjugate acid:

$$\begin{array}{c} H \\ | \\ R-C-COO^- \\ | \\ NH_3^+ \end{array} + H_3O^+ \rightleftharpoons \begin{array}{c} H \\ | \\ R-C-COOH \\ | \\ NH_3^+ \end{array} + H_2O$$

Base       Acid       Conj. acid    Conj. base

To study the acid–base properties, one can perform a simple titration. We start our titration with the amino acid being in its acidic form at a low pH:

$$\begin{array}{c} H \\ | \\ R-C-COOH \\ | \\ NH_3^+ \end{array} \quad \text{(I)}$$

As we add a base, $OH^-$, to the solution, the pH will rise. We record the pH of the solution by using a pH meter after each addition of the base. To obtain the titration curve, we plot the milliliters of NaOH added against the pH of the solution (Fig. 43.1).

**Figure 43.1**
The titration curve
of an amino acid.

Note that there are two flat portions (called legs) on the titration curve where the pH does not increase appreciably with the addition of NaOH. The midpoint of the first leg, ①, is when half of the original acidic amino acid (I) has been titrated and it becomes a zwitterion (II).

$$\begin{array}{c} H \\ | \\ R-C-COO^- \quad (II) \\ | \\ NH_3^+ \end{array}$$

The point of inflection, ②, occurs when the amino acid is entirely in the zwitterion form (II). At the midpoint of the second leg, ③, half of the amino acid is in the zwitterion form and half is in the basic form (III).

$$\begin{array}{c} H \\ | \\ R-C-COO^- \quad (III) \\ | \\ NH_2 \end{array}$$

From the pH at the midpoint of the first leg we obtain the pK value of the carboxylic acid group, since this is the group that is titrated with NaOH at this stage (the structure going from I to II). The pH of the midpoint of the second leg, ③, is equal to the pK of the $-NH_3^+$, since this is the functional group that donates its $H^+$ at this stage of the titration. The pH at the inflection point, ②, is equal to the isoelectric point. At the isoelectric point of a compound, the positive and negative charges balance each other. This occurs at the inflection point when all the amino acids are in the zwitterion form.

You will obtain a titration curve of an amino acid with a neutral side chain such as glycine, alanine, phenylalanine, leucine, or valine. If pH meters are available, you read the pH directly from the instrument after each addition of the base. If a pH meter is not available, you can obtain the pH with the aid of indicator papers. From the titration curve obtained, you can determine the pK values and the isoelectric point.

1. To study acid–base properties by titration.
2. To calculate pK values for the titratable groups.

## Procedure

1. Pipet 20 mL of 0.1 M amino acid solution (glycine, alanine, phenylalanine, leucine, or valine) that has been acidified with HCl to a pH of 1.5 into a 100-mL beaker.

2. If a pH meter is available, insert the clean and dry electrode of the pH meter into a standard buffer solution with known pH. Turn the knob of the meter to the pH mark and adjust it to read the pH of the buffer. Turn the knob of the pH meter to "Standby" position. Remove the electrode from the buffer, wash it with distilled water, and dry it. Insert the dry electrode into the amino acid solution. Turn the knob of the meter to "pH" position and record the pH of the solution. Fill a buret with 0.25 M NaOH solution. Add the NaOH solution from the buret in 1.0-mL increments to the beaker. After each increment, stir the contents with a glass rod and then read the pH of the solution. Record these on your Report Sheet. Continue the titration as described until you reach pH 12. Turn off your pH meter, wash the electrode with distilled water, wipe it dry, and store it in its original buffer.

3. If a pH meter is not available, perform the titration as above, but use pH indicator papers. After the addition of each increment and stirring, withdraw a drop of the solution with a Pasteur pipet. Touch the end of the pipet to a dry piece of the pH indicator paper. Compare the color of the indicator paper with the color on the charts supplied. Read the corresponding pH from the chart and record it on your Report Sheet.

4. Draw your titration curve. From the graph, determine your pK values and the isoelectric point of your amino acid. Record these on your Report Sheet.

### Chemicals and Equipment

1. 0.1 M amino acid solution (glycine, alanine, leucine, phenylalanine, or valine)
2. 0.25 M NaOH solution
3. pH meter and standard buffer (or pH indicator paper and Pasteur pipet)
4. 50-mL buret
5. 20-mL pipet
6. Spectroline pipet filler

## Experiment 43

**PRE-LAB QUESTIONS**

**1.** In titrating the acidic form of an amino acid with NaOH solution, at which point in the titration curve does it become a zwitterion?

**2.** If the equilibrium constant, $K_a$, for the ionization of the carboxylic acid group is $1.4 \times 10^{-4}$, what is the $pK_a$?

**3.** If in a solution of alanine, the number of negatively charged carboxylate groups, $-COO^-$, is $1 \times 10^{20}$ at the isoelectric point, what is the number of positively charged amino groups, $NH_3^+$?

## Experiment 43

## REPORT SHEET

**1.** Amino acid used for titration _____

| mL of 0.25 M NaOH added | pH | mL of 0.25 M NaOH added | pH |
|---|---|---|---|
| 0 | | 13.0 | |
| 1.0 | | 14.0 | |
| 2.0 | | 15.0 | |
| 3.0 | | 16.0 | |
| 4.0 | | 17.0 | |
| 5.0 | | 18.0 | |
| 6.0 | | 19.0 | |
| 7.0 | | 20.0 | |
| 8.0 | | 21.0 | |
| 9.0 | | 22.0 | |
| 10.0 | | 23.0 | |
| 11.0 | | 24.0 | |
| 12.0 | | 25.0 | |

**2.** Plot your data below to get the titration curve.

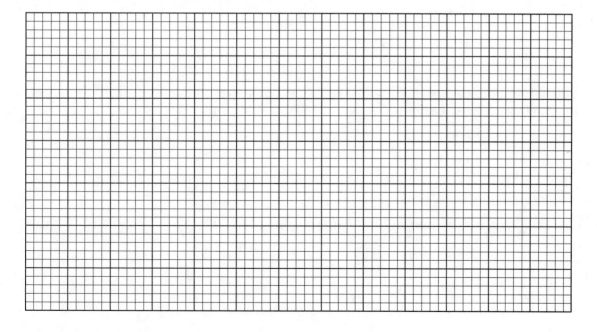

**3. a.** Indicate the positions of the midpoints of each leg and the position of the inflection point on your graph.

**b.** Record the pK values for the carboxylic acid group _____, and

for the amino group _____.

**c.** Record the pH of the isoelectric point _____.

## POST-LAB QUESTIONS

**1.** The isoelectric point of an amino acid is an intensive property.

**a.** Knowing that, would you expect to find your inflection point at a different pH value, if you had titrated 0.5 M solution of the same amino acid instead of the 0.1 M solution? Explain.

**b.** Would your result be different if you had used 50 mL of amino acid solution instead of 20 mL? Explain.

**2.** Check the pI values of the different amino acids in your textbook (Table 21.1). On the basis of the isoelectric point (pI) obtained in your experiment, how would you classify the amino acid of your experiment?

**3.** Which data can you obtain with greater accuracy from your graph—the pK values or the isoelectric point? Explain.

**4.** Write the structure of your amino acid at pH (a) 4.0 (b) 12.

## Isolation and identification of casein

Casein is the most important protein in milk. It functions as a storage protein, fulfilling nutritional requirements. Casein can be isolated from milk by acidification to bring it to its isoelectric point. At the isoelectric point, the number of positive charges on a protein equals the number of negative charges. Proteins are least soluble in water at their isoelectric points because they tend to aggregate by electrostatic interaction. The positive end of one protein molecule attracts the negative end of another protein molecule, and the aggregates precipitate out of solution.

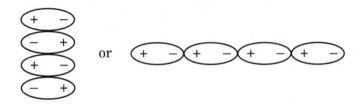

On the other hand, if a protein molecule has a net positive charge (at low pH or acidic condition) or a net negative charge (at high pH or basic condition), its solubility in water is increased.

$$\overset{+}{N}H_3 \sim COOH \xleftarrow[\text{low pH}]{H^+} \overset{+}{N}H_3 \sim COO^- \xrightarrow[\text{high pH}]{OH^-} NH_2 \sim COO^- + H_2O$$

More soluble    Least soluble    More soluble
(at isolelectric pH)

In the first part of this experiment, you are going to isolate casein from milk which has a pH of about 7. Casein will be separated as an insoluble precipitate by acidification of the milk to its isoelectric point (pH = 4.6). The fat that precipitates along with casein can be removed by dissolving it in alcohol.

In the second part of this experiment, you are going to prove that the precipitated milk product is a protein. The identification will be achieved by performing a few important chemical tests.

1. *The biuret test.* This is one of the most general tests for proteins. When a protein reacts with copper(II) sulfate, a positive test is the formation of a copper complex which has a violet color.

|              |              |                         |
| :----------: | :----------: | :---------------------: |
| Protein      | Blue color   | Protein–copper complex  |
|              |              | (violet color)          |

This test works for any protein or compound that contains two or more of the following groups:

2. *The ninhydrin test.* Amino acids with a free $-NH_2$ group and proteins containing free amino groups react with ninhydrin to give a purple-blue complex.

|              |              |
| :----------: | :----------: |
| Amino acid   | Ninhydrin    |

$+ \ RCHO \ + \ CO_2 \ + \ 3H_2O$

Purple-blue complex

3. *Heavy metal ions test.* Heavy metal ions precipitate proteins from solution. The ions that are most commonly used for protein precipitation are $Zn^{2+}$, $Fe^{3+}$, $Cu^{2+}$, $Sb^{3+}$, $Ag^+$, $Cd^{2+}$, and $Pb^{2+}$. Among these metal ions, $Hg^{2+}$, $Cd^{2+}$, and $Pb^{2+}$ are known for their notorious toxicity to humans. They can cause serious damage to proteins (especially enzymes) by denaturing them. This can result in death. The precipitation occurs because proteins become cross-linked by heavy metals as shown on the next page.

$$2NH_2 \text{~~~~~} \overset{\overset{\textstyle O}{\|}}{C}-O^- + Hg^{2+} \longrightarrow$$

Insoluble precipitate

$$2 \text{ >}SH + Hg^{2+} \longrightarrow \text{ >}S^- Hg^{2+} \text{ }^-S\text{<}$$

Insoluble precipitate

Victims swallowing $Hg^{2+}$ or $Pb^{2+}$ ions are often treated with an antidote of a food rich in proteins, which can combine with mercury or lead ions in the victim's stomach and, hopefully, prevent absorption! Milk and raw egg white are used most often. The insoluble complexes are then immediately removed from the stomach by an emetic.

4. *The xanthoprotein test.* This is a characteristic reaction of proteins that contain phenyl rings

Concentrated nitric acid reacts with the phenyl ring to give a yellow-colored aromatic nitro compound. Addition of alkali at this point will deepen the color to orange.

Tyrosine                                 Colored compound

The yellow stains on the skin caused by nitric acid are the result of the xanthoprotein reaction.

## Objectives

1. To isolate the casein from milk under isoelectric conditions.
2. To perform some chemical tests to identify proteins.

## Procedure

### Part A: Isolation of Casein

1. To a 250-mL Erlenmeyer flask, add 50.00 g of milk and heat the flask in a water bath (a 600-mL beaker containing about 200 mL of tap water; see Fig. 44.1). Stir the solution

constantly with a stirring rod. When the bath temperature has reached about 40°C, remove the flask from the water bath, and add about 10 drops of glacial acetic acid while stirring. Observe the formation of a precipitate.

**Figure 44.1**
Precipitation of casein.

2. Filter the mixture into a 100-mL beaker by pouring it through a cheese cloth which is fastened with a rubber band over the mouth of the beaker (Fig. 44.2). Remove most of the water from the precipitate by squeezing the cloth gently. Discard the filtrate in the beaker. Using a spatula, scrape the precipitate from the cheese cloth into the empty flask.

**Figure 44.2**
Filtration of casein.

3. Add 25 mL of 95% ethanol to the flask. After stirring the mixture for 5 min., allow the solid to settle. Carefully decant (pour off) the liquid that contains fats into a beaker. Discard the liquid.

4. To the residue, add 25 mL of a 1:1 mixture of diethyl ether-ethanol. After stirring the resulting mixture for 5 min., collect the solid by vacuum filtration.

> **CAUTION!**
>
> Diethyl ether is highly flammable. Make sure there is no open flame in the lab.

5. Spread the casein on a paper towel and let it dry. Weigh the dried casein and calculate the percentage of casein in the milk. Record it on your Report Sheet.

$$\% \text{ casein} = \frac{\text{weight of solid (casein)}}{50.00 \text{ g of milk}} \times 100$$

## Part B: Chemical Analysis of Proteins

1. *The biuret test.* Place 15 drops of each of the following solutions in five clean, labeled test tubes.

   **a.** 2% glycine

   **b.** 2% gelatin

   **c.** 2% albumin

   **d.** Casein prepared in Part A (one-quarter of a full spatula) + 15 drops of distilled water

   **e.** 1% tyrosine

   To each of the test tubes, add 5 drops of 10% NaOH solution and 2 drops of a dilute $CuSO_4$ solution while swirling. The development of a purplish-violet color is evidence of the presence of proteins. Record your results on the Report Sheet.

2. *The ninhydrin test.* Place 15 drops of each of the following solutions in five clean, labeled test tubes.

   **a.** 2% glycine

   **b.** 2% gelatin

   **c.** 2% albumin

   **d.** Casein prepared in Part A (one-quarter of a full spatula) + 15 drops of distilled water

   **e.** 1% tyrosine

   To each of the test tubes, add 5 drops of ninhydrin reagent and heat the test tubes in a boiling water bath for about 5 min. Record your results on the Report Sheet.

3. *Heavy metal ions test.* Place 2 mL of milk in each of three clean, labeled test tubes. Add a few drops of each of the following metal ions to the corresponding test tubes as indicated below:

   **a.** $Pb^{2+}$ as $Pb(NO_3)_2$ in test tube no. 1

   **b.** $Hg^{2+}$ as $Hg(NO_3)_2$ in test tube no. 2

   **c.** $Na^+$ as $NaNO_3$ in test tube no. 3

   Record your results on the Report Sheet.

**4.** *The xanthoprotein test.* (Perform the experiment under the hood.) Place 15 drops of each of the following solutions in five clean, labeled test tubes:

**a.** 2% glycine

**b.** 2% gelatin

**c.** 2% albumin

**d.** Casein prepared in Part A (one-quarter of a full spatula) + 15 drops of distilled water

**e.** 1% tyrosine

To each test tube, add 10 drops of concentrated $HNO_3$ while swirling. Heat the test tubes carefully in a warm water bath. Observe any change in color. Record the results on your Report Sheet.

---

**Chemicals and Equipment**

1. Hot plate
2. Büchner funnel in a no. 7 one-hole rubber stopper
3. 500-mL filter flask
4. Filter paper (Whatman no. 2, 7 cm)
5. Cheese cloth
6. Rubber band
7. Boiling chips
8. 95% ethanol
9. Diethyl ether–ethanol mixture
10. Regular milk
11. Glacial acetic acid
12. Concentrated nitric acid
13. 2% albumin
14. 2% gelatin
15. 2% glycine
16. 5% copper(II) sulfate
17. 5% lead(II) nitrate
18. 5% mercury(II) nitrate
19. Ninhydrin reagent
20. 10% sodium hydroxide
21. 1% tyrosine
22. 5% sodium nitrate

## Experiment 44

**PRE-LAB QUESTIONS**

**1.** Casein has an isoelectric point at pH 4.6. What kind of charges will be on the casein in its native environment, that is, in milk?

**2.** How do you separate the fat from the protein in the casein precipitate?

**3.** Besides tyrosine, what other amino acid would give a positive xanthoprotein test? Give the formula of the expected yellow-colored compound.

**4.** What are the three most toxic heavy metal ions?

## Experiment 44

### REPORT SHEET

#### Isolation of casein

1. Weight of milk _____ g
2. Weight of dried casein _____ g
3. Percentage of casein in milk _____ %

#### Chemical analysis of proteins

**Biuret test**

| Substance | Color formed |
|-----------|--------------|
| 2% glycine | |
| 2% gelatin | |
| 2% albumin | |
| casein + $H_2O$ | |
| 1% tyrosine | |

Which of these chemicals gives a positive test with this reagent? _____

**Ninhydrin test**

| Substance | Color formed after heating |
|-----------|----------------------------|
| 2% glycine | |
| 2% gelatin | |
| 2% albumin | |
| casein + $H_2O$ | |
| 1% tyrosine | |

Which of these chemicals gives a positive test with this reagent? _____

**Heavy metal ion test**

| Substance | Precipitates formed |
|---|---|
| Pb(NO$_3$)$_2$ | |
| Hg(NO$_3$)$_2$ | |
| NaNO$_3$ | |

Which of these metal ions gives a positive test with casein in milk?  _____

**Xanthoprotein test**

| Substance | Color formed before or after heating |
|---|---|
| 2% glycine | |
| 2% gelatin | |
| 2% albumin | |
| casein + H$_2$O | |
| 1% tyrosine | |

Which of these chemicals gives a positive test with this reagent?  _____

## POST-LAB QUESTIONS

1. According to the equation of the ninhydrin reaction given in the **Background** no. 2, one of the products is CO$_2$. Where does the CO$_2$ come from?

2. In the isolation of casein following the acidification, you removed the precipitate by filtering through a cheese cloth and squeezing the cloth. If you did not squeeze out all the liquids, would your yield of casein be different? Explain.

3. Does gelatin contain tyrosine? Explain.

**4.** If by mistake (*don't try it*) your finger touches nitric acid and you observe a yellow color on your fingers, what functional group(s) in your skin is (are) responsible for this reaction?

**5.** Why is milk or raw egg used as an antidote in cases of heavy metal ion poisoning?

**6.** According to your results, how many grams of casein are in a glass of milk (175 g)?

# Neurotransmission: an example of enzyme specificity

When our brain wants to communicate a message to the skeletal muscles or to the sympathetic ganglia, the nerve cells use acetylcholine as the neurotransmitter. This messenger is usually stored in vesicles within the neurons. Upon a signal, when the calcium ion concentration in the neurons increases, the vesicles migrate to the cell membrane. There the two membranes fuse and acetylcholine is dumped into the synapse between the neuron and the muscle cell. Acetylcholine then rapidly crosses the synapse and enters the receptor located on the muscle cell. The presence of the acetylcholine triggers a conformational change in the receptor that allows the passage of $Na^+$ and $K^+$ ions to cross the membrane of the muscle cell. The $Na^+$ flow in and the $K^+$ flow out of the muscle cell through the channels opened within the receptor. This sets up a potential across the membrane. In this manner the chemical signal of the messenger acetylcholine is converted into an electric signal. The electric signal can induce the muscles to contract. All this happens within a few milliseconds. The channel in the receptor closes again.

In special cases, the electricity is stored in columns of cells called electroplax. This happens in the electric eel, Torpedo, which can then dischange the electricity, killing or maiming its enemy.

In order that a new signal be transmitted and the channel reopen, the acetylcholine must be removed from the receptor. This is accomplished by a specific enzyme acetylcholine esterase. Acetylcholine esterase is a hydrolase and catalyzes the hydrolysis of the ester bond yielding choline and acetate:

$$CH_3-\overset{\overset{O}{\|}}{C}-O-CH_2-CH_2-\overset{\overset{CH_3}{|}}{\underset{\underset{CH_3}{|}}{N^+}}-CH_3 + H_2O \longrightarrow CH_3-\overset{\overset{O}{\|}}{C}-O^- + HO-CH_2-CH_2-\overset{\overset{CH_3}{|}}{\underset{\underset{CH_3}{|}}{N^+}}-CH_3$$

Acetylcholine        Acetate        Choline

The main feature of acetylcholine esterase is its specificity and speed of action. It can hydrolyze an acetylcholine molecule in 40 microseconds. The neurons thus can transmit 1000 signals per second, because the muscle membrane recovers its resting potential within a fraction of a millisecond.

Other esterases have much less specificity and can hydrolyze many different esters. One such enzyme is the carboxyl esterase of liver. Even though the two enzymes hydrolyze the same ester bond, they are completely different in their primary, secondary, and tertiary structure, and even in their active sites. We shall compare the specificity of these two enzymes by measuring their activity against two substrates, (a) acetylcholine and (b) o-nitrophenylacetate.

o-Nitrophenylacetate        Acetic acid        o-Nitrophenol

The way we measure the activity is a colorimetric technique. With acetylcholine as a substrate we add a yellow-colored compound, $m$-nitrophenol, to the reaction mixture. $m$-Nitrophenol absorbs light in the 400–440 nm range. As the reaction proceeds, acetic acid and choline are liberated upon hydrolysis. The increase in the intensity of the color of $m$-nitrophenol at 420 nm is in proportion to the choline concentration liberated. Thus the rate of change in absorption is a measure of the enzyme activity. We shall define the activity as the change in absorbance/min/mg enzyme.

The colorimetric technique with $o$-nitrophenyl acetate does not require an indicator. $o$-Nitrophenyl acetate is colorless and does not absorb at 420 nm. On the other hand, the liberated $o$-nitrophenol gives an intense yellow color and the absorbance at 420 nm is proportional to the $o$-nitrophenol liberated. Again, we shall define the activity as the change in absorbance/min/mg enzyme.

## Objectives

1. To study the specificity of two esterases toward different substrates.
2. To learn to measure enzyme activity by colorimetric technique.
3. To learn the use of micropipetes.

## Procedure

Before the start of experiments, your instructor will demonstrate to you the use of delivery pipets with 1–200 μL pipet tips.

1. Turn on the spectrophotometer and let it warm up for a few minutes. Turn the wavelength control knob to read 420 nm. With no sample tube in the sample compartment, adjust the amplifier control knob to read 0% transmittance or infinite absorbance. Insert the sample tube containing 2 mL distilled water into the spectrophotometer. Adjust the reading to 100% transmittance (or 0 absorbance). This zeroing must be performed every 15 min. before each enzyme activity run because some instruments have a tendency to drift.

2. Place 2.0 mL of $m$-nitrophenol solution at pH 8.0 into the sample tube. Add 0.4 mL acetylcholine solution. Mix. Read the absorbance of the solution at 420 nm. Record it on your Report Sheet (1).

3. Add 0.6 mL diluted acetylcholine esterase solution to the sample tube containing acetylcholine. Mix it. Note the exact time of addition (hour, minute, second) on your Report Sheet (2).

4. Read the absorbance of your sample 30 sec. after the addition and subsequently 1, 2, 4, 6, and 10 min. after the addition. Record these absorbances on your Report Sheet (3).

5. With 2 mL distilled water in the sample tube, zero the spectrophotometer as in step no. 1.

6. Place 2.0 mL of *m*-nitrophenol solution at pH 8.0 into the sample tube. Add 0.4 mL acetylcholine solution. Mix. Read the absorbance of the solution at 420 nm. Record it on your Report Sheet (4).

7. Add 0.51 mL of 0.15 M NaCl solution at pH 8.0 to the sample. Mix. Add 0.09 mL of the carboxyl esterase suspension to the sample tube containing acetylcholine. Note the exact time of addition (hour, minute, second) on your Report Sheet (5).

8. Read the absorbances of the sample 30 sec. after the addition. Repeat the readings at 1, 2, 4, 6 and 10 min. Record the absorbances on your Record Sheet (6).

9. Plot your data: Absorbance vs. Time for both enzymes on the graph paper provided on your Report Sheet (7).

10. With 2 mL distilled water in your sample tube, zero your spectrophotometer as in step no. 1.

11. Take 2.0 mL of *o*-nitrophenylacetate solution and read its absorbance. Record it on your Report Sheet (8).

12. Add 0.6 mL diluted acetylcholine esterase. Mix. Note the exact time of addition (hour, minute, second) on your Report Sheet (9).

13. Read the absorbance of the sample 30 sec. after addition. Repeat the readings at 1, 2, 3, 5, 7 and 10 min. Record the absorbances on your Report Sheet (10).

14. With 2 mL distilled water in your sample tube, zero your spectrophotometer as in step no. 1.

15. Take 2.0 mL of *o*-nitrophenylacetate solution. Read its absorbance and record it on your Report Sheet (11).

16. Add 0.58 mL of 0.15 M NaCl solution at pH 8.0. Mix. Add 0.02 mL carboxyl esterase suspension. Mix. Note the exact time of the carboxyl esterase addition and record it (hour, minute, second) on your Report Sheet (12).

17. Read the absorbance of the sample 30 sec. after the addition of the enzyme. Repeat the readings at 1, 2, 3, 4, and 5 min. Record the absorbances on your Report Sheet (13).

18. Plot your data on the *o*-nitrophenyl acetate experiments for both enzymes (14).

## Chemicals and Equipment

1. Spectrophotometer
2. Delivery pipet
3. Pipet tips (1–200 μL)
4. 0.02 M NaOH
5. 0.15 M NaCl at pH 8.0
6. *m*-Nitrophenol solution
7. Acetylcholine solution
8. *o*-Nitrophenyl acetate solution
9. Acetylcholine esterase
10. Carboxyl esterase

## Experiment 45

## PRE-LAB QUESTIONS

1. Acetylcholine is sold as acetylcholine chloride. Looking at the structures given in the **Background**, write the formula for the anion and the cation of this salt.

2. The molecular weight of acetylcholine chloride is 181.7. A stock solution of 1.817 g acetylcholine in 10.0 mL was prepared by the stockroom. What is the concentration of this stock solution in % w/v and in molarity? From this stock solution of acetylcholine chloride, 0.40 mL was added to the sample tube bringing the total volume in the sample tube to 3.0 mL. What is the concentration of the acetylcholine chloride in the sample tube in g/mL and in molarity?

3. Acetylcholine esterase, 3.0 mg, was dissolved in 10.0 mL solution. From this solution, 0.60 mL was added to the sample tube, bringing the volume to a total of 3.0 mL. How many mg of acetylcholine esterase did you have in the sample tube?

4. The commercial sample of carboxyl esterase contained 15.0 mg of enzyme in 0.30 mL of suspension. In your experiment, you add 0.090 mL of this suspension to the sample tube that contains a total volume of 3.0 mL. How many mg of carboxyl esterase enzyme do you have in the sample tube?

## Experiment 45

### REPORT SHEET

1. Absorbance of the acetylcholine in $m$-nitrophenol solution _____

2. Time of addition of acetylcholine esterase solution _____

3. Absorbance of the mixture

   after 30 sec. _____  after 4 min. _____

   after 1 min. _____  after 6 min. _____

   after 2 min. _____  after 10 min. _____

4. Absorbance of the acetylcholine in $m$-nitrophenol solution _____

5. Time of addition of carboxyl esterase solution _____

6. Absorbance of the mixture

   after 30 sec. _____  after 4 min. _____

   after 1 min. _____  after 6 min. _____

   after 2 min. _____  after 10 min. _____

7. Plot your data: Absorbance vs. Time for both enzyme solutions

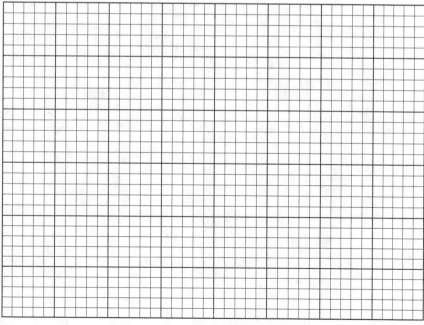

**8.** Absorbance of the a *o*-nitrophenyl acetate solution _____

**9.** Time of addition of acetylcholine esterase solution _____

**10.** Absorbance of the mixture _____

after 30 sec. _____     after 5 min. _____

after 1 min. _____     after 7 min. _____

after 2 min. _____     after 10 min. _____

after 3 min. _____

**11.** Absorbance of *o*-nitrophenyl acetate solution _____

**12.** Time of the addition of carboxyl esterase solution _____

**13.** Absorbance of the mixture _____

after 30 sec. _____     after 3 min. _____

after 1 min. _____     after 4 min. _____

after 2 min. _____     after 5 min. _____

**14.** Plot your data: Absorbance vs. Time for both enzymes

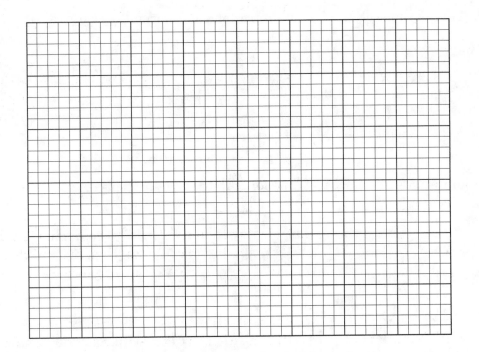

## POST-LAB QUESTIONS

1. Calculate the enzyme activities with acetylcholine as the substrate for both enzymes. Using the change in absorbencies between 30 sec. and 2 min., calculate the rate of reaction as change ($\Delta$) in absorbance/min. for

   a. acetylcholine esterase

   b. carboxyl esterase

   c. You have calculated the amount of acetylcholine esterase that you had in the sample tube (**Pre-Lab Question** no. 3). Use this number to calculate the activity of this enzyme toward acetylcholine in $\Delta$ absorbence/min/mg enzyme.

   d. You have calculated the amount of carboxyl esterase that you had in the sample tube (**Pre-Lab Question** no. 4). Use this number to calculate the activity of this enzyme toward acetylcholine in $\Delta$ absorbence/min/mg enzyme.

   e. Which enzyme catalyzes the hydrolysis of acetylcholine faster? How many times faster?

2. Calculate the enzyme activities of $o$-nitrophenyl acetate as a substrate for both enzymes. Using the change in absorbence between 30 sec. and 2 min. in the reaction, calculate the rate of reaction toward $o$-nitrophenyl acetate in $\Delta$ absorbence/min. for
   a. acetylcholine esterase

   b. carboxyl esterase

   c. You have calculated the amount of acetylcholine esterase in your sample tube (**Pre-Lab Question** no. 3). Use this number to calculate the activity of this enzyme toward $o$-nitrophenyl acetate in $\Delta$ absorbence/min/mg enzyme.

   d. You have calculated the amount of carboxyl esterase in your sample tube (**Pre-Lab Question** no. 4). Use this number to calculate the activity of this enzyme toward $o$-nitrophenyl acetate in $\Delta$ absorbence/min/mg enzyme.

   e. Which enzyme catalyzes the hydrolysis of $o$-nitrophenyl faster? How many times faster?

# Experiment 46

## Isolation and identification of DNA from onion

### Background

Hereditary traits are transmitted by genes. Genes are parts of giant deoxyribonucleic acid (DNA) molecules. In lower organisms, such as bacteria and yeast, both DNA and RNA (ribonucleic acid) occur in the cytoplasm while in higher organisms, most of the DNA is inside the nucleus and the RNA is outside the nucleus, in other organelles, and in the cytoplasm.

In this experiment, we will isolate DNA molecules from onion cells. The first task is to break up the cells. This is achieved by a combination of different techniques and agents. Grinding up the cells with sand disrupts them and the cytoplasm of many onion cells is spilled out. However, this is not a complete process. The addition of a detergent accomplishes two functions: 1) it helps to solubilize cell membranes and thereby further weakens the cell structures, and 2) it helps to inactivate the nucleic acid degrading enzymes, nucleases, that are present. Without this inhibition, the nucleases would degrade the nucleic acids to their constituent nucelotides.

Once the nucleic acids are in solution, they must be separated from the other constituents of the cell. These are proteins and small water-soluble molecules. The addition of NaCl solution minimizes the interaction between the positively charged histone (protein) molecules and the negatively charged DNA.

The addition of ethanol precipitates the large molecules (DNA, RNA, and proteins) and leaves the small molecules in solution. DNA, being the largest fibrous molecule, forms thread-like precipitates that can be spooled off onto a rod. The protein and RNA form a gelatinous precipitate that cannot be picked up by winding them on a glass rod. Thus, the spooling separates DNA from RNA and proteins. This is a crude separation and the DNA thus isolated may contain some contaminant proteins. In order that you should be able to follow the step-by-step procedure, a flow chart of the DNA isolation process is provided here.

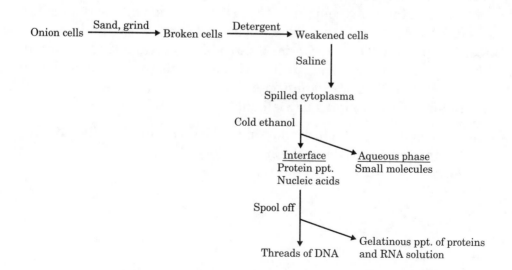

After the isolation of DNA, we will probe its identity by the diphenylamine test. The blue color of this test is specific for deoxyribose. The appearance of blue color can be used to identify the deoxyribose containing DNA molecule. The diphenylamine reagent contains small amounts of acetaldehyde. This compound is an eye irritant, although not in the concentration that exists in the reagent. Still, some stockrooms prefer not to handle acetaldehyde and prepare a less selective diphenylamine test without acetaldehyde in which the yellow-green color of deoxyribose appears very fast, while other sugars, glucose, ribose, etc., give faint coloration that evolves slowly.

## Objectives

1. To demonstrate the separation of DNA molecules from other cell constituents.
2. To prove their identity.

## Procedure

1. Cool a mortar in ice water. Add 20 to 30 g of diced yellow onion and 20 g of acid-washed sand. Grind the onion and the sand vigorously with a pestle for 5 min. to disrupt the cells. A yellowish mush will form.

2. Using a hot plate, warm 100 mL distilled water to 60°C. Add 10 g NaCl and add 10 mL of detergent. Stir to dissolve.

3. Add the ground onion and sand mush to 30 mL of the saline-detergent solution. Using a glass rod, stir the solution for 5 min., all the while maintaining the temperature at approximately 60°C.

4. Support a funnel with a ring clamp on a ring stand above a 150 mL beaker. Add cheesecloth to the funnel and decant the cell suspension onto the cheesecloth. Let the yellowish liquid filter into the beaker, while gently squeezing the cheesecloth to obtain the maximum yield, leaving the cellular debris and the sand behind. Cool the solution to room temperature.

5. To the viscous DNA containing aqueous solution, slowly add twice its volume of cold absolute ethanol, taking care that the alcohol flows along the side of the beaker, settling on top of the aqueous solution. *The next step is critical.* Insert a flame-sterilized glass rod gently into the solution reaching **just below the interface of the DNA-alcohol solution.** The DNA forms a thread-like precipitate at the interface. *Rotate (do not stir) the glass rod* in one direction (clockwise or counterclockwise). The rotation spools all the DNA precipitate onto the glass rod. Transfer the spooled DNA on the rod into a test tube containing 95% ethanol.

6. Discard the alcohol/detergent solution left in the beaker into specially labelled waste jars. Do not pour the solution down the sink.

7. Remove the rod and the spooled DNA from the test tube. Dry the DNA with a clean filter paper. Note its appearance. Dissolve the isolated crude DNA in 2 mL citrate buffer (0.15 M NaCl, 0.015 M sodium citrate).

(a) Set up 4 dry and clean test tubes. Into the test tubes add 2 mL each of the following:

| Test tube | Solution |
|---|---|
| 1 | 1% glucose |
| 2 | 1% ribose |
| 3 | 1% deoxyribose |
| 4 | crude DNA solution |

Add 5 mL diphenylamine reagent to each test tube.

---

**CAUTION!**

Diphenylamine reagent contains glacial acetic acid and concentrated sulfuric acid. Handle with care. Use gloves.

---

Mix the contents of the test tubes. Heat the test tubes in a boiling water bath for 10 min. Note the color.

(b) As an alternative procedure, one can use 1% diphenylamine dissolved in concentrated sulfuric acid. Set up the same 4 test tubes as above under (a) but add only 1 mL of solutions to each.

Keep the test tubes in a test tube rack pointed **away from your face. Using gloves,** add slowly, **dropwise**, from a pipet, a maximum of 1 mL of 1% diphenylamine solution to the first test tube. Stop adding diphenylamine reagent when coloration appears. Holding the top of the test tube between your fingers, swirl it slowly after each addition. The addition of concentrated sulfuric acid to the aqueous solution generates heat that helps the development of the color. **Do not touch the bottom of the test tubes. They are hot.** Note the color and estimate in seconds how fast it develops.

Repeat the procedure with the three other test tubes.

---

**Chemicals and Equipment**

1. Yellow onion
2. Sand
3. Mortar and pestle
4. Ivory liquid dishwashing detergent
5. NaCl
6. Absolute ethanol
7. Citrate buffer
8. Glucose solution
9. Ribose solution
10. Deoxyribose solution
11. Diphenylamine reagent
12. 95% ethanol

**Experiment 46**

## PRE-LAB QUESTIONS

Consult Chapter 24 of your textbook to answer the structural questions.

**1.** What are histones? What is a nucleosome? Where is DNA located in the nucleosome?

**2.** Draw the structures of ribose and deoxyribose.

**3.** Why must you handle the diphenylamine reagent with great care?

**4.** The most demanding part of this experiment is grinding the onion cells with sand. What does this process accomplish? Would you be able to isolate DNA without this grinding?

## Experiment 46

**REPORT SHEET**

1. Describe the appearance of the crude DNA preparation.

2. Diphenylamine test

   | *Solution* | *Color* |
   |---|---|
   | 1% glucose | _____ |
   | 1% ribose | _____ |
   | 1% deoxyribose | _____ |
   | crude DNA sample | _____ |

   Did the diphenylamine test confirm the identity of DNA?

3. How did you minimize the interaction between the DNA and histones (proteins)?

4. What compounds were left behind in the ethanol solution after spooling the DNA?

## POST-LAB QUESTIONS

1. Can the diphenylamine reagent distinguish between ribose and deoxyribose, and between DNA and RNA?

2. What was the function of the detergent in isolating DNA from onion cells?

3. Why can we isolate DNA from the precipitate, which also contains RNA and proteins, by the simple "spooling" procedure?

# Experiment 47

## Viscosity and secondary structure of DNA

In 1953, Watson and Crick proposed a three-dimensional structure of DNA which is a cornerstone in the history of biochemistry and molecular biology. The double helix they proposed for the secondary structure of DNA gained immediate acceptance, partly because it explained all known facts about DNA, and partly because it provided a beautiful model for DNA replication.

In the DNA double helix, two polynucleotide chains run in opposite directions. This means that at each end of the double helix there is one 5'-OH and one 3'-OH terminal. The sugar phosphate backbone is on the outside, and the bases point inward. These bases are paired so that for each adenine (A) on one chain a thymine (T) is aligned opposite it on the other chain. Each cytosine (C) on one chain has a guanine (G) aligned with it on the other chain. The AT and GC base pairs form hydrogen bonds with each other. The AT pair has two hydrogen bonds; the GC pair has three hydrogen bonds (Fig. 47.1).

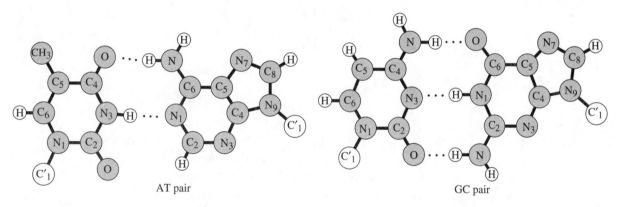

AT pair                    GC pair

**Figure 47.1** • Hydrogen bonding between base pairs.

Most of the DNA in nature has the double helical secondary structure. The hydrogen bonds between the base pairs provide the stability of the double helix. Under certain conditions the hydrogen bonds are broken. During the replication process itself, this happens and parts of the double helix unfold. Under other conditions, the whole molecule unfolds, becomes single stranded, and assumes a random coil conformation. This can happen in denaturation processes aided by heat, extreme acidic or basic conditions, etc. Such a transformation is often referred to as helix-to-coil transition. There are a number of techniques that can monitor such a transition. One of the most sensitive is the measurement of viscosity of DNA solutions.

Viscosity is the resistance to flow of a liquid. Honey has a high viscosity and gasoline a low viscosity, at room temperature. In a liquid flow, the molecules must slide past each other. The resistance to flow comes from the interaction between the molecules as they slide past each other. The stronger this interaction, i.e., hydrogen bonds vs. London dispersion forces, the greater the resistance and the higher the viscosity. Even more than the nature of the intermolecular interaction, the size and the shape of the molecules influence their viscosity. A large molecule has greater surface over which it interacts with other molecules than a small molecule. Therefore, its viscosity is greater than that of a small molecule. If two molecules have the same size and the same interaction forces but have different shapes, their viscosity will be different. For example, needle-shaped molecules, when aligned parallel by the flow of liquid, have greater surfaces of interaction than spherical molecules of the same molecular weight (Fig. 47.2). The needle-shaped molecule will have a higher viscosity than the spherical molecule. The DNA double helix is a rigid structure held together by hydrogen bonds. Its long axis along the helix exceeds by far its short axis perpendicular to it. Thus the DNA double helix has large surface area and consequently high viscosity. When the hydrogen bonds are broken and the DNA molecule becomes single stranded, it assumes a random coil shape which has much lower surface area and lower viscosity. Thus a helix-to-coil transition is accompanied by a drop in viscosity.

**Figure 47.2**

Surface area of interaction between molecules of different shapes.

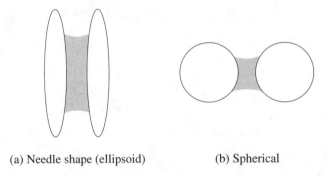

(a) Needle shape (ellipsoid)          (b) Spherical

In practice, we can measure viscosity by the efflux time of a liquid in a viscometer (Fig. 47.3). The capillary viscometer is made of two bulbs connected by a tube in which the liquid must flow through a capillary tube. The capillary tube provides a laminary flow in which concentric layers of the liquid slide past each other. Originally, the liquid is placed in the storage bulb (A). By applying suction above the capillary, the liquid is sucked up past the upper calibration mark. With a stopwatch in hand, the suction is released and the liquid is allowed to flow under the force of gravity. The timing starts when the meniscus of the liquid hits the upper calibration mark. The timing ends when the meniscus of the liquid hits the lower calibration mark of the viscometer. The time elapsed between these two marks is the efflux time.

**Figure 47.3**
Ostwald capillary viscometer.

Storage bulb A

Calibration marks

Bulb B

Capillary tube

With dilute solutions, such as the DNA in this experiment, the viscosity of the solution is compared to the viscosity of the solvent. The efflux time of the solvent, aqueous buffer, is $t_o$ and that of the solution is $t_s$. The relative viscosity of the solution is

$$\eta_{\text{rel}} = t_s/t_o$$

The viscosity of a solution also depends on the concentration; the higher the concentration, the higher the viscosity. In order to make the measurement independent of concentration, a new viscometric parameter is used which is called intrinsic viscosity, $[\eta]$. This number is

$$[\eta] = (\log \eta_{\text{rel}})/c$$

which is almost a constant for a particular solute (DNA in our case) in very dilute solutions.

In this experiment, we follow the change in the viscosity of a DNA solution when we change the pH of the solution from the very acidic (pH 2.0) to very basic (pH 12.0). At extreme pH values, we expect that the hydrogen bonds will break and the double helix will become single-stranded random coils. A change in the viscosity will tell at what pH this happens. We shall also determine whether two acid-denatured single-stranded DNA molecules can refold themselves into a double helix when we neutralize the denaturing acid.

---

### Objectives

1. To demonstrate helix-to-coil-to-helix transitions.
2. To learn how to measure viscosity.

---

*Because of the cost of viscometers the students may work in groups of 5–6.*

1. To 3 mL of a buffer solution, add 1 drop of 1.0 M HCl using a Pasteur pipet. Measure its pH with a universal pH paper. If the pH is above 2.5, add another drop of 1 M HCl. Measure the pH again. Record the pH on your Report Sheet (1).

2. Clamp one clean and dry viscometer on a stand. Pipet 3 mL of your acidified buffer solution into bulb A of your viscometer. Using a suction bulb of a Spectroline pipet filler, raise the level of the liquid in the viscometer above the upper calibration mark. Release the suction by removing the suction bulb and time the efflux time between the two calibration marks. Record this as $t_o$ on your Report Sheet (2). Remove all the liquid from your viscometer by pouring the liquid out from the wide arm. Then apply pressure with the suction bulb on the capillary arm of the viscometer and blow out any remaining liquid into the storage bulb (A); pour out this residual liquid.

3. Take 3 mL of the prepared DNA solution. Add the same amount of 1 M HCl as above (1 or 2 drops). Mix it thoroughly by shaking the solution. Test the pH of the solution with a universal pH paper and record the pH (3) and the DNA concentration of the prepared solution on your Report Sheet (4).

4. Pour the acidified DNA solution into the wide arm (bulb A) of your viscometer. Using a suction bulb, raise the level of your liquid above the upper calibration mark. Release the suction by removing the suction bulb and measure and record the efflux time of the acidified DNA solution (5).

5. Add the same amount (1 or 2 drops) as above of neutralizing 1 M NaOH solution to the liquid in the wide arm of your viscometer. With the suction bulb on the capillary arm blow a few air bubbles through the solution to mix the ingredients. Repeat the measurement of the efflux time and record it on your Report Sheet (6). For the next 100 min. or so, repeat the measurement of the efflux times every 20 min. and record the results on your Report Sheet (7–11).

6. While the efflux time measurements in viscometer no. 1 are repeated every 20 min., another dry and clean viscometer will be used for establishing the pH dependence of the viscosity of DNA solutions. First, measure the pH of the buffer solution with a universal pH paper. Record it on your Report Sheet (12). Second, transfer 3 mL of the buffer into the viscometer no. 2 and measure and record its efflux time (13). Empty the viscometer as instructed in no. 2 above. Test the pH of the DNA solution with a universal pH paper (14) and transfer 3 mL into the viscometer. Measure its efflux time and record it on your Report Sheet (15). Empty your viscometer.

7. Repeat the procedure described in step no. 6, but this time, with the aid of a Pasteur pipet, add one drop of 0.1 M HCl both to the 3-mL buffer solution, as well as to the 3-mL DNA solution. Measure the pH and the efflux times of both buffer and DNA solutions and record them (16–19) on your Report Sheet. *Make sure that you empty the viscometer after each viscosity measurement.*

8. Repeat the procedure described in step no. 6, but this time add one drop of 0.1 M NaOH solution to both the 3-mL buffer and 3-mL DNA solutions. Measure their pH and efflux times and record them on your Report Sheet (20–23).

9. Repeat the procedure described in step no. 6, but this time add 2 drops of 1 M NaOH to both buffer and DNA solutions (3 mL of each solution). Measure and record their pH and efflux times on your Report Sheet (24–27).

10. If time allows, you may repeat the procedure at other pH values; for example, by adding two drops of 1 M HCl (28–31), or two drops of 0.1 M HCl (32–35), or two drops of 0.1 M NaOH (36–39) to the separate samples of buffer and DNA solutions.

## Chemicals and Equipment

1. Viscometers, 3-mL capacity
2. Stopwatch or watch with a second hand
3. Stand with utility clamp
4. Pasteur pipets
5. Buffer at pH 7.0
6. Prepared DNA solution
7. 1 M HCl
8. 0.1 M HCl
9. 1 M NaOH
10. 0.1 M NaOH
11. Spectroline pipet fillers

## Experiment 47

**PRE-LAB QUESTIONS**

**1.** What is an efflux time?

**2.** Show the GC base pair structure before and after the addition of an acid.

**3.** Viscosity increases when the surface interaction between the molecules sliding past each other increases. Predict which will have greater viscosity:
   **a.** water or $CH_3Cl$

   **b.** an amino acid (glycine) or protein (albumin)

   **c.** tRNA (molecular weight $3 \times 10^4$) or DNA (molecular weight $1 \times 10^8$)

**4.** Show the GC base pair structure after addition of a base, $OH^-$, to the acidified DNA.

**5.** Which will have a greater surface of interaction—DNA in a double helix or the same DNA denatured, single-stranded random coil? Justify your answer with a diagram of helix-to-coil transition.

## Experiment 47

**REPORT SHEET**

1. pH of acidified buffer                           _____

2. Efflux time of acidified buffer                  _____ sec.

3. pH of acidified DNA solution                     _____

4. Concentration of DNA solution                    _____

5. Efflux time of acidified DNA solution            _____ sec.

6. Efflux time of neutralized DNA solution
   at time of neutralization                        _____ sec.

7. 20 min. later                                    _____ sec.

8. 40 min. later                                    _____ sec.

9. 60 min. later                                    _____ sec.

10. 80 min. later                                   _____ sec.

11. 100 min. later                                  _____ sec.

12. pH of neutral buffer                            _____

13. Efflux time of neutral buffer                   _____ sec.

14. pH of DNA solution in neutral buffer            _____

15. Efflux time of DNA in neutral buffer            _____ sec.

**After addition of 1 drop of 0.1 M HCl**

16. pH of buffer                                    _____

17. Efflux time of buffer                           _____ sec.

18. pH of DNA solution                              _____

19. Efflux time of DNA solution                     _____ sec.

**After addition of 1 drop of 0.1 M NaOH**

20. pH of buffer        _____

21. Efflux time of buffer        _____ sec.

22. pH of DNA solution        _____

23. Efflux time of DNA solution        _____ sec.

**After addition of 2 drops of 1 M NaOH**

24. pH of buffer        _____

25. Efflux time of buffer        _____ sec.

26. pH of DNA solution        _____

27. Efflux time of DNA solution        _____ sec.

**After addition of 2 drops of 1 M HCl**

28. pH of buffer        _____

29. Efflux time of buffer        _____ sec.

30. pH of DNA solution        _____

31. Efflux time of DNA solution        _____ sec.

**After addition of 2 drops of 0.1 M HCl**

32. pH of buffer        _____

33. Efflux time of buffer        _____ sec.

34. pH of DNA solution        _____

35. Efflux time of DNA solution        _____ sec.

**After addition of 2 drops of 0.1 M NaOH**

36. pH of buffer        _____

37. Efflux time of buffer        _____ sec.

38. pH of DNA solution        _____

39. Efflux time of DNA solution        _____ sec.

Tabulate your data on the pH dependence of relative viscosity.

| pH | $\eta_{rel}$ |
|---|---|
| (3) _____ | (5)/(2) _____ |
| (14) _____ | (15)/(13) _____ |
| (18) _____ | (19)/(17) _____ |
| (22) _____ | (23)/(21) _____ |
| (26) _____ | (27)/(25) _____ |
| (30) _____ | (31)/(29) _____ |
| (34) _____ | (35)/(33) _____ |
| (38) _____ | (39)/(37) _____ |

## POST-LAB QUESTIONS

1. Plot your tabulated data—relative viscosity on the y-axis and pH on the x-axis.

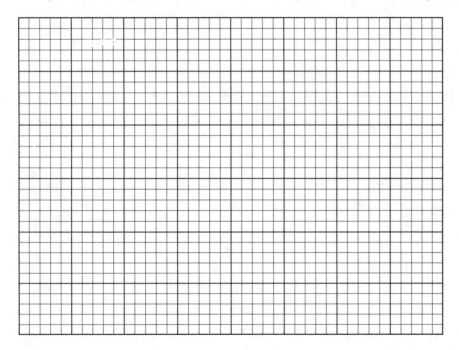

2. At what pH values did you observe helix-to-coil transitions?

**3.** Plot your data on the refolding of DNA double helix (5)–(11). Plot the time on the x-axis and the efflux times on the y-axis.

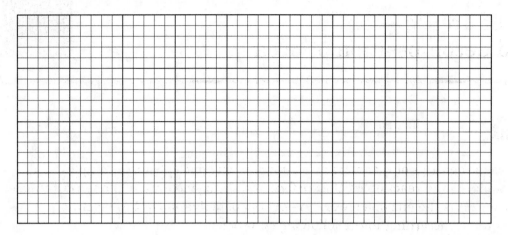

**4.** Was there any indication that, upon neutralization of the denaturing acid, the DNA did refold into a double helix? Explain.

**5.** Compare the efflux time of the neutral DNA (15) to that of the denatured DNA 100 min. after neutralization (11). What does the difference between these two efflux times tell you regarding the refolding process?

**6.** Calculate the intrinsic viscosity of your DNA at

   **a.** neutral pH = 2.3 × {log[(15)/(13)]}/(4) =

   **b.** acidic pH = 2.3 × {log[(5)/(2)]}/(4) =

   **c.** basic pH = 2.3 × {log[(27)/(25)]}/(4) =

   **d.** neutralized pH 100 min. after neutralization = 2.3 × {log[(11)/(13)]}/(4) =

**7.** A high intrinsic viscosity implies a double helix, a low intrinsic viscosity means a random coil. What do you think is the shape of the DNA after acid denaturation and subsequent neutralization? [See **6d.**] Explain your answer.

# Kinetics of urease—
catalyzed decomposition of urea

Enzymes speed up the rates of reactions by forming an enzyme-substrate complex. The reactants can undergo the reaction on the surface of the enzyme, rather than finding each other by collision. Thus the enzyme lowers the energy of activation of the reaction.

Urea decomposes according to the following equation:

$$H_2N-\overset{\overset{\displaystyle O}{\|}}{C}-NH_2 + H_2O \rightleftharpoons CO_2 + 2NH_3 \tag{1}$$

This reaction is catalyzed by a highly specific enzyme, urease. Urease is present in a number of bacteria and plants. The most common source of the enzyme is jack bean or soybean. Urease was the first enzyme that was crystallized. Sumner, in 1926, proved unequivocally that enzymes are protein molecules.

Urease is an -SH group (thiol) containing enzyme. The cysteine residues of the protein molecule must be in the reduced -SH form in order for the enzyme to be active. Oxidation of these groups will form -S-S-, disulfide bridges, and the enzyme loses its activity. Reducing agents such as cysteine or glutathione can reactivate the enzyme.

Heavy metals such as $Ag^+$, $Hg^{2+}$, or $Pb^{2+}$, which form complexes with the -SH groups, also inactivate the enzyme. For example, the poison phenylmercuric acetate is a potent inhibitor of urease.

$$\text{enzyme}-SH + CH_3-\overset{\overset{\displaystyle O}{\|}}{C}-O^-Hg^+C_6H_5 \rightleftharpoons CH_3\overset{\overset{\displaystyle O}{\|}}{C}-OH + \text{enzyme}-S-Hg-C_6H_5 \tag{2}$$

| Active | Phenylmercuric acetate | Acetic acid | Inactive |

In this experiment, we study the kinetics of the urea decomposition. As shown in equation (1), the products of the reaction are carbon dioxide, $CO_2$, and ammonia, $NH_3$. Ammonia, being a base, can be titrated with an acid, HCl, and in this way we can determine the amount of $NH_3$ that is produced.

$$NH_3(aq) + HCl(aq) \rightleftharpoons NH_4Cl(aq) \tag{3}$$

For example, a 5-mL aliquot of the reaction mixture is taken before the reaction starts. We use this as a blank. We titrate this with 0.05 M HCl to an end point. The amount of acid used was 1.5 mL. This blank then must be subtracted from all subsequent titration values. Next, we take a 5-mL sample of the reaction mixture after the reaction has proceeded for 10 min. We titrate this with 0.05 M HCl and, let's assume, get a value of 5.0 mL HCl. Therefore, $5.0 - 1.5 = 3.5$ mL of 0.05 M HCl was used to neutralize the $NH_3$ produced in a 10-min. reaction time. This means that

$$(3.5 \text{ mL} \times 0.05 \text{ moles HCl})/1000 \text{ mL} = 1.75 \times 10^{-4} \text{ moles HCl}$$

was used up. According to reaction (3), one mole of HCl neutralizes 1 mole of $NH_3$, therefore, the titration indicates that in our 5-mL sample, $1.75 \times 10^{-4}$ moles of $NH_3$ was produced in 10 min. Equation (1) also shows that for each mole of urea decomposed, 2 moles of $NH_3$ are formed. Therefore, in 10 min.

$$(1 \text{ mole urea} \times 1.75 \times 10^{-4} \text{ moles } NH_3)/2 \text{ moles } NH_3$$
$$= 0.87 \times 10^{-4} \text{ moles urea or } 8.7 \times 10^{-5} \text{ moles of urea}$$

were decomposed. Thus the rate was $8.7 \times 10^{-6}$ moles of urea per min. This is the result we obtained using a 5-mL sample in which 1 mg of urease was dissolved. This rate of reaction corresponds to $8.7 \times 10^{-6}$ moles urea/mg enzyme-min.

A *unit of activity* of urease is defined as the micromoles ($1 \times 10^{-6}$ moles) of urea decomposed in 1 min. Thus the enzyme in the preceding example had an activity of 8.7 units per mg enzyme.

In this experiment we also study the rate of the urease-catalyzed decomposition in the presence of an inhibitor. We use a dilute solution of phenylmercuric acetate to inhibit but not completely inactivate urease.

## CAUTION!

Mercury compounds are poisons. Take extra care to avoid getting the mercuric salt solution in your mouth or swallowing it.

Many of the enzymes in our body are also -SH containing enzymes, and these will be inactivated if we ingest such compounds. As a result of mercury poisoning, many body functions will be inhibited.

## Objectives

1. To demonstrate how to measure the rate of an enzyme-catalyzed reaction.
2. To investigate the effect of an inhibitor on the rate of reaction.
3. To calculate urease activity.

## Procedure

### Enzyme Kinetics in the Absence of Inhibitor

1. Prepare a 37°C water bath in a 250-mL beaker. Maintain this temperature by occasionally adding hot water to the bath. To a 100-mL Erlenmeyer flask, add 20 mL of

0.05 M Tris buffer and 20 mL of 0.3 M urea in a Tris buffer. Mix the two solutions, and place the corked Erlenmeyer flask into the water bath for 5 min. This is your *reaction vessel*.

2. Set up a buret filled with 0.05 M HCl. Place into a 100-mL Erlenmeyer flask 3 to 4 drops of a 1% $HgCl_2$ solution. This will serve to stop the reaction, once the sample is pipetted into the titration flask. Add a few drops of methyl red indicator. This Erlenmeyer flask will be referred to as the *titration vessel*.

3. Take the *reaction vessel* from the water bath. Add 10 mL of urease solution to your reaction vessel. The urease solution contains a specified amount of enzyme (e.g., 20 mg enzyme in 10 mL of solution). Note the time of adding the enzyme solution as zero reaction time. Immediately pipet a 5-mL aliquot of the urea mixture into your *titration vessel*. Stopper the reaction vessel, and put it back into the 37°C bath.

4. Titrate the contents of the titration vessel with 0.05 M HCl to an end point. The end point is reached when the color changes from yellow to pink and stays that way for 10 sec. Record the amount of acid used. This is your blank.

5. Wash and rinse your titration vessel after each titration and reuse it for subsequent titrations.

6. Take a 5-mL aliquot from the reaction vessel **every 10 min**. Pipet these aliquots into the cleaned titration vessel into which methyl red indicator and $HgCl_2$ inhibitor were already placed similar to the procedure in step no. 2 that you used in your first titration (blank). Record the time you placed the aliquots into the titration vessels and titrate them with HCl to an end point. Record the amount of HCl used in your titration. Use five samples over a period of 50 min.

## Enzyme Kinetics in the Presence of Inhibitor

> **CAUTION!**
>
> Be careful with the phenylmercuric acetate solution. Do not get it in your mouth or eyes.

1. Use the same water bath as in the first experiment. Maintain the temperature at 37°C. To a new 100-mL reaction vessel, add 19 mL of 0.05 M Tris buffer, 20 mL of 0.3 M urea solution, and 1 mL of phenylmercuric acetate ($1 \times 10^{-3}$ M). Mix the contents, and place the reaction vessel into the water bath for 5 min.

2. Ready the *titration vessel* as before by adding a few drops of $HgCl_2$ and methyl red indicator. To the *reaction vessel*, add 10 mL of urease solution. Note the time of addition as zero reaction time. Mix the contents of the *reaction vessel*. Transfer immediately a 5-mL aliquot into the *titration vessel*. This will serve as your blank.

3. Titrate it as before. Record the result. **Every 10 min.** take a 5-mL aliquot for titration. The duration of this experiment should be 40 min.

## Chemicals and Equipment

1. Tris buffer
2. 0.3 M urea
3. 0.05 M HCl
4. $1 \times 10^{-3}$ M phenylmercuric acetate
5. 1% $HgCl_2$
6. Methyl red indicator
7. Urease solution
8. 50-mL buret
9. 10-mL graduated pipets
10. 5-mL volumetric pipets
11. 10-mL volumetric pipets
12. Buret holder
13. Spectroline pipet filler

## Experiment 48

### PRE-LAB QUESTIONS

**1.** What is a blank? Read carefully the **Background** and come up with a definition.

**2.** We define the activity of urease as the number of micromoles of urea decomposed/min./mg enzyme. Yet, we measure the micromoles of ammonia produced/min./mg enzyme. Explain.

**3.** Why is phenylmercuric acetate such a dangerous poison?

**4.** What functional group is present in the active site of urease?

## Experiment 48

## REPORT SHEET

### Enzyme kinetics in the absence of inhibitor

| Reaction time (min.) | Buret readings before titration (A) | Buret readings after titration (B) | mL acid titrated (B) − (A) | mL 0.05 M HCl used up in the reaction (B) − (A) − blank |
|---|---|---|---|---|
| 0 (blank) | | | | |
| 10 | | | | |
| 20 | | | | |
| 30 | | | | |
| 40 | | | | |
| 50 | | | | |

### Enzyme kinetics in the presence of Hg salt inhibitor

| Reaction time (min.) | Buret readings before titration (A) | Buret readings after titration (B) | mL acid titrated (B) − (A) | mL 0.05 M HCl used up in the reaction (B) − (A) − blank |
|---|---|---|---|---|
| 0 (blank) | | | | |
| 10 | | | | |
| 20 | | | | |
| 30 | | | | |
| 40 | | | | |

1. Present the preceding data in the graphical form by plotting reaction time (column 1) on the x-axis and the mL of 0.5 M HCl used (column 5) on the y-axis for both reactions.

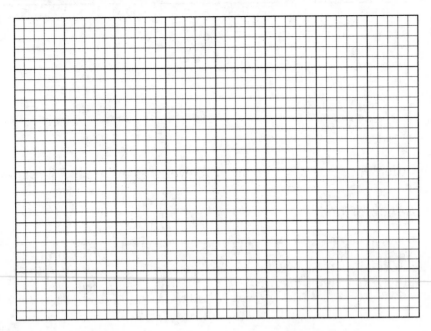

2. Calculate the urease activity only for the reaction without the inhibitor. Use the titration data from the first 10 min. of reaction (initial slope).

Urease activity

$$= \frac{X \text{ mL HCl consumed} \times 0.05 \text{ moles HCl} \times 1 \text{ mole NH}_3 \times 1 \text{ mole urea} \times 50 \text{ mL sol.}}{10 \text{ min.} \times 5 \text{ mL sol.} \times 1000 \text{ mL HCl} \times 1 \text{ mole HCl} \times 2 \text{ moles NH}_3 \times 20 \text{ mg urease}}$$

= Z units activity/mg enzyme

## POST-LAB QUESTIONS

1. What would be the urease activity if you used the slope between 40 and 50 min. instead of the initial slope from your diagram?

2. Your instructor will provide the activity of urease as it was specified by the manufacturer. Compare this activity with the one you calculated. Can you account for the difference? (Enzymes usually lose their activity in long storage.)

3. If you want to perform the urease inhibition experiment but you cannot obtain $HgCl_2$, what could you substitute as an enzyme inhibitor?

4. Is the heavy metal inhibition of urease a reversible or irreversible inhibition? Explain.

5. In studying enzyme reactions, you must work at constant temperature and pH. What steps were taken in your experiment to satisfy these requirements?

6. Your lab ran out of 0.05 M HCl. You found a bottle labeled 0.05 M $HNO_3$. Could you use it for your titration? If you do, would your calculation of urease activity be different?

7. Compare the initial rates (the first 10 min.) of the enzyme reactions with and without inhibitor. How many times slower was the reaction with inhibitor than without it?

# Isocitrate dehydrogenase—an enzyme of the citric acid cycle

## Background

The citric acid cycle is the first unit of the common metabolic pathway through which most of our food is oxidized to yield energy. In the citric acid cycle, the partially fragmented food products are broken down further. The carbons of the $C_2$ fragments are oxidized to $CO_2$, released as such, and expelled in the respiration. The hydrogens and the electrons of the $C_2$ fragments are transferred to the coenzyme, nicotinamide adenine dinucleotide, $NAD^+$, or to flavin adenine dinucleotide, FAD, which in turn become $NADH + H^+$ or $FADH_2$, respectively. These enter the second part of the common pathway, oxidative phosphorylation, and yield water and energy in the form of ATP.

The first enzyme of the citric acid cycle to catalyze both the release of one carbon dioxide and the reduction of $NAD^+$ is isocitrate dehydrogenase. The overall reaction of this step is as follows:

The reduction of the $NAD^+$ itself is given by the equation:

The enzyme has been isolated from many tissues, the best source being a heart muscle or yeast. The isocitrate dehydrogenase requires the presence of cofactors $Mg^{2+}$ or $Mn^{2+}$. As an allosteric enzyme, it is regulated by a number of modulators. ADP, adenosine diphosphate, is a positive modulator and therefore stimulates enzyme activity. The enzyme has an optimum pH of 7.0. As is the case with all enzymes of the citric acid cycle, isocitrate dehydrogenase is found in the mitochondria.

In the present experiment, you will determine the activity of isocitrate dehydrogenase extracted from pork heart muscle. The commercial preparation comes in powder form and it uses $NADP^+$ rather than $NAD^+$ as a coenzyme. The basis of the measurement of the enzyme activity is the absorption spectrum of NADPH. This reduced coenzyme has an absorption maximum at 340 nm. Therefore, an increase in the absorbance at 340 nm indicates an increase in NADPH concentration, hence the progress of the reaction. We define the unit of isocitrate dehydrogenase activity as one that causes an increase of 0.01 absorbance per min. at 340 nm.

For example, if a 10-mL solution containing isocitrate and isocitrate dehydrogenase and $NADP^+$ exhibits a 0.04 change in the absorbance in 2 min., the enzyme activity will be

$$\frac{0.04 \text{ abs.}}{2 \text{ min.} \times 10 \text{ mL}} \times \frac{1 \text{ unit}}{0.01 \text{ abs./1 min.}} = 0.2 \text{ units/mL}$$

If the 10-mL test solution contained 1 mL of isocitrate dehydrogenase solution with a concentration of 1 mg powder/1 mL of enzyme solution, then the activity will be

$$(0.2 \text{ units/mL test soln.}) \times (10 \text{ mL test soln./1 mL enzyme soln.}) \times (1 \text{ mL enzyme soln./1 mg enzyme powder})$$
$$= 2 \text{ units/mg enzyme powder.}$$

## Objectives

To measure the activity of an enzyme of the citric acid cycle, isocitrate dehydrogenase, and the effect of enzyme concentration on the rate of reaction.

## Procedure

1. Turn *on* the spectrophotometer and let it warm up for a few minutes. Turn the wavelength control knob to read 340 nm. With *no sample tube in the sample compartment*, adjust the amplifier control knob so that 0% transmittance or infinite absorbance is read.

2. Prepare a cocktail of reactants in the following manner: In a 10-mL test tube, mix 2.0 mL phosphate buffer, 1.0 mL $MgCl_2$ solution, 1.0 mL 15 mM isocitrate solution, and 5 mL distilled water.

3. To prepare a **Blank** for the spectrophotometric reading, take a sample tube and add to it 1.0 mL of reagent cocktail (prepared as above), 0.2 mL $NADP^+$ solution, and 1 mL of distilled water. Mix the solutions by shaking the sample tube. *Be careful to pipet exactly 0.2 mL $NADP^+$.*

4. Insert the sample tube with the **Blank** solution into the spectrophotometer. Adjust the reading to 100% transmittance (or 0 absorbance). This zeroing must be performed every

10 min. before each enzyme activity run because some instruments have a tendency to drift. The instrument is now ready to measure enzyme activity.

5. Prepare one sample tube for the **enzyme activity measurements**. Add 1.0 mL of reagent cocktail and 0.7 mL of distilled water. Next, add 0.2 mL $NADP^+$ solution. *Be careful to pipet exactly 0.2 mL $NADP^+$*. Mix the contents of the sample tube. Readjust the spectrophotometer with the "Blank" (prepared in step no. 3) to read 0.00 absorbance or 100.00% transmission. Remove the "Blank" and save it for future readjustments. *In the next step the timing is very important*. Take a watch, and *at a set time* (for example, 2 hr. 15 min. 00 sec.) add *exactly* 0.3 mL enzyme solution to the sample tube. Mix it thoroughly and quickly by shaking the tube. Insert the sample tube into the spectrophotometer and take a first reading 1 min. after the mixing time (i.e., 2 hr. 16 min. 00 sec.). Record the absorbancies on your Report Sheet in column 1. Thereafter, take a reading of the spectrophotometer every 30 sec. and record the readings for 5–6 min. on your Report Sheet in column 1.

6. Repeat the experiment exactly as in step no. 5: Preparing the sample solution, readjusting the instrument with the "Blank," and reading the sample solution every 30 sec. for 5 min. Record the spectrophotometric readings on your Report Sheet in column 2.

7. Prepare a new sample tube with the following contents: 1.0 mL reagent cocktail, 0.8 mL distilled water, and 0.2 mL $NADP^+$ solution. *Be careful to pipet exactly 0.2 mL $NADP^+$*. Mix it thoroughly. Readjust the spectrophotometer with the "Blank" to zero absorbance (100% transmission). *At a set time* (i.e., 2 hr. 33 min. 00 sec.), add *exactly* 0.2 mL of enzyme solution. Mix the sample tube and insert into the spectrophotometer. Take your first reading 1 min. after the mixing and every 30 sec. for 5 min. thereafter. Record the absorbancies on your Report Sheet in column 3.

8. Prepare a new sample tube with the following contents: 1.0 mL reagent cocktail, 0.6 mL distilled water, and *exactly* 0.2 mL $NADP^+$ solution. Mix it thoroughly. Readjust the spectrophotometer with the "Blank" to zero absorbance (100% transmission). *At a set time* (i.e., 2 hr. 45 min. 00 sec.), add *exactly* 0.4 mL enzyme solution to the sample tube. Mix it thoroughly. Insert the sample tube into the spectrophotometer. Take a first reading 1 min. after mixing and every 30 sec. for 5 min. thereafter. Record the absorbancies on your Report Sheet in column 4.

9. Plot the numerical data you recorded in the four columns on graph paper. Note that somewhere *between 3 and 5 min.* your graphs are linear. Obtain the slopes of these linear portions and record them on your Report Sheet. Calculate the activities of your enzyme first as (a) units per mL sample solution and second as (b) units per mg enzyme powder.

## Experiment 49

### PRE-LAB QUESTIONS

1. Explain the difference between the structures of citrate and isocitrate.

2. Identify the reactions in the citric acid cycle that (a) precedes and (b) follows the isocitrate oxidation (dehydrogenation).

3. What other reactant or product concentration could be used to measure the isocitrate dehydrogenase activity?

4. What is the difference between $NAD^+$ and $NADP^+$? (Consult your textbook.)

SECTION

DATE

GRADE

## Experiment 49

## REPORT SHEET

| Time (sec.) after mixing | Absorbance of sample | | | |
| --- | --- | --- | --- | --- |
| | Column 1 | 2 | 3 | 4 |
| 60 | | | | |
| 90 | | | | |
| 120 | | | | |
| 150 | | | | |
| 180 | | | | |
| 210 | | | | |
| 240 | | | | |
| 270 | | | | |
| 300 | | | | |
| 330 | | | | |
| 360 | | | | |

1. Plot your data: Absorbance versus time.

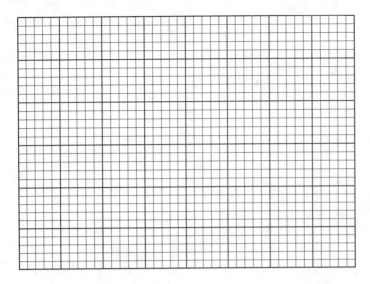

2. Calculate the enzyme activity:

   a. Units of enzyme activity/mL reaction mixture: The slope of the plot is usually a straight line. If so, read the value of change in absorbance per min. Divide it by 0.01. This gives you the number of enzyme activity units/reaction mixture. (One unit of enzyme activity is 0.01 absorbance/min.) Your reaction mixture had a volume of 2.2 mL. Thus dividing by 2.2 will give you the activities in units/mL reaction mixture.

|  | (1) | (2) | (3) | (4) |
|---|---|---|---|---|
| Units/reaction mixture | | | | |
| Units/mL reaction mixture | | | | |

   b. Calculate the isocitrate dehydrogenase activity per mg powder extract. For example, your enzyme solution contained 0.2 mg powder extract/mL solution. If you added 0.2 mL of enzyme solution, it contained 0.04 mg powder extract. Dividing the units/reaction mixture (obtained above) by the number of mg of powder extract added gives you the units/mg powder extract.

|  | (1) | (2) | (3) | (4) |
|---|---|---|---|---|
| Units of enzyme activity/mg powder extract | | | | |

## POST-LAB QUESTIONS

1. You ran two experiments with the same enzyme concentration in (1) and (2). Calculate the average activity for this concentration of enzyme in units/mg powder. Does the reproducibility fall within ±5%?

2. Does your enzyme activity (units/mg powder) give you the same number for the different enzyme concentrations employed?

3. If your powder extract contained 80% protein, what would be the average isocitrate dehydrogenase activity per mg protein (enzyme)?

4. In step 2 of your procedure you added $MgCl_2$ to the reaction mixture. What was the purpose of this addition? If you would have forgotten to add this reagent, would the activity of the enzyme be different from that you obtained? If so, in what way?

5. Assume that because of a mix-up, you started to collect data on absorbances only after 100 sec. had elapsed. Would the calculated enzyme activity be (a) the same, (b) greater, or (c) smaller than if you had performed the experiment correctly? Explain.

# Quantitative analysis of vitamin C contained in foods

## Background

Ascorbic acid is commonly known as vitamin C. It was one of the first vitamins that played a role in establishing the relationship between a disease and its prevention by proper diet. The disease *scurvy* has been known for ages, and a vivid description of it was given by Jacques Cartier, a 16th century explorer of the American continent: "Some did lose their strength and could not stand on their feet. . . . Others . . . had their skin spotted with spots of blood . . . their mouth became stinking, their gums so rotten that all the flesh did fall off." Prevention of scurvy can be obtained by eating fresh vegetables and fruits. The active ingredient in fruits and vegetables that helps to prevent scurvy is ascorbic acid. It is a powerful biological antioxidant (reducing agent). It helps to keep the iron in the enzyme, prolyl hydroxylase, in the reduced form and, thereby, it helps to maintain the enzyme activity. Prolyl hydroxylase is essential for the synthesis of normal collagen. In scurvy, the abnormal collagen causes skin lesions and broken blood vessels.

Vitamin C cannot be synthesized in the human body and must be obtained from the diet (e.g., citrus fruits, broccoli, turnip greens, sweet peppers, tomatoes) or by taking synthetic vitamin C (e.g., vitamin C tablets, "high-C" drinks, and other vitamin C-fortified commercial foods). The minimum recommended adult daily requirement of vitamin C to prevent scurvy is 60 mg. Some people, among them the late Linus Pauling, twice Nobel Laureate, suggested that very large daily doses (250 to 10,000 mg) of vitamin C could help prevent the common cold, or at least lessen the symptoms for many individuals. No reliable medical data support this claim. At present, the human quantitative requirement for vitamin C is still controversial and requires further research.

In this experiment, the amount of vitamin C is determined quantitatively by titrating the test solution with a water-soluble form of iodine $I_3^-$:

$$:\!\ddot{\underset{\cdot\cdot}{I}}\!:\!\ddot{\underset{\cdot\cdot}{I}}\!: + :\!\ddot{\underset{\cdot\cdot}{I}}\!:^- \longrightarrow \left[:\!\ddot{\underset{\cdot\cdot}{I}}\!: \; \ddot{\underset{\cdot\cdot}{I}}\!: : \ddot{\underset{\cdot\cdot}{I}}\!:\right]^- \text{(tri-iodode ion)}$$

Expanded octets

Vitamin C is oxidized by $I_2$ (as $I_3^-$) according to the following chemical reaction:

Vitamin C     (MW 254)        Oxidized product
(MW 176)                         (MW 174)

As vitamin C is oxidized by iodine, $I_2$ becomes reduced to $I^-$. When the end point is reached (no vitamin C is left), the excess of $I_2$ will react with a starch indicator to form a starch-iodine complex which is blackish-blue in color.

$$I_2 + starch \rightarrow iodine - starch\ complex\ (blackish-blue)$$

It is worthwhile to know that although vitamin C is very stable when dry, it is readily oxidized by air (oxygen) when in solution; therefore, a solution of vitamin C should not be exposed to air for long periods. The amount of vitamin C can be calculated by using the following conversion factor:

$$1\ mL\ of\ I_2\ (0.01\ M) = 1.76\ mg\ vitamin\ C$$

## Objective

To determine the amount of vitamin C that is present in certain commercial food products by the titration method.

## Procedure

1. Pour about 60 mL of a fruit drink that you wish to analyze into a clean, dry 100-mL beaker. The fruit drink should be light colored, apple, orange, or grapefruit, but not dark colored, such as grape. Record the kind of drink on the Report Sheet (1).

2. If the fruit drink is cloudy or contains suspended particles, it can be clarified by the following procedure: Add Celite, used as a filter aid, to the fruit drink (about 0.5 g). After swirling it thoroughly, filter the solution through a glass funnel, bedded with a large piece of cotton. Collect the filtrate in a 50-mL Erlenmeyer flask (Fig. 50.1).

3. Using a 10-mL volumetric pipet and a Spectroline pipet filler, transfer 10.00 mL of the fruit drink into a 125-mL Erlenmeyer flask. Then add 20 mL of distilled water, 5 drops of 3 M HCl (as a catalyst), and 10 drops of 2% starch solution to the flask.

4. Clamp a clean, dry 50-mL buret onto the buret stand. Rinse the buret twice with 5-mL portions of iodine solution. Let the rinses run through the tip of the buret and discard them. Fill the buret slightly above the zero mark with a standardized iodine solution. (A dry funnel may be used for easy transfer.) Air bubbles should be removed by turning

the stopcock several times to force the air bubbles out of the tip. Record the molarity of standardized iodine solution (2). Record the initial reading of standardized iodine solution to the nearest 0.02 mL (3a).

**Figure 50.1**
Clarification of fruit drinks.

Cotton

5. Place the flask that contains the vitamin C sample under the buret and add the iodine solution dropwise, while swirling, until the indicator just changes to dark blue. This color should persist for at least 20 sec. Record the final buret reading (3b). Calculate the total volume of iodine solution required for the titration (3c), the weight of vitamin C in the sample (4), and percent (w/v) of vitamin C in the drink (5). Repeat this titration procedure twice more, except using 20- and 30-mL portions of the same fruit drink instead of 10 mL. Record the volumes of iodine solution that are required for each titration.

---

### Chemicals and Equipment

1. 50-mL buret
2. Buret clamp
3. Spectroline pipet filler
4. 10-mL volumetric pipet
5. 50-mL Erlenmeyer flask
6. Cotton
7. Filter aid
8. Hi-C apple drink
9. Hi-C orange drink
10. Hi-C grapefruit drink
11. 0.01 M iodine in potassium iodide
12. 3 M HCl
13. 2% starch solution

## Experiment 50

### PRE-LAB QUESTIONS

**1.** What foods contain appreciable amounts of vitamin C?

**2.** Vitamin C is ascorbic acid. Yet, in the structure given in the **Background**, one does not see a -COOH group. Instead, one sees a lactone in the structure. Circle the lactone. Explain from what functional groups it is formed, and how it is formed.

**3.** What is the minimum daily requirement of vitamin C to prevent scurvy in adults?

**4.** What enzyme is oxidized in the absence of vitamin C and causes the symptoms of scurvy? Which natural product is synthesized by this enzyme?

## Experiment 50

**REPORT SHEET**

**1.** The kind of fruit drink _____

**2.** Molarity of iodine solution _____

**3.** Titration results

|  | Sample 1 (10.0 mL) | Sample 2 (20.0 mL) | Sample 3 (30.0 mL) |
|---|---|---|---|
| **a.** Initial buret reading | _____ mL | _____ mL | _____ mL |
| **b.** Final buret reading | _____ mL | _____ mL | _____ mL |
| **c.** Total volume of iodine solution used: (b − a) | _____ mL | _____ mL | _____ mL |

**4.** The weight of vitamin C in the fruit drink sample: [(3c) × 1.76 mg/mL]

_____ mg     _____ mg     _____ mg

**5.** Concentration of vitamin C in the fruit drink (mg/100 mL): [(4)/volume of drink] × 100

_____     _____     _____

**6.** Average concentration of vitamin C in the fruit drink _____ mg/100 mL

## POST-LAB QUESTIONS

**1.** Why is HCl added for the titration of vitamin C?

**2.** What gives the blue color in your titration?

**3.** What volume of fruit drink would satisfy your minimum daily vitamin C requirement?

**4.** Why was it necessary to use Celite (Filter Aid)?

**5.** You have analyzed a 15.0-mL sample of orange juice for vitamin C. Using a 0.005 M iodine solution for titration, your initial buret reading was 1.0 mL and the final reading was 7.4 mL. What was the concentration of vitamin C (mg/100 mL) in your orange juice? Show your work.

## Analysis of vitamin A in margarine

Vitamin A, or retinol, is one of the major fat-soluble vitamins. It is present in many foods; the best natural sources are liver, butter, margarine, egg yolk, carrots, spinach, and sweet potatoes. Vitamin A is the precursor of retinal, the essential component of the visual pigment rhodopsin.

Vitamin A (All-*trans*-retinol)                    11-*cis*-retinal

When a photon of light penetrates the eye, it is absorbed by the 11-*cis*-retinal. The absorption of light converts the 11-*cis*-retinal to all-*trans*-retinal:

11-*cis*-retinal                    All-*trans*-retinal

This isomerization converts the energy of a photon into an atomic motion which in turn is converted into an electrical signal. The electrical signal generated in the retina of the eye is transmitted through the optic nerve into the brain's visual cortex.

Even though part of the all-*trans*-retinal is regenerated in the dark to 11-*cis*-retinal, for good vision, especially for night vision, a constant supply of vitamin A is needed. The recommended daily allowance of vitamin A is 750 $\mu$g. Deficiency in vitamin A results in night blindness and keratinization of epithelium. The latter compromises the integrity of healthy skin. In young animals, vitamin A is also required for growth. On the other hand, large doses of vitamin A, sometimes recommended in faddish diets, can be harmful. A daily dose above 1500 $\mu$g can be toxic

To analyze the vitamin A content of margarine by spectrophotometric method.

## Procedure

The analysis of vitamin A requires a multistep process. In order that you should be able to follow the step-by-step procedure, a flow chart is provided here:

1. Margarine is largely fat. In order to separate vitamin A from the fat in margarine, first the sample must be saponified. This converts the fat to water-soluble products, glycerol and potassium salts of fatty acids. Vitamin A can be extracted by diethyl ether from the products of the saponification process. To start, weigh a cover glass to the nearest 0.1 g. Report this weight on the Report Sheet (1). Add approximately 10 g of margarine to the watch glass. Record the weight of watch glass plus sample to the nearest 0.1 g on your

Report Sheet (2). Transfer the sample from the watch glass into a 250-mL Erlenmeyer flask with the aid of a glass rod, and wash it in with 75 mL of 95% ethanol. Add 25 mL of 50% KOH solution. Cover the Erlenmeyer flask loosely with a cork and put it on an electric hot plate. Bring it gradually to a boil. Maintain the boiling for 5 min. with an occasional swirling of the flask using tongs. The stirring should aid the complete dispersal of the sample. Remove the Erlenmeyer from the hot plate and let it cool to room temperature (approximately 20 min.).

> **CAUTION!**
>
> 50% KOH solution can cause burns on your skin. Handle the solution with care; do not spill it. If a drop gets on your skin, wash it immediately with copious amounts of water. Use gloves when working with this solution.

2. While the sample is cooling, prepare a chromatographic column. Take a 25-mL buret. Add a small piece of glass wool. With the aid of a glass rod, push it down near the stopcock. Add 15–16 mL of petroleum ether to the buret. Open the stopcock slowly, and allow the solvent to fill the tip of the buret. Close the stopcock. You should have 12–13 mL of petroleum ether above the glass wool. Weigh about 20 g of alkaline aluminum oxide (alumina) in a 100-mL beaker. Place a small funnel on top of your buret. Pour the alumina slowly, in small increments, into the buret. Allow it to settle to form a 20-cm column. Drain the solvent but do not allow the column to run dry. **Always have at least 0.5 mL clear solvent on top of the column.** If the alumina adheres to the walls of the buret, wash it down with more solvent.

> **CAUTION!**
>
> Diethyl ether is very volatile and flammable. Make certain that there are no open flames, not even a hot electrical plate in the vicinity of the operation.

3. Transfer the solution (from your reaction in step no. 1) from the Erlenmeyer flask to a 500-mL separatory funnel. Rinse the flask with 30 mL of distilled water and add the rinsing to the separatory funnel. Repeat the rinsing two more times. Add 100 ml of diethyl ether to the separatory funnel. Close the separatory funnel with the glass stopper. Shake the separatory funnel vigorously. (See Exp. 35, Fig. 35.1 for technique.) Allow it to separate into two layers. Drain the bottom aqueous layer into an Erlenmeyer flask. Add the top (diethyl ether) layer to a second clean 250-mL Erlenmeyer flask. Pour back the aqueous layer into the separatory funnel. Add another 100-mL portion of diethyl ether. Shake and allow it to separate into two layers. Drain again the bottom (aqueous) layer and discard. Combine the first diethyl ether extract with the residual diethyl ether extract in the separatory funnel. Add 100 mL of distilled water to the combined diethyl ether extracts in the separatory funnel. Agitate it gently and allow the water to drain. Discard the washing.

4. Transfer the diethyl ether extracts into a clean 300-mL beaker. Add 3–5 g of anhydrous $Na_2SO_4$ and stir it gently for 5 min. to remove traces of water. Decant the diethyl ether extract into a clean 300-mL beaker. Add a boiling chip or a boiling stick. Evaporate the diethyl ether solvent to about 25 mL volume by placing the beaker **in the hood** on a steam bath. Transfer the sample to a 50-mL beaker and continue to evaporate on the steam bath until an oily residue forms. Remove the beaker from the steam bath. Cool it in an ice bath for 1 min. Add 5 mL of petroleum ether and transfer the liquid (without the boiling chip) to a 10-mL volumetric flask. Add sufficient petroleum ether to bring it to volume.

5. Add 5 mL of extracts in petroleum ether to your chromatographic column. By opening the stopcock drain the sample into your column, but **take care not to let the column run dry**. (Always have about 0.5 mL liquid on top of the column.) Continue to add solvent to the top of your column. Collect the eluents in a beaker. First you will see the orange-colored carotenes moving down the column. With the aid of a UV lamp, you can also observe a fluorescent band following the carotenes. This fluorescent band contains your vitamin A. Allow all the orange color band to move to the bottom of your column and into the collecting beaker. When the fluorescent band reaches the bottom of the column, close the stopcock. By adding petroleum ether on the top of the column continuously, elute the fluorescent band from the column into a 25-mL graduated cylinder. Continue the elution until all the fluorescent band has been drained into the graduated cylinder. Close the stopcock, and record the volume of the eluate in the graduated cylinder on your Report Sheet (4). Add the vitamin A in the petroleum ether eluate to a dry and clean 50-mL beaker. Evaporate the solvent in the hood on a steam bath. The evaporation is complete when an oily residue appears in the beaker. Add 5 mL of absolute ethanol to the beaker. Transfer the sample into a 10-mL volumetric flask and bring it to volume by adding absolute ethanol.

6. Place your sample in a 1-cm length quartz spectroscopic cell. The control (blank) spectroscopic cell should contain absolute ethanol. Read the absorbance of your sample against the blank, according to the instructions of your spectrophotometer, at 325 nm. Record the absorption at 325 nm on your Report Sheet (5).

7. Calculate the amount of margarine that yielded the vitamin A in the petroleum ether eluate. Remember that you added only half (5 mL) of the extract to the column. Report this value on your Report Sheet (6). Calculate the grams of margarine that would have yielded the vitamin A in 1 mL absolute ethanol by dividing (6)/10 mL. Record it on your Report Sheet (7). Calculate the vitamin A in a pound of margarine by using the following formula:

$$\mu g \text{ vitamin A/lb of margarine} = \text{Absorption} \times 5.5 \times [454/(7)].$$

Record your value on the Report Sheet (8).

## Chemicals and Equipment

1. Separatory funnel (500 mL)
2. Buret (25 mL)
3. UV lamp
4. Spectrophotometer (near UV)
5. Margarine
6. Petroleum ether (30–60°C)
7. 95% ethanol
8. Absolute ethanol
9. Diethyl ether
10. Glass wool
11. Alkaline aluminum oxide (alumina)

## Experiment 51

### PRE-LAB QUESTIONS

1. The structure of β-carotene is given below. What is the difference between β-carotene and vitamin A?

β-carotene

2. Why must you work in the hood when dealing with diethyl ether solutions? Why should you make sure that there is no lit Bunsen burner in the lab during this experiment?

3. In the saponification process, you hydrolyzed fat in the presence of KOH. Write an equation of a reaction in which the fat is hydrolyzed in the presence of HCl. What is the difference between the products of the saponification and that of the acid hydrolysis?

4. In the flowchart provided in the **Procedure,** there are two steps in sequence: (a) washing the extract with water, and (b) getting rid of water by drying the anyhydrous sodium sulfate. Why are these contradicting steps necessary?

## Experiment 51

**REPORT SHEET**

1. Weight of watch glass           _____ g

2. Weight of watch glass + margarine      _____ g

3. Weight of margarine: (2) − (1)       _____ g

4. Volume of petroleum ether eluate      _____ mL

5. Absorption at 325 nm            _____

6. Grams margarine in 1 mL of petroleum ether eluate: 2 × [(3)/(4)]    _____ g

7. Grams of margarine in 1 mL of absolute ethanol: (6)/10 mL    _____ g

8. $\mu$g vitamin A/lb margarine: (5) × 5.5 × [454/(7)]    _____

## POST-LAB QUESTIONS

1. How did you detect the fluorescent band of vitamin A during the chromatography? Was it easy to see?

2. In your separation scheme the fatty acids of the margarine ended up in the aqueous wash, which was discarded. Could you have removed the fatty acids, similarly, if, instead of saponification, you used acid hydrolysis? Explain.

3. What chemical processes are needed to convert vitamin A to the 11-*cis*-retinal?

4. The label on a commercial margarine sample states that 1 g of it contains 15% of the daily recommended allowance. Was your sample richer or poorer in vitamin A than the above mentioned commercial sample?

## Urine analysis

The kidney is an important organ that filters materials from the blood that are harmful, or in excess, or both. These materials are excreted in the urine. A number of tests are routinely run in clinical laboratories on urine samples. These involve the measurements of glucose or reducing sugars, ketone bodies, albumin, specific gravity, and pH.

Normal urine contains little or no glucose or reducing sugars; the amount varies from 0.05 to 0.15%. Higher concentrations may occur if the diet contains a large amount of carbohydrates or if strenuous work was performed shortly before the test. Patients with diabetes or liver damage have chronically elevated glucose content in the urine. A semiquantitative test of glucose levels can be performed with the aid of test papers such as Clinistix. This is a quick test that uses a paper containing two enzymes—glucose oxidase and peroxidase. In the presence of glucose, the glucose oxidase catalyzes the formation of gluconic acid and hydrogen peroxide. The hydrogen peroxide is decomposed with the aid of peroxidase and yields atomic oxygen.

α-D-glucose          Gluconic acid          Hydrogen peroxide

$$H_2O_2 \xrightarrow{\text{Peroxidase}} H_2O + [O]$$

In Clinistix, the atomic oxygen reacts with an indicator, *o*-tolidine, and produces a purple color. The intensity of the purple color is proportional to the glucose concentration.

*o*-Tolidine

This test is specific for glucose only. No other reducing sugar will give positive results.

Normal urine contains no albumin or only a trace amount of it. In cases of kidney failure or malfunction, the protein passes through the glomeruli and is not reabsorbed in the tubule. So, albumin and other proteins end up in the urine. The condition known as proteinuria may be symptomatic of kidney disease. The loss of albumin and other blood proteins will decrease the osmotic pressure of blood. This allows water to flow from the blood into the tissues, and creates swelling (edema). Renal malfunction is usually accompanied by swelling of the tissues. The Albustix test is based on the fact that a certain indicator at a certain pH changes its color in the presence of proteins.

Albustix contains the indicator, tetrabromophenol blue, in a citrate buffer at pH 3. At this pH, the indicator has a yellow color. In the presence of protein, the color changes to green. The higher the protein concentration, the greener the indicator will be. Therefore, the color produced by the Albustix can be used to estimate the concentration of protein in urine.

Three substances that are the products of fatty acid catabolism—acetoacetic acid, $\beta$-hydroxybutyric acid, and acetone—are commonly called ketone bodies. These are normally present in the blood in small amounts and can be used as an energy source by the cells. Therefore, no ketone bodies will normally be found in the urine. However, when fats are the only energy source, excess production of ketone bodies will occur. They will be filtered out by the kidney and appear in the urine. Such abnormal conditions of high fat catabolism take place during starvation or in diabetes mellitus when glucose, although abundant, cannot pass through the cell membranes to be utilized inside where it is needed. Acetoacetic acid ($CH_3COCH_2COOH$), and to a lesser extent acetone ($CH_3COCH_3$) and $\beta$-hydroxybutyric acid ($CH_3CHOHCH_2COOH$), react with sodium nitroprusside $\{Na_2[Fe(CN)_5NO]\}\cdot 2H_2O$ to give a maroon-colored complex. In Ketostix, the test area contains sodium nitroprusside and a sodium phosphate buffer to provide the proper pH for the reaction. The addition of lactose to the mixture in the Ketostix enhances the development of the color.

Some infants are born with a genetic defect known as phenylketonuria (PKU). They lack the enzyme phenylalanine oxidase, which converts phenylalanine to tyrosine. Thus phenylalanine accumulates in the body and it is degraded to phenylpyruvate by transamination:

| Phenylalanine | Pyruvate | Phenylpyruvate | Alanine |

Phenylpyruvate is excreted in the urine. Normal urine does not contain any phenylpyruvate. People suffering from PKU have varying amounts of phenylpyruvate in their urine. PKU causes severe mental retardation in infants if it is not treated immediately after birth, which is done by restricting the phenylalanine content of the diet. In many states, the law requires that every newborn be tested for phenylpyruvate in the urine. The test is based on the reaction of the iron (III) ion with the phenylpyruvate, producing a gray-green color.

Phenistix strips which are coated with $Fe(NH_4)(SO_4)_2$ and a buffer can detect as little as 8 mg of phenylpyruvate in 100 mL of urine. Some drugs such as aspirin produce metabolites (salicylic acid) that are excreted in the urine and give color with an iron(III) ion. However, this produces a deep purple color and not the gray-green of PKU. The purple color that is given by the Phenistix can be used to diagnose an overdose of aspirin. Other drugs, such as phenylthiazine, in an overdose, give a gray-purple color with Phenistix. For PKU diagnosis only the appearance of the gray-green color means a positive test.

Urobilinogen and other bile pigments are normally minor components of urine (2 to 50 $\mu$g/100 mL). They are the products of hemoglobin breakdown. Bile pigments are usually excreted in the feces. In case of obstruction of the bile ducts (gallstones, obstructive jaundice), the normal excretion route through the small intestines is blocked and the excess bilirubin is filtered out of the blood by the kidneys and appears in the urine. Urobilistix is a test paper that can detect the presence of urobilinogen, because it is impregnated with p-dimethylaminobenzaldehyde. In strongly acidic media, this reagent gives a yellow-brownish color with urobilinogen.

p-Dimethylaminobenzaldehyde

The specific gravity of normal urine may range from 1.008 to 1.030. After excessive fluid intake (like a beer party), the specific gravity may be on the low side; after heavy exercise and perspiration, it may be on the high side. High specific gravity indicates excessive dissolved solutes in the urine.

The pH of normal urine can vary between 4.7 through 8.0. The usual value is about 6.0. High-protein diets and fever can lower the pH of urine. In severe acidosis, the pH may be as low as 4.0. Vomiting and respiratory or metabolic alkalosis can raise the pH above 8.0.

## Objectives

1. To perform quick routine analytical tests on urine samples.
2. To compare results obtained on "normal" and "pathological" urine samples.

## Procedure

Each student must analyze her (his) own urine. A fresh urine sample will be collected prior to the laboratory period. The stockroom will provide paper cups for sample collection. *While handling body fluids, such as urine, plastic gloves should be worn.* The used body fluid will be collected in a special jar and disposed of collectively. The plastic gloves worn will be collected and autoclaved before disposal. In addition, the stockroom will provide one "normal" and two "pathological" urine samples.

Place 5 mL of the urine sample from each source into four different test tubes. These will be tested with the different test papers.

## Glucose Test

For the glucose test, use for comparison two test tubes, each half-filled, one with 0.25% and the other with 1% glucose solutions. Take six strips of Clinistix from the bottle. Replace the cap immediately. Dip the test area of the Clinistix into one of the samples. Tap the edge of the strip against a clean, dry surface to remove the excess urine. Compare the test area of the strip to the color chart supplied on the bottle exactly 10 sec. after the wetting. Do not read the color changes that occur after 10 sec. Record your observation: The "light" on the color chart means 0.25%, or less, glucose; the "medium" means 0.4%; and the "dark" means 0.5%, or more. Repeat the test with the other five samples.

## Protein Test

For the protein test, take four Albustix strips, one for each of the four urine samples, from the bottle. Replace the cap immediately. Dip the Albustix strips into the test solutions, making certain that the reagent area of the strip is completely immersed. Tap the edge of the strip against a clean, dry surface to remove the excess urine. Compare the color of the Albustix test area with the color chart supplied on the bottle. The time of the comparison is not critical; you can do it immediately or any time within 1 min. after wetting. Read the color from yellow (negative) to different shades of green, indicating trace amounts of albumin, up to 0.1%.

## Ketone Bodies

To measure the ketone bodies' concentration in the urine, take four Ketostix strips from the bottle. Replace the cap immediately. Dip a Ketostix in each of the urine samples and remove the strips at once. Tap them against a clean, dry surface to remove the excess urine. Compare the color of the test area of the strips to the color chart supplied on the bottle. Read the colors exactly 15 sec. after wetting. A buff pink color indicates the absence of ketone bodies. A progression to a maroon color indicates increasing concentration of ketone bodies from 52 to 160 mg/L of urine.

## Test for PKU

To test for PKU disease, use four Phenistix strips, one for each urine sample. Immerse the test area of the strips into the urine samples and remove them immediately. Remove the excess urine by tapping the strips against a clean, dry surface. Read the color after 30 sec. of wetting, and match them against the color charts provided on the bottle. Record your estimated phenylpyruvate content: negative or 0.015 to 0.1%.

## Urobilinogen

To measure the urobilinogen content of the urine samples, use Urobilistix strips, one for each urine sample. Dip them into the samples and remove the excess urine by tapping the strips against a clean, dry surface. Sixty seconds after wetting, compare the color of the test area of the strips to the color chart provided on the bottle. Estimate the urobilinogen content and record it in Ehrlich units.

## pH in Urine

To measure the pH of each urine sample, use pH indicator paper such as pHydrion test paper within the pH range of 3.0 to 9.0.

For the preceding tests, you may use a multipurpose strip such as Labstix that contains test areas for all these tests, except for the test for phenylpyruvate, on one strip. The individual test areas are separated from each other and clearly marked. The time requirements to read the colors are also indicated on the chart. The results of five tests can be read in 1 min.

## Specific Gravity of Urine

The specific gravity of your urine samples will be measured with the aid of a hydrometer (urinometer; see Fig. 52.1). Place the bulb in a cylinder. Add sufficient urine to the cylinder to make the bulb float. Read the specific gravity of the sample from the stem of the hydrometer where the meniscus of the urine intersects the calibration lines. Be sure the hydrometer is freely floating and does not touch the walls of the cylinder. In order to use as little urine as possible, the instructor may read the normal and two pathological urine samples for the whole class. If so, you will measure the specific gravity of your own urine sample only.

**Figure 52.1**
A urinometer.

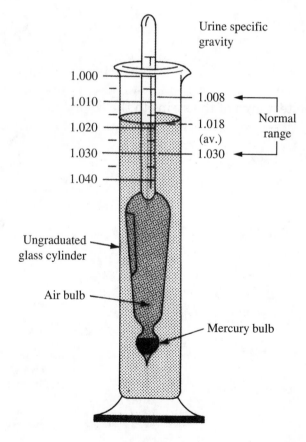

## Chemicals and Equipment

1. 0.25 and 1% glucose solutions
2. Clinistix
3. Albustix
4. Ketostix
5. Urobilistix
6. pH paper in the 3.0 to 9.0 range
7. A multipurpose Labstix (instead of these test papers)
8. Phenistix
9. 3 M NaOH
10. Hydrometer (urinometer)

## Experiment 52

### PRE-LAB QUESTIONS

1. Can a color-blind person perform the pH measurement of urine samples with the method given? If not, what can she/he do?

2. In what tests do we use the following reagents?

   (a) $Fe^{3+}$

   (b)

   (c)

3. What does a high specific gravity reading on a urine sample indicate?

4. A patient's urine shows a high specific gravity, 1.04. The pH is 7.8, and the Phenistix test indicates a purple color that is not characteristic of PKU. The patient has had a high fever for a few days and has been given aspirin. Do these tests indicate any specific disease, or are they symptomatic of recovering from a high fever? Explain.

## Experiment 52

### REPORT SHEET

*Urine Samples*

| Test | Normal | Pathological A | Pathological B | Your own | Remarks |
|------|--------|----------------|----------------|----------|---------|
| Glucose | | | | | |
| Ketone bodies | | | | | |
| Albumin | | | | | |
| Urobilinogen | | | | | |
| pH | | | | | |
| Phenylpyruvate | | | | | |
| Specific gravity | | | | | |

### POST-LAB QUESTIONS

1. Did you find any indication that your urine is not normal? If so, what may be the reason?

2. Why is the phenylpyruvate test mandatory with newborns in many states?

**3.** If a urine sample shows unusually high protein content, what disease is suggested by this test?

**4.** A patient's urine was tested with Clinistix, and the color was read 60 sec. after wetting the strip. It showed 1.0% glucose in the urine. Is the patient diabetic? Explain.

**5.** Assume that you did not have enough urine to completely immerse the urinometer. Would your specific gravity readings be meaningful? Explain your answer.

# Appendix

## List of Apparatus and Equipment in Student's Locker

**Amount and description**

(1) Beaker, 50 mL
(1) Beaker, 100 mL
(1) Beaker, 250 mL
(1) Beaker, 400 mL
(1) Beaker, 600 mL
(1) Bunsen burner
(3) Clamps, extension
(3) Clamps, holder
(1) Cylinder, graduated by 0.1 mL, 10 mL
(1) Cylinder, graduated by 1 mL, 100 mL
(1) Dropper, medicine with rubber bulb
(1) Evaporating dish
(2) Flasks, Erlenmeyer, 50 mL
(2) Flasks, Erlenmeyer, 125 mL
(1) Flask, Erlenmeyer, 250 mL
(1) Flask, Erlenmeyer, 500 mL
(1) File, triangular
(1) Forceps
(1) Funnel, short stem
(1) Gauze, wire
(1) Spatula, stainless steel
(1) Sponge
(1) Striker (or box of matches)
(12) Test tubes, approximately 125 × 16 mm
(12) Test tubes, approximately 100 × 13 mm
(1) Test tube brush
(1) Test tube holder
(1) Test tube rack
(1) Thermometer, 150°C
(1) Tongs, crucible
(1) Wash bottle, plastic
(1) Watch glass, 3 in.
(1) Watch glass, 5 in.

# Appendix 2

## List of Common Equipment and Materials in the Laboratory

Each laboratory should be equipped with hoods and safety-related items such as fire extinguisher, fire blankets, safety shower, and eye wash fountain. The equipment and materials listed here for 25 students should be made available in each laboratory.

Acid tray
Aspirators (splashgun type) on sink faucet
Balances, single pan, triple beam (or Centogram) or top-loading
Barometer
Clamps, extension
Clamps, thermometer
Clamps, utility
Containers for solid chemical waste disposal
Containers for liquid organic waste disposal
Corks
Detergent for washing glassware
Drying oven
Filter paper
Glass rods, 4 and 6 mm OD
Glass tubing, 6 and 8 mm OD
Glycerol (glycerine) in dropper bottles
Hot plates
Ice maker
Paper towel dispensers
Pasteur pipets
Rings, support, iron, 76 mm OD
Ring stands
Rubber tubing, pressure
Rubber tubing, latex (0.25 in. OD)
Water, deionized or distilled
Weighing dishes, polystyrene, disposable, 73 × 73 × 25 mm
Weighing paper

# Appendix 3

## Special Equipment and Chemicals

In the instructions below every time a solution is to be made up in "water" you must use *distilled water*.

**Experiment 1**   **Laboratory techniques: use of the laboratory gas burner; basic glassworking**

### Special Equipment
| | |
|---|---|
| (25) | Wing tops |
| (25) | Crucible tongs |
| (25) | Wire gauze |
| (50) | Glass tubing (6 mm OD), 25-cm segments |
| (25) | Solid glass rod, 25-cm segments |

**Experiment 2**   **Laboratory measurements**

### Special Equipment
| | |
|---|---|
| (25) | 50-mL graduated beakers |
| (25) | 50-mL graduated Erlenmeyer flasks |
| (25) | Metersticks or rulers, with both English and metric scale |
| (25) | Hot plates |
| (25) | Single pan, triple beam balances (Centogram) |
| (5) | Platform triple beam balances |
| (2) | Top-loading balances |

**Experiment 3**   **Density determination**

### Special Equipment
| | |
|---|---|
| (4) | 250-mL beakers (labeled for unknown metals) |
| (25) | Magnetic stir-bars, small (1/2 × 5/16 in.; must be small enough to fit into a 50-mL graduated cylinder) |
| (25) | Magnetic stirrers |
| (25) | Solid wood blocks, rectangular or cubic |
| (25) | Spectroline pipet fillers |
| (25) | 10-mL volumetric pipets |
| (50) | Polyethylene plastic chips, 2–4 mm dia. |

### Chemicals
| | |
|---|---|
| (1 L) | Acetone, reagent |

### *Unknown metals*
| | |
|---|---|
| (100 g) | Aluminum, pellets or rod |
| (100 g) | Tin, pellets or cut strips |
| (100 g) | Zinc, pellets |
| (100 g) | Lead, shot |

**Experiment 4**     **The separation of the components of a mixture**

### Special Equipment

| | |
|---|---|
| (2) | Top-loading balances (weigh to 0.001 g) |
| (15) | Single pan, triple beam (Centogram) balances (weigh to 0.01 g) as an alternative |
| (25) | Evaporating dishes, porcelain, 6 cm dia. |
| (25) | Rubber policeman |
| (1 box) | Filter paper, 15 cm, fast flow |
| (25) | Mortar and pestle |

### Chemicals

| | |
|---|---|
| (30 g) | Unknown mixture: mix 3.0 g naphthalene (10%), 15 g sodium chloride (50%), 12 g sea sand (40%) |
| (1 jar) | Boiling stones (silicon carbide chips if available) |

**Experiment 5**     **Resolution of a mixture by distillation**

### Special Equipment

| | |
|---|---|
| (25) | Distillation kits with 19/22 standard taper joints: 100-mL round bottom flasks (2); distilling head; thermometer adapter; 110°C thermometer; condenser; vacuum adapter |
| (25) | Nickel wires |

### Chemicals

| | |
|---|---|
| (1 jar) | Boiling chips |
| (1 jar) | Silicone grease |
| (2 L) | 5% sodium chloride solution: dissolve 100 g NaCl in enough water to make 2 L |
| (100 mL) | 0.5 M silver nitrate: dissolve 8.5 g $AgNO_3$ in enough water to make 100 mL solution |
| (50 mL) | Concentrated nitric acid, $HNO_3$ |

**Experiment 6**     **The empirical formula of a compound: the Law of Constant Composition**

### Special Equipment

| | |
|---|---|
| (1 box) | Filter papers (Whatman no. 2), 7.0 cm |
| (25) | Hot plates |
| (25) | Rubber policemen |
| (25) | Vacuum filtration setup: (2) 250-mL filter flasks; rubber stopper (no. 6, 1 hole) with glass tubing inserted (10-cm length × 7 mm OD); (2) vacuum tubing, 2-ft. lengths; Büchner funnel (65 mm OD) in a no. 6 1-hole rubber stopper |

### Chemicals

| | |
|---|---|
| (25) | Aluminum wires, no. 16 or 18 (45 cm; approx. 1.5 g) |
| (500 mL) | Acetone (use dropper bottles) |
| (500 mL) | 6 M aqueous ammonia, 6 M $NH_3$ (aq): add 200 mL concentrated $NH_3$ (28%) into a 500-mL volumetric flask and add enough water to bring to the mark. Place the prepared solution in dropper bottles. **Prepare in the hood using a face shield, rubber apron, and rubber gloves.** |

|          | (200 g) | Copper(II) chloride, $CuCl_2$ |

**Experiment 7**   **Determination of the formula of a metal oxide**

**Special Equipment**

| (25) | Porcelain crucibles and covers |
| (25) | Clay triangles |
| (25) | Crucible tongs |
| (25) | Eye droppers |

**Chemicals**

| (25) | Magnesium ribbon, 12-cm strip |
| (500 mL) | 6 M HCl: take 250 mL 12 M HCl and add to enough ice cold water to make 500 mL of solution. **Wear a face mask, rubber gloves, and a rubber apron during the preparation. Do in the hood.** |

**Experiment 8**   **Classes of chemical reactions**

**Special Equipment**

| (1 box) | Wood splints |

**Chemicals**

| (25 pieces) | Aluminum foil (2 × 0.5 in. each) |
| (50 pieces) | Aluminum wire (1 cm each) |
| (25 pieces) | Copper foil (2 × 0.5 in. each) |
| (25) | Pre-1982 copper penny (optional) |
| (20 g) | Ammonium carbonate, $(NH_4)_2CO_3$ |
| (20 g) | Potassium iodide, KI |
| (40 g) | Potassium iodate, $KIO_3$ |
| (20 g) | Calcium turnings, Ca |
| (20 g) | Magnesium ribbon, Mg |
| (20 g) | Mossy zinc, Zn |
| (20 g) | Lead shot, Pb |

**All the following solutions should be placed in dropper bottles:**

| (100 mL) | 3 M hydrochloric acid, 3 M HCl: 25 mL 12 M HCl diluted with ice cold water to 100 mL |
| (100 mL) | 6 M hydrochloric acid, 6 M HCl: 50 mL 12 M HCl diluted with ice cold water to 100 mL |
| (100 mL) | 3 M sulfuric acid, 3 M $H_2SO_4$: 16.7 mL 18 M $H_2SO_4$ is slowly added to 60 mL ice cold water; stir slowly and dilute with water to 100 mL |
| (100 mL) | 3 M sodium hydroxide, 3 M NaOH: dissolve 12 g NaOH per 100 mL solution |

**In preparing the above solutions, rubber gloves, a rubber apron, and a face shield should be worn. Do all preparations in the hood.**

| (100 mL) | 0.1 M silver nitrate, 0.1 M $AgNO_3$: dissolve 1.70 g $AgNO_3$ per 100 mL solution |
| (100 mL) | 0.1 M sodium chloride, 0.1 M NaCl: dissolve 0.58 g NaCl per 100 mL solution |
| (100 mL) | 0.1 M sodium nitrate, 0.1 M $NaNO_3$: dissolve 0.85 g $NaNO_3$ per 100 mL solution |

(100 mL)  0.1 M sodium carbonate, 0.1 M $Na_2CO_3$: dissolve 1.24 g $Na_2CO_3 \cdot H_2O$ per 100 mL solution

(100 mL)  0.1 M potassium nitrate, 0.1 M $KNO_3$: dissolve 1.01 g $KNO_3$ per 100 mL solution

(100 mL)  0.1 M potassium chromate, 0.1 M $K_2CrO_4$: dissolve 1.94 g $K_2CrO_4$ per 100 mL solution

(100 mL)  0.1 M barium chloride, 0.1 M $BaCl_2$: dissolve 2.08 g $BaCl_2$ per 100 mL solution

(100 mL)  0.1 M copper(II) nitrate, 0.1 M $Cu(NO_3)_2$: dissolve 1.88 g $Cu(NO_3)_2$ per 100 mL solution

(100 mL)  0.1 M copper(II) chloride, 0.1 M $CuCl_2$: dissolve 1.70 g $CuCl_2 \cdot 2H_2O$ per 100 mL solution

(100 mL)  0.1 M lead(II) nitrate, 0.1 M $Pb(NO_3)_2$: dissolve 3.31 g $Pb(NO_3)_2$ per 100 mL solution

(100 mL)  0.1 M iron(III) nitrate, 0.1 M $Fe(NO_3)_3$: dissolve 4.04 g $Fe(NO_3)_3 \cdot 9H_2O$ per 100 mL solution

## Experiment 9  Chemical properties of consumer products

### Special Equipment

(1 roll)   Copper wire
(2 vials)  Litmus papers, blue
(2 vials)  Litmus papers, red

### Chemicals

**All solutions should be placed in dropper bottles. In preparing all acid and base solutions, observe personal safety practices. Use a face shield, rubber gloves, and a rubber apron. Do preparations in the hood.**

(100 mL)  Commercial ammonia solution, $NH_3$ (2.8% ammonia solution can be substituted: 10 mL 28% $NH_3$ solution diluted to 100 mL with water)

(100 mL)  Commercial bleach containing sodium hypochlorite, NaOCl

(100 g)   Commercial baking soda, $NaHCO_3$

(100 g)   Commercial detergent containing sodium phosphate, $Na_3PO_4$

(100 g)   Garden fertilizer containing ammonium phosphate, $(NH_4)_3PO_4$

(100 g)   Epsom salt, $MgSO_4 \cdot 7H_2O$

(100 g)   Table salt, sodium chloride, NaCl

(100 g)   Ammonium chloride, $NH_4Cl$

(100 g)   Potassium iodide, KI

(100 mL)  Ammonium molybdate reagent: preparation
*solution 1*: dissolve 100 g of ammonium molybdate, $(NH_4)_6Mo_7O_{24}$, in 400 mL water; add slowly, with stirring, 80 mL concentrated $NH_3$ (15 M $NH_3$).
*solution 2*: add 400 mL concentrated $HNO_3$ (16 M $HNO_3$), slowly, with stirring, to 600 mL ice cold water.
Mix *solution 1* and *solution 2* in the proportion of 1:2.

(100 mL)  1 M barium chloride, 1 M $BaCl_2$: dissolve 24.42 g $BaCl_2 \cdot 2H_2O$ per 100 mL solution

(100 mL)  5% barium hydroxide, 5% $Ba(OH)_2$: dissolve 5.0 g $Ba(OH)_2 \cdot 8H_2O$ per 100 mL solution

(200 mL)  3 M nitric acid, 3 M $HNO_3$: 50 mL concentrated $HNO_3$ (12 M $HNO_3$) diluted to 200 mL solution with ice cold water

(200 mL)  6 M nitric acid, 6 M $HNO_3$: 100 mL concentrated $HNO_3$ (12 M $HNO_3$) diluted to 200 mL solution with ice cold water.

(100 mL)  0.1 M silver nitrate, 0.1 M $AgNO_3$: dissolve 1.70 g $AgNO_3$ per 100 mL solution

(100 mL)  6 M sodium hydroxide, 6 M NaOH: dissolve 24 g NaOH per 100

mL solution

(100 mL)     1 M sodium phosphate, 1 M $Na_3PO_4$: dissolve 38 g
$Na_3PO_4 \cdot 12H_2O$ per 100 mL solution

(100 mL)     6 M sulfuric acid, 6 M $H_2SO_4$: pour 33.3 mL concentrated $H_2SO_4$
(18 M $H_2SO_4$) into 50 mL ice cold water. Stir slowly. Dilute to 100
mL volume.

(100 mL)     0.01% $p$-nitrobenzene-azoresorcinol: dissolve 0.01 g
$p$-nitrobenzene-azoresorcinol in 100 mL 0.025 M NaOH

(100 mL)     Hexane, $CH_3CH_2CH_2CH_2CH_2CH_3$

## Experiment 10    Water Analysis

### Special Equipment

| | |
|---|---|
| (25) | 50-mL burets |
| (25) | Buret clamps |
| (25) | 50-mL volumetric pipets |
| (25) | 25-mL volumetric pipets |
| (25) | 5-mL volumetric pipets |
| (25) | 5-mL graduated pipets |
| (25) | 1-mL graduated pipets |
| (25) | Magnetic stirrers |
| (25) | Magnetic stir-bars |

### Chemicals

**The following chemicals should be placed in dropper bottles.**

(1 L)     6 N nitric acid, 6 N $HNO_3$: add slowly 500 mL concentrated $HNO_3$
(12 M $HNO_3$) to 400 mL ice cold water, with stirring, and bring to 1 L
volume with water. **Do preparation in the hood. Use a face
shield, rubber apron, and rubber gloves.**

(1.5L)     0.01 N silver nitrate, 0.01 N $AgNO_3$: dissolve 2.55 g $AgNO_3$ in
enough water to bring to 1.5 L volume

(1 L)     0.01 N ammonium thiocyanate, 0.01 N $NH_4SCN$: add 5 mL of
standard volumetric concentrate ammonium thiocyanate solution
[obtainable from chemical suppliers, for example, from Thomas
Scientific as "ACCULUATE" (Anachemia)] to enough water to
bring to 1 L volume

(200 mL)     40% iron(III) nitrate, 40% $Fe(NO_3)_3$: add 80 g
$Fe(NO_3)_3 \cdot 9H_2O$ to 100 mL water, stir, and bring to 200 mL
volume

(2 L)     Brackish water: dissolve 300 mg NaCl in 1 L water and bring to 2
L volume with water

## Experiment 11    Calorimetry: the determination of the specific heat of a metal

### Special Equipment

| | |
|---|---|
| (25) | Wire loops for stirring |
| (25) | Rubber rings (cut from latex tubing) |
| (50) | Styrofoam cups (8 oz.) |
| (25) | Lids for styrofoam cups |
| (25) | Thermometers, 110°C |
| (25) | Stopwatches |
| (25) | Thermometer clamps |
| (25) | Test tubes, 25 x 150 mm |

### Chemicals

| | |
|---|---|
| (1 kg) | Lead shot, no. 8 |
| (1 kg) | Aluminum metal, turnings or wire |
| (1 kg) | Iron metal |

| (1 kg) | Tin metal, granular (mossy) |
| (1 kg) | Zinc shot |

**Experiment 12**    **Boyle's Law: the pressure–volume relationship of a gas**

**Special Equipment**

| (25) | Boyle's Law apparatus: the apparatus can be constructed as follows. A piece of glass tubing, 30 cm in length, 3 mm OD, is sealed at one end. A Pasteur disposable pipet is drawn out to form a capillary; the capillary needs to be only small enough to be inserted into the 3-mm OD glass tubing and long enough to reach half the length of the tubing (approx. 15 cm). Mercury is transferred with the pipet into the tubing; enough mercury is placed in the tube to give a 10-cm long column. The tube is attached to a 1-ft. ruler by means of rubber bands. The ruler should read with both English (to nearest 1/16 in.) and metric (to nearest mm) scales. |
| (25) | 30°-60°-90° plastic triangles |

**Experiment 13**    **Charles's Law: the volume–temperature relationship of a gas**

**Special Equipment**

| (25) | Bunsen burner (or hot plate) |
| (50) | 250-mL Erlenmeyer flask |
| (50) | 800-mL beaker |
| (50) | Clamp |
| (25) | Glass tubing (5- to 8-cm length; 7 mm OD) |
| (25) | Marking pencil |
| (25) | One-hole rubber stopper (size no. 6) |
| (25) | Premade stopper assembly for 250-mL Erlenmeyer flask (optional alternative) |
| (25) | Ring stand |
| (25) | Ring support |
| (25) | Rubber tubing, latex (2-ft. length) |
| (25) | Thermometer, 110°C |
| (25) | Wire gauze |

**Chemicals**

| (1 jar) | Boiling stones |

**Experiment 14**    **Properties of gases: determination of the molecular weight of a volatile liquid**

**Special Equipment**

| (25) | Aluminum foil, 2.5 × 2.5 in. |
| (25) | Aluminum foil, 3 × 3 in. |
| (1 roll) | Copper wire |
| (25) | Beaker tongs |
| (25) | Crucible tongs |
| (25) | Rubber bands |
| (25) | Hot plates |
| (25) | Lead sinkers |

**Chemicals**

| (1 jar) | Boiling chips |

The following liquids should be placed in dropper bottles.

| | |
|---|---|
| (100 mL) | Pentane |
| (100 mL) | Acetone |
| (100 mL) | Methanol (methyl alcohol) |
| (100 mL) | Hexane |
| (100 mL) | Ethanol (ethyl alcohol) |
| (100 mL) | 2-Propanol (isopropyl alcohol) |

**Experiment 15**

## Physical properties of chemicals: melting point, sublimation, and boiling point

### Special Equipment

| | |
|---|---|
| (1 roll) | Aluminum foil |
| (1 bottle) | Boiling chips |
| (1) | Commercial melting point apparatus (if available) |
| (25) | Glass tubing, 20-cm segments |
| (25) | Hot plates |
| (100) | Melting point capillary tubes |
| (50) | Rubber rings (cut 0.25-in. rubber tubing into narrow segments) |
| (25) | Thermometer clamps |
| (25) | Thiele tube melting point apparatus |

### Chemicals

| | |
|---|---|
| (20 g) | Acetamide |
| (20 g) | Acetanilide |
| (20 g) | Adipic acid |
| (20 g) | Benzophenone |
| (20 g) | Benzoic acid |
| (20 g) | p-Dichlorobenzene |
| (20 g) | Naphthalene, pure |
| (50 g) | Naphthalene, impure: mix 47.5 g (95%) naphthalene and 2.5 g (5%) charcoal powder |
| (20 g) | Stearic acid |

**The following liquids should be placed in dropper bottles.**

| | |
|---|---|
| (200 mL) | Acetone |
| (200 mL) | Cyclohexane |
| (200 mL) | Ethyl acetate |
| (200 mL) | Hexane |
| (200 mL) | 2-Propanol (isopropyl alcohol) |
| (200 mL) | Methanol (methyl alcohol) |
| (200 mL) | 1-Propanol |

**Experiment 16**

## Solubility and solutions

### Special Equipment

| | |
|---|---|
| (12) | Electrical conductivity apparatus (one for each pair of students) |
| (50) | Beakers, 30 mL |
| (25) | Beaker tongs |
| (25) | Hot plates |

### Chemicals

| | |
|---|---|
| (2 lb) | Granulated table sugar, sucrose |
| (10 g) | Table salt, NaCl |
| (10 g) | Naphthalene |
| (10 g) | Iodine |
| (500 mL) | Ethanol (ethyl alcohol) |

| (500 mL) | Acetone |
| (500 mL) | Petroleum ether (b.p. 30–60°C) |
| (500 mL) | 1 M NaCl: dissolve 29.22 g of NaCl in water and bring to 500 mL volume |
| (500 mL) | 0.1 M NaCl: take 50 mL of 1 M NaCl and add enough water to make 500 mL |
| (500 mL) | 1 M sucrose: dissolve 171.15 g sucrose in water and bring to 500 mL volume |
| (500 mL) | 0.1 M sucrose: take 50 mL of 1 M sucrose and add enough water to make 500 mL |
| (500 mL) | 1 M HCl: add 41.7 mL concentrated HCl (12 M HCl) to 200 mL of ice cold water; add water to bring to 500 mL volume. **Use a face shield, rubber gloves, and a rubber apron during the preparation. Do in the hood.** |
| (500 mL) | 0.1 M HCl: add 50 mL of 1 M HCl to enough water to make 500 mL **(use the same precautions as in the above preparation)** |
| (500 mL) | Glacial acetic acid |
| (500 mL) | 0.1 M acetic acid: take 3 mL glacial acetic acid and add water to bring to 500 mL volume |

## Experiment 17  Water of hydration

### Special Equipment

| (25) | Crucibles and covers |
| (25) | Clay triangles |
| (25) | Crucible tongs |
| (25) | Ring stands |

### Chemicals

| (25 g) | Calcium chloride, anhydrous, $CaCl_2$ |
| (100 g) | Copper(II) sulfate pentahydrate, $CuSO_4 \cdot 5H_2O$ |

## Experiment 18  Colligative properties: freezing point depression and osmotic pressure

### Special Equipment

| (150) | Capillary melting point tubes |
| (50) | Narrow rubber bands |
| (25) | Test tubes ($25 \times 200$ mm) |
| (25) | Hot plates |
| (5) | Student microscopes |
| (250) | Microscope slides |
| (250) | Cover slides |
| (5) | Razor blades or dissecting knives |
| (3) | Mortars |
| (3) | Pestles |
| (25) | Thermometer clamps |

### Chemicals

| (5 g) | Lauric acid |
| (5 g) | Benzoic acid |
| (100 mL) | 0.1 M glucose solution: dissolve 1.8 g glucose in water and dilute to 100 mL solution |
| (100 mL) | 0.5 M glucose solution: dissolve 9.0 g glucose in water and dilute to 100 mL solution |

| (250 mL) | 0.89% NaCl solution: dissolve 2.23 g NaCl in water and dilute to 250 mL solution |
| --- | --- |
| (250 mL) | 3.0% NaCl solution: dissolve 7.5 g NaCl in distilled water and dilute to 250 mL solution |
| (2 each) | Fresh carrot, scallion, celery |
| (50 mL) | Fresh (not more than 1 week old) whole bovine blood. Obtainable from a slaughterhouse. *Refrigerate.* **Even though bovine blood is not a source of HIV or hepatitis virus, prudence requires handling of the blood samples with care, using plastic gloves. The blood samples after the experiment should be collected in a special jar. Both the blood and the gloves should be autoclaved before disposal.** |

## Experiment 19    Factors affecting rate of reactions

### Special Equipment

| (5) | Mortars |
| --- | --- |
| (5) | Pestles |
| (25) | 10-mL graduated pipets |
| (25) | 5-mL volumetric pipets |

### Chemicals

**Solutions should be put into dropper bottles. In preparing the solutions, wear a face shield, rubber gloves, and a rubber apron. Do in the hood.**

| (100 mL) | 3 M $H_2SO_4$; dissolve 16.7 mL concentrated $H_2SO_4$ (18 M $H_2SO_4$) in 60 mL ice cold water. Stir gently and bring to 100 mL volume. |
| --- | --- |
| (500 mL) | 6 M HCl: add 250 mL concentrated HCl (12 M HCl) to 200 mL ice cold water. Mix and bring it to 500 mL volume. |
| (100 mL) | 2 M $H_3PO_4$: add 13.3 mL concentrated $H_3PO_4$ (15 M $H_3PO_4$) to 50 mL ice cold water. Mix and bring it to 100 mL volume. |
| (100 mL) | 6 M $HNO_3$: add 50.0 mL concentrated $HNO_3$ (12 M $HNO_3$) to 50 mL ice cold water. Mix and bring it to 100 mL volume. |
| (100 mL) | 6 M acetic acid: add 34.4 mL glacial acetic acid (99–100%) to 50 mL water. Mix and bring it to 100 mL volume. |
| (500 mL) | 0.1 M $KIO_3$: **Caution!** *This solution must be fresh. Prepare it on the day of the experiment.* Dissolve 10.7 g $KIO_3$ in 500 mL water. |
| (250 mL) | 4% starch indicator: add 10 g soluble starch to 50 mL cold water. Stir it to make a paste. Bring 200 mL water to a boil in a 500-mL beaker. Pour the starch paste into the boiling water. Stir and cool to room temperature. |
| (500 mL) | 0.01 M $NaHSO_3$: dissolve 0.52 g $NaHSO_3$ in 100 mL water. Add slowly 2 mL concentrated sulfuric acid (18 M $H_2SO_4$). Stir and bring it to 500 mL volume. |
| (250 mL) | 3% hydrogen peroxide: take 25 mL concentrated $H_2O_2$ (30%) and bring it to 250 mL volume with water. |
| (150) | Mg ribbons, 1 cm long |
| (25) | Zn ribbons, 1 cm long |
| (25) | Cu ribbons, 1 cm long |
| (25 g) | $MnO_2$ |

**Law of chemical equilibrium and Le Chatelier's principle**

### Special Equipment

| | |
|---|---|
| (2 rolls) | Litmus paper, blue |
| (2 rolls) | Litmus paper, red |

### Chemicals

| | |
|---|---|
| (50 mL) | 0.1 M copper(II) sulfate: dissolve 0.80 g $CuSO_4$ (or 1.25 g $CuSO_4 \cdot 5H_2O$) in 50 mL water |
| (50 mL) | 1 M ammonia: dilute 3.3 mL concentrated $NH_3$ (28%) with water to 50 mL volume. **In the preparation wear a face shield, rubber gloves, and a rubber apron. Do in the hood.** |
| (25 mL) | Concentrated HCl (12 M HCl) |
| (100 mL) | 1 M hydrochloric acid: add 8.5 mL concentrated HCl (12 M HCl) to 50 mL ice water; add enough water to bring to volume. **In the preparation wear a face shield, rubber gloves and a rubber apron. Do in the hood.** |
| (150 mL) | 0.1 M phosphate buffer: dissolve 1.74 g $K_2HPO_4$ in 100 mL water. Also dissolve 1.36 g $KH_2PO_4$ in 100 mL water. Mix 100 mL $K_2HPO_4$ with 50 mL of $KH_2PO_4$ solution. |
| (100 mL) | 0.1 M potassium thiocyanate: dissolve 0.97 g KSCN in 100 mL water |
| (100 mL) | 0.1 M iron(III) chloride: dissolve 2.7 g $FeCl_3 \cdot 6H_2O$ (or 1.6 g $FeCl_3$) in 100 mL water |
| (100 mL) | Saturated saline solution: add 290 g NaCl to warm (60°C) water. Stir until dissolved. Cool to room temperature. |
| (50 mL) | 1.0 M cobalt chloride: dissolve 11.9 g $CoCl_2 \cdot 6H_2O$ in 50 mL water. |

**pH and buffer solutions**

### Special Equipment

| | |
|---|---|
| (5) | pH meters |
| (12 rolls) | pHydron paper (pH range 0 to 12) |
| (5 boxes) | Kimwipes |
| (5) | Wash bottles |
| (100) | 10-mL graduated pipets |
| (25) | Spot plates |
| (25) | 10-mL beakers |

### Chemicals

| | |
|---|---|
| (250 mL) | 0.1 M acetic acid, 0.1 M $CH_3COOH$: dissolve 1.4 mL glacial acetic acid in water to make 250 mL volume |
| (500 mL) | 0.1 M sodium acetate, 0.1 M $CH_3COONa$: dissolve 6.8 g $CH_3COONa \cdot 3H_2O$ in water to make 500 mL volume |
| (1 L) | 0.1 M hydrochloric acid, 0.1 M HCl: add 8.5 mL concentrated HCl (12 M HCl) to 100 mL ice cold water with stirring; dilute with water to 1 L. **Prepare in the hood; wear a face shield, rubber gloves, and a rubber apron.** |
| (1 L) | 0.1 M sodium bicarbonate, 0.1 M $NaHCO_3$: dissolve 8.2 g $NaHCO_3$ in water to make 1 L volume |
| (500 mL) | Saturated carbonic acid, $H_2CO_3$: use a bottle of Club Soda or Seltzer water; these solutions are approximately 0.1 M carbonic acid. |

**The following solutions should be placed in dropper bottles.**

| | |
|---|---|
| (100 mL) | 0.1 M HCl prepared above |
| (100 mL) | 0.1 M ammonia, 0.1 M $NH_3$: dilute 0.7 mL concentrated $NH_3$ (28%) with water to make 100 mL volume |

|  |  |
|---|---|
| (100 mL) | 0.1 M sodium hydroxide, 0.1 M NaOH: dissolve 0.4 g NaOH in water to make 100 mL volume |

**Experiment 22**    **Analysis of vinegar by titration**

### Special Equipment

| | |
|---|---|
| (25) | 25-mL (or 50-mL) burets |
| (25) | Buret clamps |
| (25) | Ring stands |
| (25) | 5-mL volumetric pipets |
| (25) | Small funnels |

### Chemicals

| | |
|---|---|
| (500 mL) | Vinegar |
| (2L) | 0.2 M NaOH standardized solution: dissolve 16.8 g NaOH in water to make 2 L volume. Standardize the solution as follows: place approximately 1 g potassium hydrogen phthalate, $KC_8H_5O_4$, in a tared weighing bottle. Weigh it to the nearest 0.001 g. Dissolve it in 20 mL water. Add a few drops of phenolphthalein indicator and titrate with the NaOH solution prepared above. The molarity of NaOH is calculated as follows: M = mass of phthalate/(0.2043 × mL NaOH used in titration). Write the calculated molarity of the NaOH on the bottle of the standardized NaOH solution. |
| (100 mL) | Phenolphthalein indicator: dissolve 0.1 g phenolphthalein in 60 mL 95% ethanol and bring it to 100 mL volume with water |

**Experiment 23**    **Analysis of antacid tablets**

### Special Equipment

| | |
|---|---|
| (25) | 25-mL (or 50-mL) burets |
| (25) | 100-mL burets |
| (25) | Buret clamps |
| (25) | Ring stands |
| (5) | Balances to read to 0.001 g |

### Chemicals

| | |
|---|---|
| (5 bottles) | Commercial antacids such as Alka-Seltzer, Gelusil, Maalox, Rolaids, Di-Gel, Tums, etc. Have at least two different kinds available. |
| (1 L) | 0.2 M NaOH, sodium hydroxide, standardized: dissolve 8.4 g NaOH in 1 L water. Standardize as follows: accurately weigh to the nearest 0.001 g approximately 1 g potassium hydrogen phthlate, $KC_8H_5O_4$, MW = 204.3 g/mole, and dissolve it in 20 mL water. Add a few drops of phenolphthalein and titrate the potassium hydrogen phthalate with the prepared NaOH solution. The molarity (M) of the NaOH solution is calculated as follows: M = mass of phthalate/(0.2043 × mL NaOH used in the titration). Write the calculated molarity on the bottle of the standardized NaOH solution. |
| (1 L) | 0.2 M HCl, hydrochloric acid: add 16.7 mL concentrated HCl (12 M HCl) to 100 mL ice cold water; dilute with water to 1 L volume. **(Prepare in the hood; wear a face shield, rubber gloves, and a rubber apron.)** Standardize the acid solution by titration against the standardized 0.2 M NaOH solution. Write the calculated molarity on the bottle of the standardized HCl solution. |

(100 mL)  Thymol blue indicator: dissolve 0.1 g thymol blue in 50 mL 95% ethanol and dilute with water to 100 mL volume. Put in a dropper bottle.

(100 mL)  Phenolphthalein indicator: dissolve 0.1 g phenolphthalein in 60 mL 95% ethanol and bring to 100 mL volume with water. Put in a dropper bottle.

## Experiment 24   Structure in organic compounds: use of molecular models. I

### Special Equipment

**(Color of spheres may vary depending on the set; substitute as necessary.)**

| | |
|---|---|
| (50) | Black spheres—4 holes |
| (300) | Yellow spheres—1 hole |
| (50) | Colored spheres (e.g., green)—1 hole |
| (25) | Blue spheres—2 holes |
| (400) | Sticks |
| (25) | Protractors |
| (75) | Springs (optional) |

## Experiment 25   Stereochemistry: use of molecular models. II

### Special Equipment

Commercial molecular model kits vary in style, size, material composition, and the color of the components. The set which works best in this exercise is the *Molecular Model Set for Organic Chemistry* available from Allyn and Bacon, Inc. (Newton, MA). Wood ball and stick models work as well. For 25 students, 25 of these sets should be provided. If you wish to make up your own kit, you would need the following for 25 students:

(25)  Cyclohexane model kits: each consisting of the following components:
8 carbons—black, 4 hole
18 hydrogens—white, 1 hole
2 substituents—red, 1 hole
24 connectors—bonds

(25)  Chiral model kits: each consisting of the following components:
8 carbons—black, 4 hole
32 substituents—8 red, 1 hole; 8 white, 1 hole; 8 blue, 1 hole; 8 green, 1 hole
28 connectors—bonds

(5)  Small hand mirrors

## Experiment 26   Identification of hydrocarbons

### Special Equipment

| | |
|---|---|
| (2 vials) | Litmus paper, blue |
| (250) | $100 \times 13$ mm test tubes |

### Chemicals

(1 g)  Anhydrous aluminum chloride, $AlCl_3$

**The following solutions should be placed in dropper bottles.**

| | |
|---|---|
| (100 mL) | Concentrated $H_2SO_4$ (18 M $H_2SO_4$) |
| (100 mL) | Cyclohexene |

| (100 mL) | Hexane |
|---|---|
| (100 mL) | Ligroin (b.p. 90–110°C) |
| (100 mL) | Toluene |
| (100 mL) | 1% $Br_2$ in cyclohexane (**wear a face shield, rubber gloves, and a rubber apron; prepare under hood**): mix 1.0 mL $Br_2$ with enough cyclohexane to make 100 mL. *Prepare fresh solutions prior to use; keep in a dark-brown dropper bottle; do not store.* |
| (100 mL) | 1% aqueous $KMnO_4$: dissolve 1.0 g potassium permanganate in 50 mL distilled water by gently heating for 1 hr.; cool and filter; dilute to 100 mL. Store in a dark-brown dropper bottle. |
| (100 mL) | Unknown A = hexane |
| (100 mL) | Unknown B = cyclohexene |
| (100 mL) | Unknown C = toluene |

**Experiment 27**  **Column and paper chromatography; separation of plant pigments**

### Special Equipment

| (50) | Melting point capillaries open at both ends |
|---|---|
| (25) | 25-mL burets |
| (1 jar) | Glass wool |
| (25) | Filter papers (Whatman no.1), 20 × 10 cm |
| (3) | Heat lamp (optional) |
| (25) | Ruler with both English and metric scale |
| (1) | Stapler |
| (15) | Hot plates with or without water bath |

### Chemicals

| (1 lb) | Tomato paste |
|---|---|
| (500 g) | Aluminum oxide (alumina) |
| (500 mL) | 95% ethanol |
| (500 mL) | Petroleum ether, b.p. 30–60°C |
| (500 mL) | Eluting solvent: mix 450 mL petroleum ether with 10 mL toluene and 40 mL acetone. |
| (10 mL) | 0.5% $\beta$-carotene solution: dissolve 50 mg in 10 mL petroleum ether. Wrap the vial in aluminum foil to protect from light and keep in refrigerator until used. |
| (150 mL) | Saturated bromine water: mix 5.5 g bromine with 150 mL water. **Prepare in hood; wear a face shield, rubber gloves, and a rubber apron.** |
| (500 mg) | Iodine crystals |

**Experiment 28**  **Identification of alcohols and phenols**

### Special Equipment

| (125) | Corks (for test tubes 100 × 13 mm) |
|---|---|
| (125) | Corks (for test tubes 150 × 18 mm) |
| (25) | Hot plate |
| (5 rolls) | Indicator paper (pH 1−12) |

### Chemicals

**The following solutions should be placed in dropper bottles.**

| (100 mL) | Acetone (reagent grade) |
|---|---|
| (100 mL) | 1-Butanol |
| (100 mL) | 2-Butanol |
| (100 mL) | 2-Methyl-2-propanol (*t*-butyl alcohol) |

| | |
|---|---|
| (100 mL) | 1, 2-Dimethoxyethane |
| (200 mL) | 20% aqueous phenol: dissolve 80 g of phenol in 20 mL distilled water; dilute to 400 mL. |
| (100 mL) | Lucas reagent **(prepare under hood; wear a face shield, rubber gloves, and a rubber apron):** cool 100 mL of concentrated HCl (12 M HCl) in an ice bath; with stirring, add 150 g anhydrous $ZnCl_2$ to the cold acid. |
| (150 mL) | Chromic acid solution **(prepare under hood; wear a face shield, rubber gloves, and a rubber apron):** dissolve 20 g potassium dichromate, $K_2Cr_2O_7$, in 100 mL concentrated sulfuric acid (18 M $H_2SO_4$). Carefully add this solution to enough ice cold water to bring to 1 L. |
| (100 mL) | 2.5% iron(III) chloride solution: dissolve 2.5 g anhydrous $FeCl_3$ in 50 mL water; dilute to 100 mL. |
| (100 mL) | Iodine in KI solution: mix 20 g of KI and 10 g of $I_2$ in 100 mL water |
| (250 mL) | 10% sodium hydroxide, 10% NaOH: dissolve 25.00 g NaOH in 100 mL water. Dilute to 250 mL with water. |
| (100 mL) | Unknown A = 1-butanol |
| (100 mL) | Unknown B = 2-butanol |
| (100 mL) | Unknown C = 2-methyl-2-propanol (*t*-butyl alcohol) |
| (100 mL) | Unknown D = 20% aqueous phenol |

**Experiment 29**    **Identification of aldehydes and ketones**

**Special Equipment**

| | |
|---|---|
| (250) | Corks (to fit 100 × 13 mm test tube) |
| (125) | Corks (to fit 150 × 18 mm test tube) |
| (1 box) | Filter paper (students will need to cut to size) |
| (25) | Hirsch funnels |
| (25) | Hot plates |
| (25) | Neoprene adapters (no. 2) |
| (25) | Rubber stopper assemblies: a no. 6 one-hole stopper fitted with glass tubing (15 cm in length × 7 mm OD) |
| (25) | 50-mL side-arm filter flasks |
| (25) | 250-mL side-arm filter flasks |
| (50) | Vacuum tubing, heavy-walled (2-ft. lengths) |

**Chemicals**

| | |
|---|---|
| (50 g) | Hydroxylamine hydrochloride |
| (100 g) | Sodium acetate |

**The following solutions should be placed in dropper bottles.**

| | |
|---|---|
| (100 mL) | Acetone (reagent grade) |
| (100 mL) | Benzaldehyde (freshly distilled) |
| (100 mL) | *Bis*(2-ethoxymethyl) ether |
| (100 mL) | Cyclohexanone |
| (100 mL) | 1,2-Dimethoxyethane |
| (500 mL) | Ethanol (absolute) |
| (500 mL) | Ethanol (95%) |
| (100 mL) | Isovaleraldehyde |
| (500 mL) | Methanol |
| (100 mL) | Pyridine |
| (150 mL) | Chromic acid reagent: dissolve 20 g potassium dichromate, $K_2Cr_2O_7$, in 100 mL concentrated sulfuric acid (18 M $H_2SO_4$). Carefully add this solution to enough ice cold water to bring to 1 L. **Wear a face shield, rubber gloves, and a rubber apron during the preparation. Do in the hood.** |

*Tollens' reagent*

(100 mL)     Solution A: dissolve 9.0 g silver nitrate in 90 mL of water; dilute to 100 mL.

(100 mL)     Solution B: 10 g NaOH dissolved in enough water to make 100 mL

**The following reagents should be placed in dropper bottles.**

(100 mL)     10% ammonia water: 35.7 mL of concentrated (28%) $NH_3$ diluted to 100 mL

(100 mL)     10% sodium hydroxide, 10% NaOH: dissolve 10.00 g NaOH in enough water to make 100 mL

(500 mL)     Iodine-KI solution: mix 100 g of KI and a 50 g of iodine in enough distilled water to make 500 mL

(100 mL)     2,4-dinitrophenylhydrazine reagent: dissolve 3.0 g of 2,4-dinitrophenylhydrazine in 15 mL concentrated $H_2SO_4$ (18 M $H_2SO_4$). In a beaker, mix together 10 mL water and 75 mL 95% ethanol. With vigorous stirring slowly add the 2,4-dinitrophenylhydrazine solution to the aqueous ethanol mixture. After thorough mixing, filter by gravity through a fluted filter paper. **Wear a face shield, rubber gloves, and a rubber apron during the preparation. Do in the hood.**

(100 mL)     Semicarbazide reagent: dissolve 22.2 g of semicarbazide hydrochloride in 100 mL of distilled water

(100 mL)     Unknown A = isovaleraldehyde

(100 mL)     Unknown B = benzaldehyde

(100 mL)     Unknown C = cyclohexanone

(100 mL)     Unknown D = acetone

Additional compounds for use as unknowns:

### Aldehydes

(100 mL)     2-Butenal (crotonaldehyde)

(100 mL)     Octanal (caprylaldehyde)

(100 mL)     Pentanal (valeraldehyde)

### Ketones

(100 mL)     Acetophenone

(100 mL)     Cyclopentanone

(100 mL)     2-Pentanone

(100 mL)     3-Pentanone

**Experiment 30**     **Properties of carboxylic acids and esters**

### Special Equipment

(5 rolls)     pH paper (range 1–12)

(100)     Disposable Pasteur pipets

(5 vials)     Litmus paper, blue

(25)     Hot plates

### Chemicals

(10 g)     Salicylic acid

(10 g)     Benzoic acid

**The following solutions are placed in dropper bottles.**

(75 mL)     Acetic acid

(50 mL)     Formic acid

(25 mL)     Benzyl alcohol

(50 mL)     Ethanol (ethyl alcohol)

(25 mL)     2-Methyl-1-propanol (isobutyl alcohol)

(25 mL)     3-Methyl-1-butanol (isopentyl alcohol)

(50 mL)     Methanol (methyl alcohol)

| (25 mL) | Methyl salicylate |
| (250 mL) | 6 M hydrochloric acid, 6 M HCl: take 125 mL of concentrated HCl (12 M HCl) and add to 50 mL of ice cold water; dilute with enough water to 250 mL. **Wear a face shield, rubber gloves, and a rubber apron during the preparation. Do in the hood.** |
| (100 mL) | 3 M hydrochloric acid, 3 M HCL: take 50 mL 6 M HCl and bring to 100 mL; **follow the same precautions as above.** |
| (300 mL) | 6 M sodium hydroxide, 6 M NaOH: dissolve 72.00 g NaOH in enough water to bring to 300 mL; **follow the same precautions as above.** |
| (150 mL) | 2 M sodium hydroxide, 2 M NaOH: take 50 mL 6 M NaOH and bring to 150 mL; **follow the same precautions as above.** |
| (25 mL) | Concentrated sulfuric acid (18 M $H_2SO_4$). |

## Experiment 31    Properties of amines and amides

### Special Equipment

| (2 rolls) | pH paper (range 0 to 12) |
| (25) | Hot plates |

### Chemicals

| (20 g) | Acetamide |

**The following chemicals and solutions should be placed in dropper bottles.**

| (25 mL) | Triethylamine |
| (25 mL) | Aniline |
| (25 mL) | N,N-Dimethylaniline |
| (100 mL) | Diethyl ether (ether) |
| (100 mL) | 6 M ammonia solution, 6 M $NH_3$: add 40 mL concentrated $NH_3$ (28%) to 50 mL water; then add enough water to 100 mL volume. **Do in the hood.** |
| (100 mL) | 6 M hydrochloric acid, 6 M HCl: add 50 mL concentrated HCl (12 M HCl) to 40 mL ice cold water; then add enough water to 100 mL volume. **Wear a face shield, rubber gloves and a rubber apron when preparing. Do in the hood.** |
| (50 mL) | Concentrated hydrochloric acid (12 M HCl) |
| (250 mL) | 6 M sulfuric acid, 6 M $H_2SO_4$: pour 83.4 mL concentrated $H_2SO_4$ (18 M $H_2SO_4$) into 125 mL ice cold water. Stir slowly. Then add enough water to 250 mL volume. **Wear a face shield, rubber gloves, and a rubber apron when preparing. Do in the hood.** |
| (250 mL) | 6 M sodium hydroxide, 6 M NaOH: dissolve 60.00 g NaOH in 100 mL water. Then add enough water to 250 mL volume. **Do in the hood.** |

## Experiment 32    Polymerization reactions

### Special Equipment

| (25) | Hot plates |
| (25) | Cylindrical paper rolls or sticks |
| (25) | Bent wire approximately 10 cm long |
| (25) | 10-mL pipets or syringes |
| (25) | Spectroline pipet fillers |
| (25) | Test tubes, 16/18 x 150 mm, Pyrex |
| (25) | Beaker tongs |

## Chemicals

| | |
|---|---|
| (1 kg) | Sea sand |

**The following chemicals and solutions should be placed in dropper bottles.**

| | |
|---|---|
| (75 mL) | Styrene |
| (250 mL) | Xylene |
| (10 mL) | *t*-butyl peroxide benzoate (also called *t*-butyl benzoyl peroxide); store at 4°C. |
| (75 mL) | 20% sodium hydroxide: dissolve 15.00 g NaOH in enough water to make 75 mL |
| (300 mL) | 5% adipoyl chloride: dissolve 15.00 g adipoyl chloride in enough cylohexane to make 300 mL |
| (300 mL) | 5% hexamethylene diamine: dissolve 15.00 g hexamethylene diamine in enough water to make 300 mL |
| (200 mL) | 80% formic acid: add 40 mL water to 160 mL formic acid |

**Experiment 33**   ## Preparation of acetylsalicylic acid (aspirin)

### Special Equipment

| | |
|---|---|
| (25) | Büchner funnels (65 mm OD) |
| (25) | Filtervac or no. 2 neoprene adapters |
| (1 box) | Filter paper (5.5 cm, Whatman no. 2) |
| (25) | 250-mL filter flasks |
| (25) | Hot plates |

### Chemicals

| | |
|---|---|
| (1 jar) | Boiling chips |
| (25) | Commercial aspirin tablets |
| (100 mL) | Concentrated phosphoric acid (15 M $H_3PO_4$) (in a dropper bottle) |
| (100 mL) | 1% iron(III) chloride: dissolve 1 g $FeCl_3 \cdot 6H_2O$ in enough distilled water to make 100 mL (in a dropper bottle) |
| (100 mL) | Acetic anhydride, freshly opened bottle |
| (300 mL) | Ethyl acetate |
| (100 g) | Salicylic acid |

**Experiment 34**   ## Measurement of the active ingredient in aspirin pills

### Special Equipment

| | |
|---|---|
| (25) | Büchner funnels (65 mm OD) |
| (25) | Filtervac or no. 2 neoprene adapters |
| (25) | 250-mL filter flasks |
| (1) | Drying oven at 110°C |
| (25) | Mortars, 100-mL capacity |
| (25) | Pestles |
| (1 box) | Filter paper (5.5 cm, Whatman no. 2) |
| (1 box) | Microscope slides, 3 × 1 in., plain |
| (25) | 25-mL beakers |
| (25) | Rubber policeman |

### Chemicals

| | |
|---|---|
| (1.5 L) | 95% ethanol |
| (300 g) | Commercial aspirin tablets |
| (100 mL) | Hanus iodine solution: dissolve 1.2 g KI in 80 mL water. Add 0.25 g $I_2$. Stir until the iodine dissolves. Add enough water to make 100 mL volume. Store in dark dropper bottle. |

**Experiment 35**  **Isolation of caffeine from tea leaves**

**Special Equipment**

| | |
|---|---|
| (25) | Cold finger condensers (115 mm long × 15 mm OD) |
| (1 box) | Filter paper; 7.0 cm, fast flow (Whatman no.1) |
| (25) | Hot plates |
| (50) | Latex tubing, 2-ft. lengths |
| (1 vial) | Melting point capillaries |
| (25) | No. 2 neoprene adaptors |
| (25) | Rubber stopper (no. 6, 1-hole) with glass tubing inserted (10 cm length × 7 mm OD) |
| (25) | 125-mL separatory funnels |
| (25) | 25-mL side-arm filter flasks |
| (25) | 250-mL side-arm filter flasks |
| (25) | Small sample vials |
| (1) | Stapler |
| (50) | Vacuum tubing, 2-ft. lengths |
| (1 box) | Weighing paper |

**Chemicals**

| | |
|---|---|
| (1 jar) | Boiling chips |
| (500 mL) | Dichloromethane, $CH_2Cl_2$ |
| (25 g) | Sodium sulfate, anhydrous, $Na_2SO_4$ |
| (50 g) | Sodium carbonate, anhydrous, $Na_2CO_3$ |
| (50) | Tea bags |

**Experiment 36**  **Carbohydrates**

**Special Equipment**

| | |
|---|---|
| (50) | Medicine droppers |
| (125) | Microtest tubes or 25 depressions white spot plates |
| (2 rolls) | Litmus paper, red |

**Chemicals**

| | |
|---|---|
| (20 g) | Boiling chips |
| (400 mL) | Fehling's reagent (solutions A and B, from Fisher Scientific Co.) |
| (200 mL) | 3 M NaOH: dissolve 24.00 g NaOH in 100 mL water and then add enough water to 200 mL volume |
| (200 mL) | 2% starch solution: place 4 g soluble starch in a beaker. With vigorous stirring, add 10 mL water to form a thin paste. Boil 190 mL water in another beaker. Add the starch paste to the boiling water and stir until the solution becomes clear. Store in a dropper bottle. |
| (200 mL) | 2% sucrose: dissolve 4 g sucrose in 200 mL water |
| (50 mL) | 3 M sulfuric acid: add 8.5 mL concentrated $H_2SO_4$ (18 M $H_2SO_4$) to 30 mL ice cold water; **pour the sulfuric acid slowly along the walls of the beaker, this way it will settle on the bottom without much mixing;** stir slowly in order not to generate too much heat; when fully mixed bring the volume to 50 mL. **Wear a face shield, rubber gloves, and a rubber apron when preparing. Do in the hood.** |
| (100 mL) | 2% fructose: dissolve 2 g fructose in 100 mL water. Store in a dropper bottle. |
| (100 mL) | 2% glucose: dissolve 2 g glucose in 100 mL water. Store in a dropper bottle. |

| (100 mL) | 2% lactose: dissolve 2 g lactose in 100 mL water. Store in a dropper bottle. |
|---|---|
| (100 mL) | 0.01 M iodine in KI: dissolve 1.2 g KI in 80 mL water. Add 0.25 g $I_2$. Stir until the iodine dissolves. Dilute the solution to 100 mL volume. Store in a dark dropper bottle. |

## Experiment 37     Fermentation of a carbohydrate: ethanol from sucrose

### Special Equipment

| (50) | Extension clamps and holders |
|---|---|
| (50) | Ring stands |
| (25) | Heating mantles (or Thermowells) |
| (25) | Variacs |
| (1 jar) | Boiling chips |
| (1 jar) | Celite (Filter Aid) |
| (1 jar) | Silicone grease |
| (3 pads) | Steel wool, coarse (1 package contains 12 pads) |
| (25) | Kits for fractional distillation with 19/22 standard taper joints: 250-mL round bottom flask; condensers (2); distilling head; thermometer adapter; 11°C thermometer; vacuum adapter; latex tubing (2 lengths, 2 ft. each). |

### Chemicals

| (1 L) | Calcium hydroxide solution, $Ca(OH)_2$, saturated: mix 75 g of calcium hydroxide with 1 L of water and let stand for several days. |
|---|---|
| (25 g) | Disodium hydrogenphosphate, $Na_2HPO_4$ |
| (50 g) | Potassium phosphate, $K_3PO_4$ |
| (600 g) | Sucrose |
| (50 mL) | Mineral oil, in a dropper bottle |
| (75 g) | Dried baker's yeast (Fleischmann's All Natural Yeast comes in individual packets of 7 g each) |

## Experiment 38     Preparation and properties of a soap

### Special Equipment

| (25) | Büchner funnels (85 mm OD) |
|---|---|
| (25) | No. 7 one-hole rubber stoppers |
| (1 box) | Filter paper (7.0 cm, Whatman no. 2) |
| (1 roll) | pHydrion paper (pH range 0 to 12) |

### Chemicals

| (1 jar) | Boiling chips |
|---|---|
| (1 L) | 95% ethanol |
| (1 L) | Saturated sodium chloride (sat. NaCl): dissolve 360 g NaCl in 1 L water |
| (1 L) | 25% sodium hydroxide (25% NaOH): dissolve 250 g NaOH in 1 L water |
| (1 L) | Vegetable oil |
| (100 mL) | 5% iron(III) chloride (5% $FeCl_3$): dissolve 5 g $FeCl_3 \cdot 6H_2O$ in 100 mL water. Store in a dropper bottle. |
| (100 mL) | 5% calcium chloride (5% $CaCl_2$): dissolve 5 g $CaCl_2 \cdot H_2O$ in 100 mL water. Store in a dropper bottle. |
| (100 mL) | Mineral oil. Store in a dropper bottle. |
| (100 mL) | 5% magnesium chloride (5% $MgCl_2$): dissolve 5 g $MgCl_2$ in 100 mL water. Store in a dropper bottle. |

**Experiment 39**     **Preparation of a hand cream**

**Special Equipment**

(25)          Bunsen burners or hot plates

**Chemicals**

(100 mL)      Triethanolamine
(40 mL)       Propylene glycol (1,2-propanediol)
(500 g)       Stearic acid
(40 g)        Methyl stearate (ethyl stearate may be substituted)
(400 g)       Lanolin
(400 g)       Mineral oil

**Experiment 40**     **Extraction and identification of fatty acids from corn oil**

**Special Equipment**

(12)          Water baths
(2)           Heat lamps or hair dryers
(25)          15 × 6.5 cm silica gel TLC plates
(25)          Rulers, metric scale
(25)          Polyethylene, surgical gloves
(150)         Capillary tubes, open on both ends
(1)           Drying oven, 110°C

**Chemicals**

(50 g)        Corn oil
(5 mL)        Methyl palmitate solution: dissolve 25 mg methyl palmitate in 5 mL petroleum ether
(5 mL)        Methyl oleate solution: dissolve 25 mg methyl oleate in 5 mL petroleum ether
(5 mL)        Methyl linoleate solution: dissolve 25 mg methyl linoleate in 5 mL petroleum ether
(100 mL)      0.5 M KOH: dissolve 2.81 g KOH in 25 mL water and add 75 ml of 95% ethanol
(500 g)       Sodium sulfate, $Na_2SO_4$, anhydrous, granular
(100 mL)      Concentrated hydrochloric acid (12 M HCl)
(1 L)         Petroleum ether (b.p. 30–60°C)
(300 mL)      Methanol: perchloric acid mixture: mix 285 mL methanol with 15 mL $HClO_4 \cdot 2H_2O$ (73% perchloric acid)
(400 mL)      Hexane:diethyl ether mixture: mix 320 mL hexane with 80 mL diethyl ether
(10 g)        Iodine crystals, $I_2$

**Experiment 41**     **Analysis of lipids**

**Special Equipment**

(25)          Hot plates
(25)          Cheesecloth 3 × 3 in.

**Chemicals**

(3 g)         Cholesterol (ash free) 95–98% pure from Sigma Co.
(3 g)         Lecithin (prepared from dried egg yolk) 60% pure from Sigma Co.
(10 g)        Glycerol
(10 g)        Corn oil
(10 g)        Butter
(1)           Egg yolk obtained from one fresh egg before the lab period. Stir and mix.

| (250 mL) | Molybdate solution: dissolve 0.8 g $(NH_4)_6Mo_7O_{24} \cdot 4H_2O$ in 30 mL water. Put in an ice bath. Pour slowly 20 mL concentrated sulfuric acid (18 M $H_2SO_4$) into the solution and stir slowly. After cooling to room temperature bring the volume to 250 mL. **Wear a face shield, rubber gloves, and a rubber apron during the preparation. Do in the hood.** |
|---|---|
| (50 mL) | 0.1 M ascorbic acid solution: dissolve 0.88 g ascorbic acid (vitamin C) in water and bring it to 50 mL volume. This must be prepared fresh every week and stored at 4°C. |
| (250 mL) | 6 M sodium hydroxide, 6 M NaOH: dissolve 60.00 g NaOH in water and bring the volume to 250 mL |
| (250 mL) | 6 M nitric acid, 6 M $HNO_3$: into a 250-mL volumetric flask containing 100 mL ice cold water, pipet 125 mL concentrated nitric acid (12 M $HNO_3$); add enough water to bring to 250 mL. **Wear a face shield, rubber gloves, and a rubber apron during the preparation. Do in the hood.** |
| (200 mL) | Chloroform |
| (75 mL) | Acetic anhydride |
| (50 mL) | Concentrated sulfuric acid (18 M $H_2SO_4$) |
| (75 g ) | Potassium hydrogen sulfate, $KHSO_4$ |

**Experiment 42**   Separation of amino acids by paper chromatography

### Special Equipment

| (1) | Drying oven, 105–110°C |
|---|---|
| (2) | Heat lamps or hair dryers |
| (25) | Whatman no. 1 chromatographic paper, 15 x 8 cm |
| (25) | Rulers, metric scale |
| (25) | Polyethylene, surgical gloves |
| (150) | Capillary tubes, open on both ends |
| (1 roll) | Aluminum foil |
| (2) | Wide-mouth jars |

### Chemicals

| (25 mL) | 0.12% aspartic acid solution: dissolve 30 mg aspartic acid in 25 mL distilled water |
|---|---|
| (25 mL) | 0.12% phenylalanine solution: dissolve 30 mg phenylalanine in 25 mL distilled water |
| (25 mL) | 0.12% leucine solution: dissolve 30 mg leucine in 25 mL distilled water |
| (25 mL) | Aspartame solution: dissolve 150 mg Equal sweetener powder in 25 mL distilled water |
| (50 mL) | 3 M HCl solution: place 10 mL ice cold distilled water into a 50-mL volumetric flask. Add slowly 12.5 mL of concentrated HCl (12 M HCl) and bring it to volume with distilled water. **Wear a face shield, rubber gloves, and a rubber apron when preparing. Do in the hood.** |
| (1 L) | Solvent mixture: mix 600 mL 1-butanol with 150 mL acetic acid and 250 mL distilled water. **Solvent mixture *must* be freshly prepared the day it is to be used.** |
| (1 can) | Ninhydrin spray reagent (0.2% ninhydrin in ethanol or acetone). Do not use any reagent older than 6 months. |
| (1 can) | Diet Coke |
| (4 packets) | Equal or NutraSweet, sweeteners |
| (10 g) | Iodine crystals, $I_2$ |

**Acid–base properties of amino acids**

**Special Equipment**

| | |
|---|---|
| (10) | pH meters or |
| (5 rolls) | pHydrion short-range papers, from each range: pH: 0.0 to 3.0; 3.0 to 5.5; 5.2 to 6.6; 6.0 to 8.0; 8.0 to 9.5 and 9.0 to 12.0 |
| (25) | 20-mL pipets |
| (25) | 50-mL burets |
| (25) | Spectroline pipet fillers |
| (25) | Pasteur pipets |

**Chemicals**

| | |
|---|---|
| (500 mL) | 0.25 M NaOH: dissolve 5.00 g NaOH in 100 mL water and then add enough water to 500 mL volume |
| (750 mL) | 0.1 M alanine solution: dissolve 6.68 g L-alanine in 500 mL; add sufficient 1 M HCl to bring the pH to 1.5. Add enough water to 750 mL volume. |

**or**

Do as above but use either 5.63 g glycine or 9.84 g leucine or 12.39 g phenylalanine or 8.79 g valine.

---

**Isolation and identification of casein**

**Special Equipment**

| | |
|---|---|
| (25) | Hot plates |
| (25) | 600-mL beakers |
| (25) | Büchner funnels (O.D. 85 mm) in no. 7 1-hole rubber stopper |
| (7 boxes) | Whatman no. 2 filter paper, 7 cm |
| (25) | Rubber bands |
| (25) | Cheese cloths (6 × 6 in.) |

**Chemicals**

| | |
|---|---|
| (1 jar) | Boiling chips |
| (1 L) | 95% ethanol |
| (1 L) | Diethyl ether:ethanol mixture (1:1) |
| (0.5 gal) | Regular milk |
| (500 mL) | Glacial acetic acid |

**The following solutions should be placed in dropper bottles:**

| | |
|---|---|
| (100 mL) | Concentrated nitric acid (12 M $HNO_3$) |
| (100 mL) | 2% albumin suspension: dissolve 2 g albumin in 100 mL water |
| (100 mL) | 2% gelatin: dissolve 2 g gelatin in 100 mL water |
| (100 mL) | 2% glycine: dissolve 2 g glycine in 100 mL water |
| (100 mL) | 5% copper(II) sulfate: dissolve 5 g $CuSO_4$ (or 7.85 g $CuSO_4 \cdot 5H_2O$) in 100 mL water |
| (100 mL) | 5% lead(II) nitrate: dissolve 5 g $Pb(NO_3)_2$ in 100 mL water |
| (100 mL) | 5% mercury(II) nitrate: dissolve 5 g $Hg(NO_3)_2$ in 100 mL water |
| (100 mL) | Ninhydrin reagent: dissolve 3 g ninhydrin in 100 mL acetone. Do not use a reagent older than 6 months. |
| (100 mL) | 10% sodium hydroxide: dissolve 10 g NaOH in 100 mL water |
| (100 mL) | 1% tyrosine: dissolve 1 g tyrosine in 100 mL water |
| (100 mL) | 5% sodium nitrate: dissolve 5 g $NaNO_3$ in 100 mL water |

**Experiment 45**    **Neurotransmission: an example of enzyme specificity**

### Special Equipment

| | |
|---|---|
| (5) | Spectrophotometers |
| (5) | Delivery pipets (1-200 μL) |
| (20) | Natural pipet tips (1-200 μL) (USA Scientific Co. # 1111-0810) |

### Chemicals

(1L)      0.01 M Acetic acid: take 0.6 mL glacial acetic acid and add enough water to bring to 1 L.

(500 mL)   0.02 M NaOH: dissolve 0.40 g NaOH in enough water to make 500 mL of solution.

(1 L)      0.150 M NaCl solution: dissolve 8.766 g NaCl in 800 mL `water. Adjust the pH to 8.0 using 0.02 M NaOH. Add enough water to bring to 1.00 L. **The pH of the solution must be adjusted to 8.0 before every experiment.**

(100 mL)   $m$-Nitrophenol solution: dissolve 500 mg $m$-nitrophenol in 5 mL absolute ethanol. Bring the solution to 100 mL by adding 0.15 M NaCl solution, adjust to pH 8.0.

(10 mL)    Acetylcholine esterase (EC.3.1.1.7) from electric eel (Sigma C-3389): dissolve 3 mg. approximately 300 unit/mg, in 10 mL 0.15 M NaCl solution, adjust to pH 8.0

(10 mL)    Acetylcholine substrate solution: take 1.817 g acetylcholine and dissolve completely in enough 0.15 M NaCl solution, adjust to pH 8.0, to make 10 mL.

(40 mL)    $o$-Nitrophenyl acetate solution: dissolve 181.1 mg $o$-nitrophenyl acetate in 5 mL dimethylsulfoxide. Just before the start of the experiment, add 35 mL of 0.15 M NaCl solution, adjust to pH 8.0. **Mix the solution thoroughly.**

(10 mL)    Carboxyl esterase (EC 3.1.1.1) from porcine liver, available as a 0.3 mL suspension (conc. is 15 mg protein/mL, 250 units/mg protein): add ammonium sulfate buffer to get 10 mL of solution, adjust to pH 8.0.

**Experiment 46**    **Isolation and identification of DNA from onion**

### Special equipment

| | |
|---|---|
| (25) | Mortars |
| (25) | Pestles |

### Chemicals

| | |
|---|---|
| (12) | Yellow onions |
| (500 g) | Acid washed sand |
| (1 bottle) | Ivory concentrated liquid dishwashing detergent, 12.6 oz. |
| (250 g) | Sodium chloride, NaCl |
| (100 mL) | Citrate buffer: dissolve 0.88 g NaCl and 0.39 g sodium citrate in 100 mL water. |
| (2 L) | Absolute ethanol |
| (200 mL) | 95% ethanol |
| (50 mL) | 1% glucose solution: dissolve 0.5 g D-glucose in 50 mL water. |
| (50 mL) | 1% ribose solution: dissolve 0.5 g D-ribose in 50 mL water. |
| (50 mL) | 1% deoxyribose solution: dissolve 0.5 g 2-deoxy-D-ribose in 50 mL water. |
| (500 mL) | Diphenylamine reagent: dissolve 7.5 g reagent grade diphenylamine (Sigma D3409) in 50 mL glacial acetic acid. |

Add 7.5 mL concentrated sulfuric acid, $H_2SO_4$ slowly, with stirring. Prior to use, add 2.5 mL 1.6% acetaldehyde (made by dissolving 0.16 g acetaldehyde in 10 mL water); then add to enough water to make 500 mL. **This reagent must be prepared shortly before use in the laboratory. Wear a face shield, rubber gloves, and a rubber apron when preparing the reagent. Do in the hood.**

As an alternative:

(200 mL)    Diphenylamine solution without acetaldehyde: dissolve 2.0 g reagent grade diphenylamine (Sigma D3409) in 200 mL concentrated sulfuric acid, $H_2SO_4$. Swirl slowly with a glass rod until compeltely dissolved. **Prepare this reagent fresh each week. Wear a face shield, rubber gloves, and a rubber apron when preparing the reagent. Do in the hood.**

| | |
|---|---|
| **Experiment 47** | **Viscosity and secondary structure of DNA** |

### Equipment

(5)    Ostwald (or Cannon-Ubbelhode) capillary viscometers; 3-mL capacity, approximate capillary diameter 0.2 mm; **efflux time of water = 40–50 sec.**

(5)    Stopwatches (Wristwatches can also time the efflux with sufficient precision.)

(5)    Stands with utility clamps

(25)    Pasteur pipets

(10)    Spectroline pipet fillers

### Chemicals

(500 mL)    Buffer solution: dissolve 4.4 g sodium chloride, NaCl, and 2.2 g sodium citrate, $Na_3C_6H_5O_7 \cdot 2H_2O$ in 450 mL distilled water. Adjust the pH with either 0.1 M HCl or 0.1 M NaOH to pH 7.0. Add enough water to bring to 500 mL volume.

(200 mL)    DNA solution: dissolve 20 mg of calf thymus Type I highly polymerized DNA (obtainable from Sigma as well as from other companies) in 200-mL buffer solution at pH 7.0. The purchased DNA powder should be kept in the freezer. The DNA solutions should be prepared fresh or maximum 2–3 hr. in advance of the experiment. The solution should be kept at 4°C; 1–2 hr. before the experiment; the solution should be allowed to come to room temperature. Label the solution as 0.01 g/dL concentration.

(100 mL)    1 M hydrochloric acid, 1 M HCl: add 8.3 mL concentrated HCl (12 M HCl) to 50 mL ice cold water; add enough water to bring to 100 mL volume. **Wear a face shield, rubber gloves, and a rubber apron during preparation. Do in the hood.**

(100 mL)    0.1 M hydrochloric acid: add 10.0 mL 1 M HCl to 50 mL water; add enough water bring to 100 mL volume. **Follow safety procedure described above.**

(100 mL)    1 M sodium hydroxide: dissolve 4.00 g NaOH in 50 mL water; add enough water to bring to 100 mL volume.

(100 mL)    0.1 M sodium hydroxide: dissolve 0.40 g NaOH in 50 mL water; add enough water to bring to 100 mL volume.

**Experiment 48**　　**Kinetics of urease catalyzed decomposition of urea**

### Special Equipment

| | |
|---|---|
| (25) | 5-mL pipets |
| (25) | 10-mL graduated pipets |
| (25) | 10-mL volumetric pipets |
| (25) | 50-mL burets |
| (25) | Buret holders |
| (25) | Spectroline pipet fillers |

### Chemicals

(3.5 L)　0.05 M Tris buffer: dissolve 21.05 g Tris buffer in water (3 L). Adjust the pH to 7.2 with 1 M HCl solution; add sufficient water to make 3.5 L. Portions of buffer solution will be used to make urea and enzyme solutions.

(2.5 L)　0.3 M urea solution: dissolve 45 g urea in 2.5 L Tris buffer

(50 mL)　$1 \times 10^{-3}$ M phenylmercuric acetate: dissolve 16.5 mg phenylmercuric acetate in 40 mL water; add enough water to bring the volume to 50 mL.

**CAUTION! Phenylmercuric acetate is a poison. Do not touch the chemical with your hands. Do not swallow the solution. Wear rubber gloves in the preparation.**

(50 mL)　1% $HgCl_2$ solution: dissolve 0.5 g $HgCl_2$ in enough water to make 50 mL solution

(100 mL)　0.04% methyl red indicator: dissolve 40 mg methyl red in 100 mL distilled water

(500 mL)　Urease solution: prepare the enzyme solution on the week of the experiment and store at 4°C. Take 1.0 g urease, dissolve in 500 mL Tris buffer. (One can buy urease with 5 to 6 units activity, for example, from Nutritional Biochemicals, Cleveland, Ohio.) The activity of the enzyme printed on the label should be checked by the stockroom personnel or instructor.

(1.0 L)　0.05 N HCl: add 4.2 mL concentrated HCl (12 M HCl) to 100 mL ice cold water; add enough water to bring to 1.0 L volume. **Wear a face shield, rubber gloves, and a rubber apron during the preparation. Do in the hood.**

**Experiment 49**　　**Isocitrate dehydrogenase—an enzyme of the citric acid cycle**

### Special Equipment

| | |
|---|---|
| (15) | Spectrophotometers with 5 cuvettes each |
| (25) | 1-mL graduated pipets |

### Chemicals

(40 mL)　Phosphate buffer at pH 7.0: mix together 25 mL 0.1 M $KH_2PO_4$ and 15 mL 0.1 M NaOH. To prepare 0.1 M NaOH, add 0.20 g NaOH to 20 mL water in a 50-mL volumetric flask; stir to dissolve; add enough water to bring to 50 mL volume. To prepare 0.1 M $KH_2PO_4$, add 0.68 g potassium dihydrogen phosphate to 40 mL water in a 50-mL volumetric flask; stir to dissolve; add enough water to bring to 50 mL volume.

(20 mL)　0.1 M $MgCl_2$: add 0.19 g magnesium chloride to 20 mL water; stir to dissolve.

|  |  |
|---|---|
| (50 mL) | Isocitrate dehydrogenase: commercial preparations from porcine heart are obtainable from companies such as Sigma, etc. (EC 1.1.1.42) (activity about 8 units per mg of solid). Dissolve 10 mg of the enzyme in 50 mL water. This solution should be made fresh before the lab period and kept in a refrigerator until used. |
| (20 mL) | 6.0 mM $\beta$-nicotinamide adenine dinucleotide, $\beta$-NADP$^+$, solution: dissolve 92 mg NADP$^+$ in 20 mL water |
| (50 mL) | 15 mM sodium isocitrate solution: dissolve 160 mg sodium isocitrate in 50 mL water |

**Experiment 50**

## Quantitative analysis of vitamin C contained in foods

### Special Equipment

|  |  |
|---|---|
| (25) | 50-mL burets |
| (25) | Buret clamps |
| (25) | Ring stands |
| (25) | Spectroline pipet fillers |
| (25) | 10-mL volumetric pipets |
| (1 box) | Cotton |

### Chemicals

|  |  |
|---|---|
| (500 g) | Celite, filter aid |
| (1 can) | Hi-C orange drink |
| (1 can) | Hi-C grapefruit drink |
| (1 can) | Hi-C apple drink |
| (2 L) | 0.01 M iodine solution: add 32 g KI to 800 mL water; stir to dissolve. Add 5 g I$_2$; stir to dissolve. Add enough water to bring to 2 L volume. Store in dark bottle. **Caution! Iodine is poisonous if taken internally.** |
| (100 mL) | 3 M HCl: add 25 mL concentrated HCl (12 M HCl) to 50 mL ice cold water; add enough water to bring to 100 mL volume. **Wear a face shield, rubber gloves, and a rubber apron during the preparation. Do in the hood.** |
| (100 mL) | 2% starch solution: place 2 g soluble starch in a 50-mL beaker. Add 10 mL water. Stir vigorously to form a paste. Boil 90 mL water in a second beaker. Add the starch paste to the boiling water. Stir until the solution becomes translucent. Cool to room temperature. |

**Experiment 51**

## Analysis of vitamin A in margarine

### Special Equipment

|  |  |
|---|---|
| (1) | UV spectrophotometer with suitable UV light source. Preferably it should be able to read down to 200 nm. |
| (1 pair) | Matched quartz cells, with 1-cm internal path length. |
| (2) | Long-wavelength UV lamp. The lamp should provide radiation in the 300-nm range (for example, #UVSL-55; LW 240 from Ultraviolet Products Inc.) |
| (12) | 500-mL separatory funnels |
| (12) | 25-mL (or 50-mL) burets |
| (12) | Hot plates, each with a water bath |
| (12) | Beaker tongs |

### Chemicals

| | |
|---|---|
| (0.5 lb) | Margarine |
| (400 mL) | 50% KOH: weigh 200 g KOH and add 200 mL water, with constant stirring; add enough water to make 400 mL |
| (1 L) | 95% ethanol |
| (150 mL) | Absolute ethanol |
| (2 L) | Petroleum ether, 30–60°C |
| (3 L) | Diethyl ether |
| (350 g) | Alkaline aluminum oxide (alumina) |

**Experiment 52**

## Urine analysis

### Special Equipment

| | |
|---|---|
| (12) | Hydrometers (urinometers) from 1.00 to 1.40 specific gravity |
| (3 bottles) | Clinistix (50 reagent strips/bottle) |
| (3 bottles) | Urobilistix (50 reagent strips/bottle) |
| (3 bottles) | Phenistix (50 reagent strips/bottle) |
| (1 bottle) | Albustix (100 reagent strips/bottle) |
| (1 bottle) | Ketostix (100 reagent strips/bottle) |

These are obtainable from Ames Co., Division Miles Lab. Inc., Elkhart, Indiana, 46515. Instead of the individual "Stix," you may purchase 4 bottles of multipurpose Labstix (100 reagent strips/bottle).

### Chemicals

| | |
|---|---|
| (500 mL) | Normal urine. This and all other urine samples must be kept at 4°C until 30 min. prior to the lab period. Alternatively, you may ask each student to provide fresh urine samples for analysis. |
| (500 mL) | "Pathological urine A": add 4 g glucose, 2 mL acetone and 2 g citric acid to 500 mL water |
| (500 mL) | "Pathological urine B": add 50 mg phenylpyruvate, and 500 mg sodium phoshate, $Na_3PO_4$, to 500 mL water |
| (250 mL) | 1% glucose: dissolve 2.5 g glucose in 250 mL water |
| (200 mL) | 0.25% glucose: dilute 50 mL of 1% glucose solution with water to 200 mL volume |